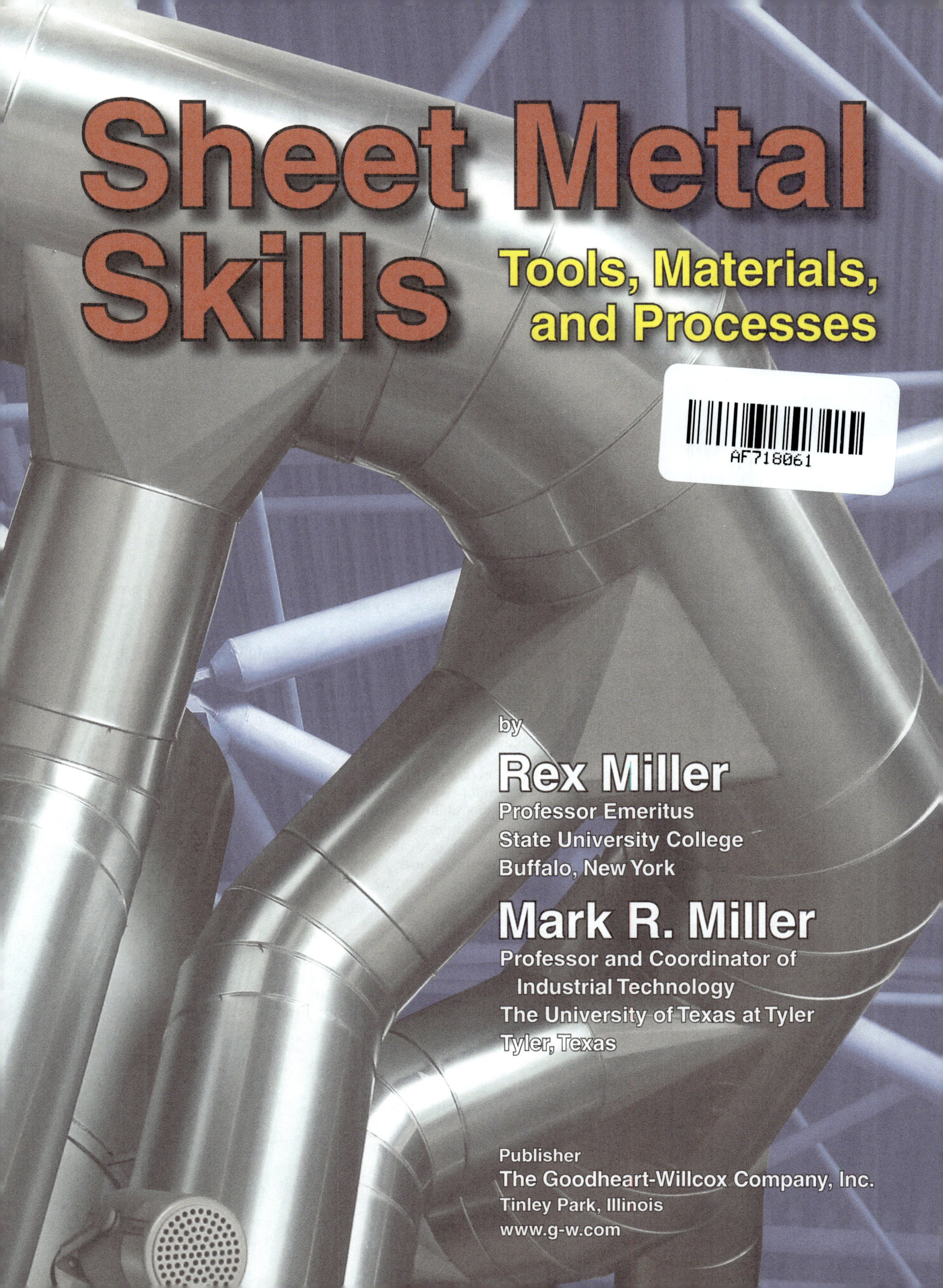

Sheet Metal Skills

Tools, Materials, and Processes

by

Rex Miller
Professor Emeritus
State University College
Buffalo, New York

Mark R. Miller
Professor and Coordinator of Industrial Technology
The University of Texas at Tyler
Tyler, Texas

Publisher
The Goodheart-Willcox Company, Inc.
Tinley Park, Illinois
www.g-w.com

Manufactured in the United States of America.

Library of Congress Catalog Card Number 2010003693

ISBN 978-1-60525-325-1

2 3 4 5 6 7 8 – 11 – 22 21 20 19 18

The Goodheart-Willcox Company, Inc. Brand Disclaimer: Brand names, company names, and illustrations for products and services included in this text are provided for educational purposes only and do not represent or imply endorsement or recommendation by the author or the publisher.

The Goodheart-Willcox Company, Inc. Safety Notice: The reader is expressly advised to carefully read, understand, and apply all safety precautions and warnings described in this book or that might also be indicated in undertaking the activities and exercises described herein to minimize risk of personal injury or injury to others. Common sense and good judgment should also be exercised and applied to help avoid all potential hazards. The reader should always refer to the appropriate manufacturer's technical information, directions, and recommendations; then proceed with care to follow specificequipmentoperatinginstructions.Thereadershouldunderstandthesenoticesandcautionsarenotexhaustive.

The publisher makes no warranty or representation whatsoever, either expressed or implied, including but not limited to equipment, procedures, and applications described or referred to herein, their quality, performance, merchantability, or fi tness for a particular purpose. The publisher assumes no responsibility for any changes, errors, or omissions in this book. The publisher specificallydisclaimsanyliabilitywhatsoever,includinganydirect,indirect,incidental,consequential, special, or exemplary damages resulting, in whole or in part, from the reader's use or reliance upon the information, instructions, procedures, warnings, cautions, applications, or other matter contained in this book. The publisher assumes no responsibility for the activities of the reader.

Library of Congress Cataloging-in-Publication Data

Miller, Rex
Sheet metal skills : tools, materials, and processes / by Rex Miller, Mark R. Miller.
p. cm.
Includes index.
ISBN 978-1-60525-325-1
1. Sheet-metal work. I. Miller, Mark R. II. Title.
TS250.M544 2011
671.8'23--dc22
2010003693

Introduction

Sheet Metal Skills: Tools, Materials, and Processes is designed for students interested in the study of sheet metal work. The text provides a comprehensive view of the sheet metal industry and the work involved in it. This write-in text includes up-to-date information on the tools, materials, and processes used in the industry.

Sheet Metal Skills begins with an overview of sheet metal history, sheet metal uses, and the types of work available to well-trained sheet metal workers. An in-depth safety chapter addresses the many safety precautions necessary when working with sheet metal. The tools and methods used in measurement and layout are explained, along with how to read prints.

Sheet Metal Skills focuses on the specific tools and tasks used in sheet metal work. Typically used hand tools and machinery are explained. Processes covered include joining, drilling, punching, and fastening metal pieces into finished products. Drawing and developing patterns is also discussed in detail.

Additional topics covered include materials and materials testing, specialized sheet metal forming processes, and sheet metal working careers.

Sheet metal projects and experiments are included in the text to provide hands-on activities for students. These activities are intended to increase students' skill and confidence levels. In the Appendix of the text, there is helpful information regarding math skills needed to do sheet metal work, conversion charts between the U.S. Customary and metric systems, and sheet metal and wire gage designations.

Sheet Metal Skills: Tools, Materials, and Processes covers all aspects of sheet metal working and the industry, providing students with an excellent foundation for further learning.

Table of Contents

About the Authors

Rex Miller is professor emeritus of industrial technology at State University College at Buffalo. Dr. Miller taught technical curriculums at the technical school, high school, and college levels for more than 40 years. He has authored or co-authored more than 100 texts for vocational and industrial arts programs, covering a number of technical areas. He was presented with the State University of New York Chancellor's Award for Excellence in Teaching and the Distinguished Service Award from the Technology Education Association.

Mark R. Miller is a professor of technology and the coordinator for the industrial technology and industrial management programs in the Department of Human Resource Development & Technology at The University of Texas at Tyler. Dr. Miller teaches technology courses for future middle managers and has authored or co-authored more than 40 technical books and numerous technical articles. He has taught technical courses at the high school and university levels for over 25 years.

Chapter 1 Introduction to Sheet Metal

Objectives

After studying this chapter, you will be able to:

- Define sheet metal.
- Explain how sheet metal is measured.
- Discuss the different types of metal.
- Identify various uses for sheet metal.
- List the requirements for becoming an apprentice.

Technical Terms

alloy
aluminum
American Standard Wire Gage
apprenticeship
brass
cold-rolled steel
copper sheet
ductwork
fitting
gage
galvanized steel
galvanizing
heating, ventilation, and air-conditioning (HVAC) system
nonferrous metal
plate
sheet metal
sheet steel
stainless steel
tin plate
United States Standard Gage for Sheet and Plate Iron and Steel

Sheet Metal History

In ancient days, many workers toiled many hours to hammer a single sheet of metal into its final form. Today, sheet metal can be fabricated into its final form in a matter of hours. The history of sheet metal is lengthy, beginning in prehistoric times and continuing until today.

The Bronze Age marks the widespread use of metals in everyday life. Ancient people discovered copper and bronze could be made into tools, shields, swords, and decorative items. Even though this discovery was most likely accidental, it revolutionized civilization. Metal production created rapid growth in populations and in trade among nations. Towns and cities developed, as people moved to metal-producing areas. Workers were needed to mine and process the raw materials, and to create the tools and weapons. Workers were also needed to produce food, build housing, and transport the finished products. The use of metal weapons also meant these civilizations were able to defeat less-advanced civilizations in war. The Bronze Age lasted for several thousand years and spread throughout the Middle East, Asia, and Europe. It was followed by the Iron Age.

Like the Bronze Age, the Iron Age was also a time of discovery and growth. Iron was now the metal being used to make tools, weapons, farm machinery, and other items. Civilizations evolved further, wars were fought with new weapons and vehicles, and trade flourished. These people built on the knowledge gained during the Bronze Age to develop a more advanced society.

The Industrial Revolution changed the way products were made and work was done. Machines such as the rolling mill transformed the production of sheet metal. Sheets could now be made by running metal through rollers, creating flat sheets or rolls. Metal no longer had to be processed into sheets by hand. It could now be formed into thin, strong, flat sheets and strips in a much shorter amount of time. Readily available, sheet metal was used in a variety of products.

The use of computers continues to revolutionize sheet metal work. Computers now do jobs that were previously done using calculators, hand tools, or hands. For example, computers software is used to design the complex curves and bends found on aircraft bodies and wings, and automobile bodies. The computer does the calculations needed for the tools to shape the body parts. Skyscrapers contain complex ductwork to heat, cool, and ventilate buildings. Computer software is used to design the ducts and the fittings to install them. Most sheet metal cutting and punching operations are done by computer numerical control (CNC) lasers or punch presses.

Sheet metal work has evolved into a well-established, skilled craft. Today's sheet metal workers use many processes to fabricate, install, and maintain a variety of products. So many products are made of sheet metal, it is nearly impossible to count them. These products include heating, ventilating, and air-conditioning systems, automobile and aircraft bodies, buildings, and appliances.

Basics of Sheet Metal

Sheet metal is any type of metal produced in flat sheets or rolls, **Figure 1-1.** Sheets and rolls each have advantages and disadvantages. These are important to consider when selecting material for a project. Sheet metal thickness varies, but generally is less than 1/4″ (7 mm) thick. Metal thicker than 1/4″ is called ***plate***. Extremely thin metal used for decoration is called foil.

sheet metal: Any type of metal produced in flat sheets or rolls less than 1/4″ (7 mm) thick.

plate: Any type of metal produced in flat sheets or rolls more than 1/4″ (7 mm) thick.

Sheet Size

Sheet metal can be ordered directly from the mill in any size, dimension, or gage. If specific conditions exist, then the metal can be shaped or cut to fit the requirements of that particular object. For instance, the B-47 bomber had very long wings and required flexible metal to cover them. A plant was built to produce the long, tapered lengths of metal. There are no limits on available sheet metal sizes. However, commonly used sizes include 96″ or 120″ lengths. Most smaller sheet metal shops make 30″ x 96″ and 36″ x 120″ sizes.

Gage

The thickness of sheet metal is its ***gage***. Each gage represents a specific metal thickness. The lower the gage number, the thicker the metal. For example, 18-gage sheet metal is thicker than 26-gage sheet metal. Gage is also the name of the tool used to measure gage, **Figure 1-2.**

gage: The thickness of sheet metal. Each gage represents a specific metal thickness

Types of Sheet Metal				
Shapes	**Standard Length**	**Measurement Method**	**Form of Purchase**	**Additional Information**
 Sheet less than 1/4″ thick	To 144″	Thickness × width Widths to 72″	Weight, foot, or piece	Available in coils of much longer lengths.
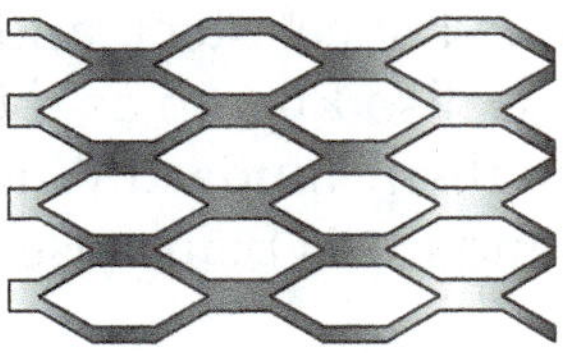 Expanded Sheet	To 96″	Gage number (U.S. Standard)	36 × 96″ and size of openings	Metal is pierced and expanded (stretched) to diamond shape; also available rolled to thickness after it has been expanded.
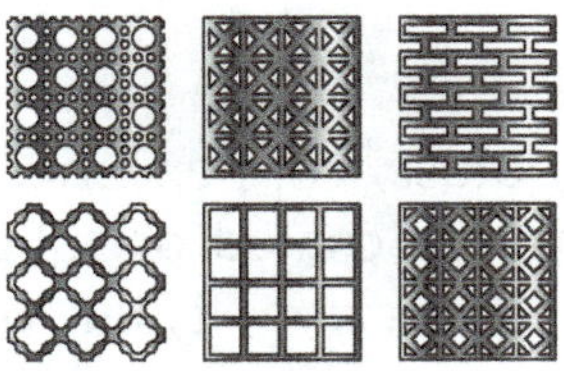 Perforated Sheet	To 96″	Gage number (U.S. Standard)	30 × 36″ 36 × 48″ 36 × 96″	Design is cut in sheet. Many designs available.

Figure 1-1. Common shapes and sizes of sheet metal.

Figure 1-2. Sheet metal gage. (L.S. Starrett)

American Standard Wire Gage: The standard measurement used for nonferrous metals and for wire.

United States Standard Gage for Sheet and Plate Iron and Steel: The standard measurement used for plate iron, steel, and uncoated steel. It is based on weight in ounces per square foot.

nonferrous metal: Metal other than iron.

The ***American Standard Wire Gage,*** or Brown and Sharpe, is the standard measurement used for ***nonferrous metals*** (metals other than iron) and for wire. The recognized commercial standard in the United States for plate iron, steel, and uncoated steel is the ***United States Standard Gage for Sheet and Plate Iron and Steel.*** It is commonly referred to as the U.S. Standard Gage. It is based on weight in ounces per square foot.

To use a gage, insert the sheet metal edge in the slot of the gage. See **Figure 1-3.** Find the smallest slot the metal easily slides into. The slot has a number above it. That is the gage. Coated materials have a thin coating on both sides, to protect from corrosion. Therefore, coated metals are usually one gage thinner than what they measure. In other words, a 26-gage metal sheet will measure 25-gage.

There are several other standards in addition to the American Standard Wire Gage and U.S. Standard Gage, **Figure 1-4.** The English Standard Wire Gage is used to gage iron wire, hot- and cold-rolled sheet steel, and, in some cases, sheet iron. English Standard Wire Gage is also known as the Birmingham Wire Gage or Stubs' Iron Wire Gage. An English Standard Wire Gage is shown in **Figure 1-5.** Always specify which standard is being used in measurements.

Types of Metals

There are many types of metal: steel, aluminum, brass, copper. Sheet steel is commonly used in school shops. Sheet steel can be coated or solid (uncoated). Coating reduces rust and corrosion on steel. For example, tar- or asphalt-based products are used to coat the undersides of automobiles to reduce corrosion. The most commonly used coated sheet steels are galvanized steel

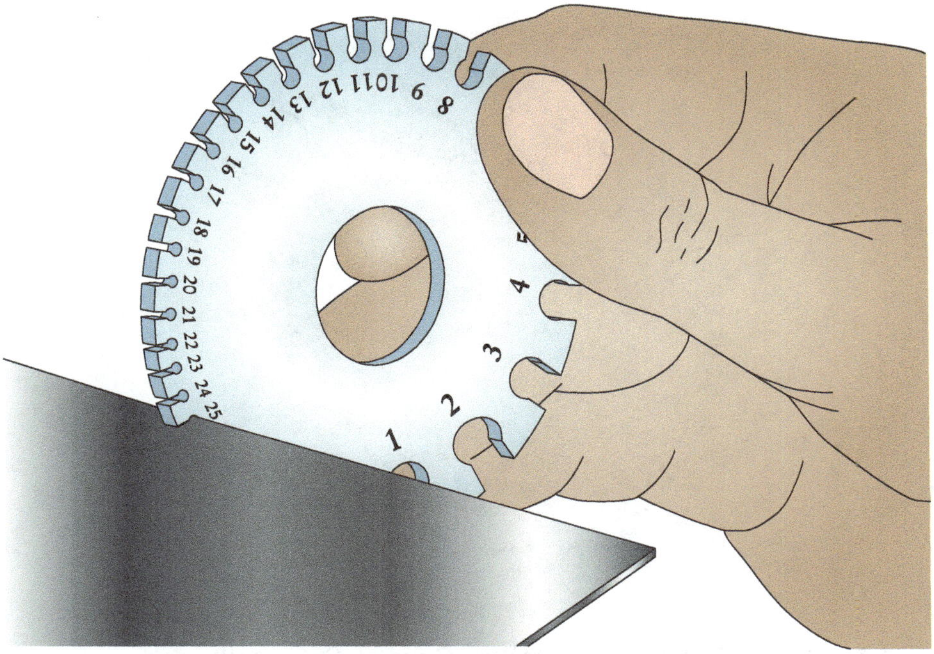

Figure 1-3. Measuring a piece of sheet metal with a gage.

Standards for Sheet and Wire Gages						
Number of Wire Gage	American or Brown & Sharpe	Birmingham or Stubs' Iron Wire	Washburn & Moen, (Worcester, MA)	American S. & W. Co's. Music Wire Gage	Stubs' Steel Wire	U.S. Standard Gage for Sheet and Plate Iron and Steel
1	.289279	.300	.2830	.010	.227	.28125
2	.257626	.284	.2625	.011	.219	.265625
3	.229423	.259	.2437	.012	.212	.250
4	.204307	.238	.2253	.013	.207	.234375
5	.181941	.220	.2070	.014	.204	.21875
6	.162023	.203	.1920	.016	.201	.203125

Figure 1-4. A comparison of various standards for sheet and wire gages. The dimensions are in decimal parts of an inch.

and tin plate. The most commonly used solid sheet steel is stainless steel. Stainless steel is an ***alloy***, a combination of two or more metals. Combining metals is done to overcome the less desirable properties of individual metals. Other well known alloys include bronze and brass.

alloy: A material with metallic properties that is a mixture of two or more metals.

Sheet Steel

Uncoated ***sheet steel*** has a bluish black surface. Because uncoated steel easily corrodes, after assembly, products are coated. Sheet steel applications include stovepipes, safety guards, and parts for electrical appliances.

sheet steel: An uncoated sheet of steel having a bluish-black surface. After being assembled into a finished product, the steel is coated to prevent corrosion.

Figure 1-5. A typical English Standard Wire Gage. (L.S. Starrett)

galvanized steel: Sheet steel that is coated with a thin layer of zinc to prevent corrosion. Commonly used to make roofing, siding, gutters, and downspouts.

galvanizing: Process that coats steel with a thin layer of zinc to prevent corrosion.

cold-rolled steel: Steel that is not processed until it is cooled. Its finished appearance is silver gray, with a smooth surface.

copper sheet: Reddish brown metal. An excellent conductor of electricity that contains a small quantity of alloy.

tin plate: Sheet metal coated with a layer of pure tin, making it corrosion resistant.

Galvanized Steel

Galvanized steel is used often in the sheet metal shop. ***Galvanizing*** is the process of coating sheet steel with hot, liquid zinc. Galvanized steel has a shaded appearance. The zinc coating protects the metal from corrosion. The most common uses for galvanized steel are in heating and air-conditioning ducts and in gutters and downspouts. Other products made with galvanized steel include garbage cans and storage or grain bins. Gages for galvanized sheet are shown in **Figure 1-6.**

Cold-Rolled Steel

Cold-rolled steel is not processed into its final shape and thickness until it is cooled. It is silver gray and has a smooth finish. School lockers and cabinets are made of cold-rolled steel.

Copper Sheet

Copper sheet is easily recognized by its reddish brown color. It is an excellent conductor of electricity and contains a small quantity of alloy. It is more than 99 percent copper, with less than 0.4 percent alloy. Copper sheet is commonly used to make pots and pans, electronic parts (printed circuit boards), and roofing.

Tin Plate

Tin plate is sheet metal coated with a layer of pure tin, making it corrosion resistant. Tin coating is food safe, meaning it can safely come in contact with surfaces that touch food. It was used in kitchen utensils, tin food cans, dairy equipment, and other food industry products. However, it has been replaced in many products with less-expensive stainless steel.

Galvanized Sheet Gages			
Gage Number	Thickness in Inches	Gage Number	Thickness in Inches
10	0.1382	22	0.0336
11	0.1233	23	0.0306
12	0.1084	24	0.0276
13	0.0934	25	0.0247
14	0.0785	26	0.0217
15	0.0710	27	0.0202
16	0.0635	28	0.0187
17	0.0575	29	0.0172
18	0.0516	30	0.0157
19	0.0456	31	0.0142
20	0.0396	32	0.0134
21	0.0366		

Figure 1-6. Gages for galvanized sheet steel.

Aluminum

Aluminum is used in alloys with other metals, such as magnesium or manganese, allowing for easy forming and bending. Typical uses include automobile trim, ductwork, and electronic parts.

Stainless Steel

Stainless steel is steel that contains 12–25 percent chromium. It is a durable metal that resists rust and corrosion. Stainless steel is more useful than coated metals because coating can chip, wear off, or crack. This does not happen with stainless steel. Stainless steel is used in restaurant and cafeteria equipment, dairy equipment, escalators, and household appliances.

Brass

Brass is an alloy made of copper and zinc. The zinc content is not more than 40 percent of the total. Some lead, tin, or aluminum may also be used in very small amounts. Brass has many uses in jewelry, and plumbing fixtures. Sheet brass is used in a variety of applications, from belt buckles to decorative plaques.

aluminum: A lightweight, inexpensive, easily worked metal. It is used in alloys with other metals, such as magnesium or manganese.

stainless steel: A durable, somewhat expensive metal that contains 12–25 percent chromium and resists rust and corrosion.

brass: An alloy of copper and zinc, available in various degrees of hardness. Colors range from copper to nearly white.

Sheet Metal Uses

Look around the classroom. Find one or two products made from sheet metal. These products are everywhere. Sheet metal is an economical material. It is easy to form. It can also be decorative, **Figure 1-7.** From dishwashers to airplanes, sheet metal is a versatile material.

Figure 1-7. This portion of the Chrysler Building in New York City is covered in stainless steel.

Heating, Ventilation, and Air-Conditioning Systems

heating, ventilation, and air-conditioning (HVAC) system: System that circulates either warm or cooled air throughout a building.

ductwork: Combination of pipes and fittings in the HVAC system that contains the circulating air.

fitting: Hardware used to change airflow direction, the size of the duct ends, or both.

Many sheet metal workers in the building trades work on ***heating, ventilation, and air-conditioning*** (*HVAC*) ***systems***. These systems circulate air, either warming or cooling homes and businesses, **Figure 1-8.** Some furnace parts and the ductwork are made by sheet metal workers.

Ductwork is a combination of pipes and fittings used to circulate air. Ducts can be square, rectangular, or round pieces. They are open at both ends and are one size, **Figure 1-9.** ***Fittings*** are pieces of hardware used to change airflow direction, the size of the duct ends, or both, **Figure 1-10.**

Approximately 30 percent of sheet metal work involves fittings, and about 70 percent involves ducts. Various types and sizes of ducts and fittings can be purchased from manufacturers, rather than building custom pieces. These premade pieces are used more in residential work than in commercial buildings. Commonly used premade pieces in commercial work include access doors, dampers, stationary and moving louvers, and penthouses.

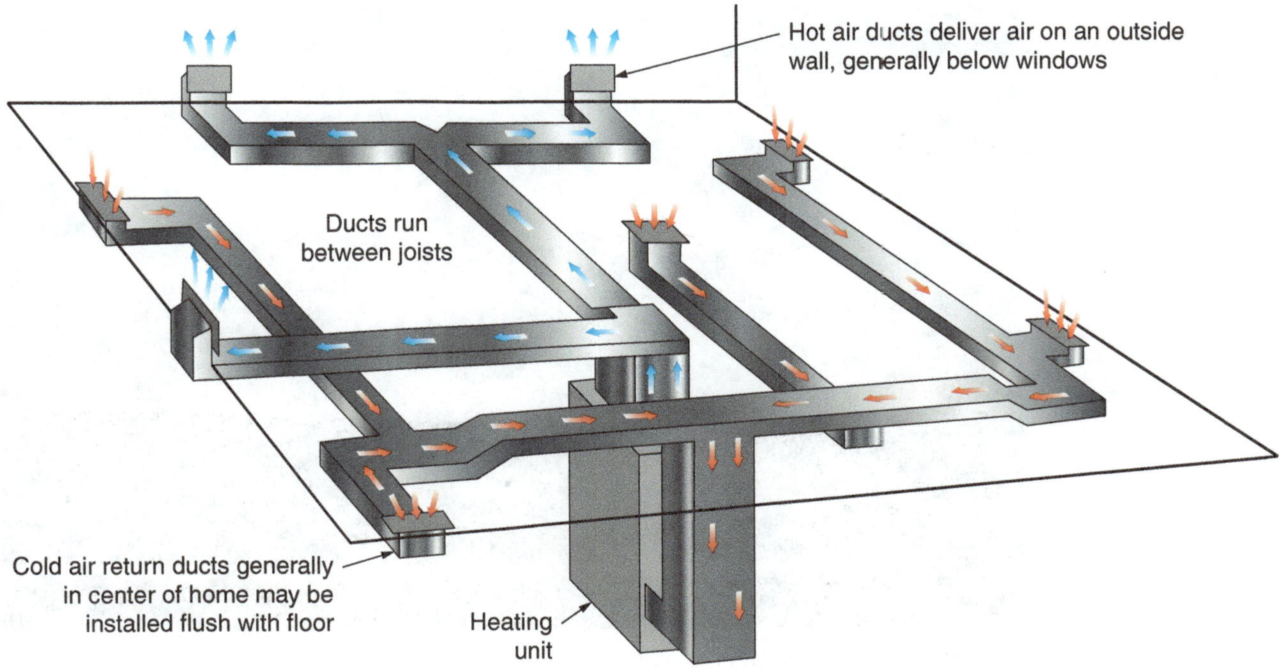

Figure 1-8. Ductwork for a forced-air heating system.

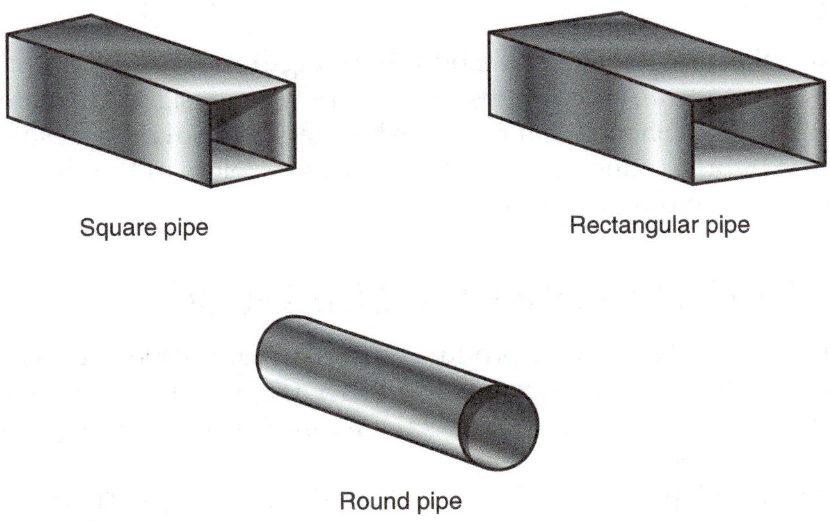

Figure 1-9. Pipes are open on both ends.

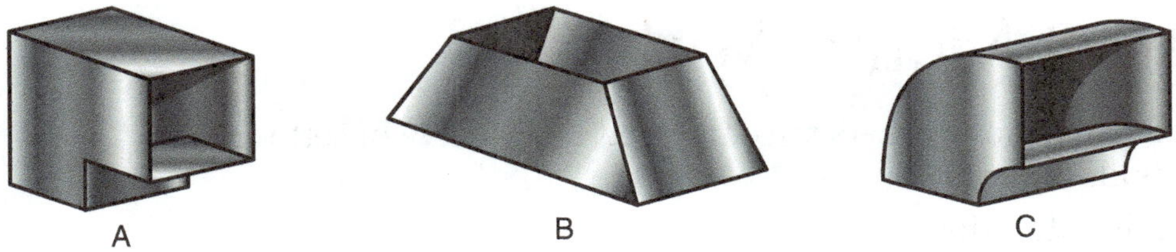

Figure 1-10. Fittings for ductwork. A—Changes direction of airflow. B—Acts as transition between two differently sized pipes. C—Changes direction and acts as transition.

Figure 1-11. Sheet metal building.

Buildings

Sheet metal is also used to make buildings, **Figure 1-11.** Roofing material may also be sheet metal. Many times a roof is specially designed to match the style of the building. Sheet metal roofs are becoming popular because seams can be sealed tightly, reducing the chance of leaks. Colors in the sheet material can be baked on during production.

Mass-Produced Sheet Metal Goods

Mass-produced sheet metal products include many types of consumer goods. Dishwashers, clothes dryers, and other major appliances found in today's home are made from sheet metal. Automobiles are mass produced and made of sheet metal.

Mass produced parts are also found in the housing industry. Metal hangers are often used to reinforce joints or increase the strength of junction point, **Figure 1-12.**

Sheet Metal Workers

Sheet metal workers are employed in a variety of settings:

- Building trades.
- Aircraft industry.
- Automobile industry.
- Shipbuilding industry.
- Railroad industry.

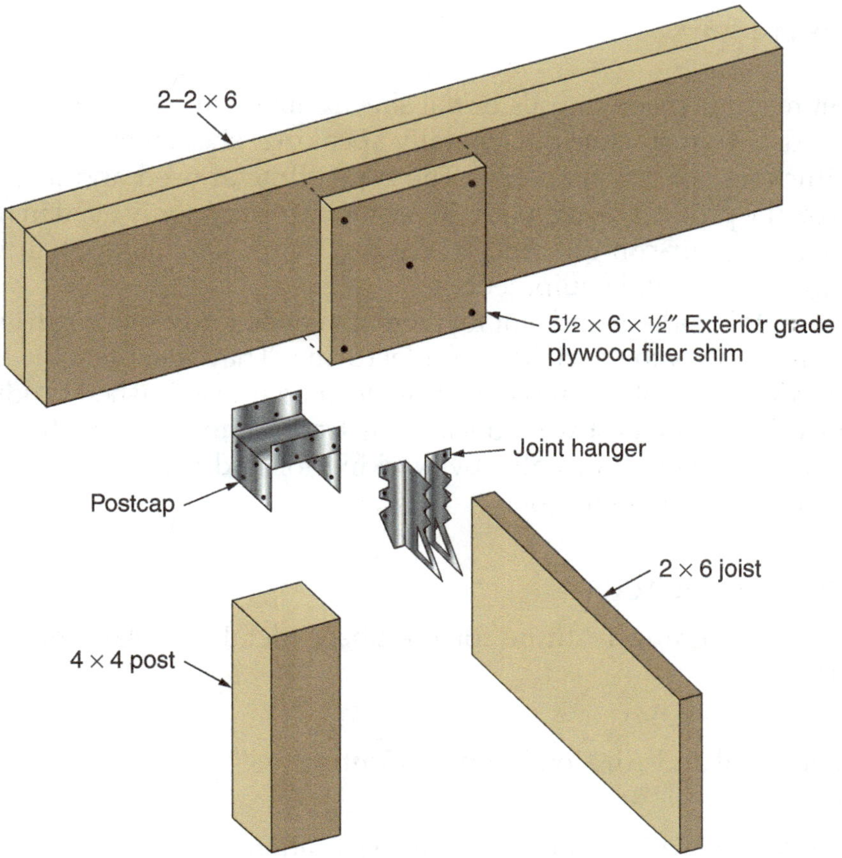

Figure 1-12. Hangers and brackets used in construction.

Others may work in shops that manufacture specialty products, such as custom kitchen equipment or electrical generating and distributing machinery.

Sheet metal workers must be able to lay out, fabricate, install, and service their products. Each step in this process requires specific skills. For instance, computers play an increasingly important role in the layout of a pattern. Therefore, computer skills are becoming a requirement for sheet metal workers.

Apprenticeships are considered the best way to learn the sheet metal trade. In order to be accepted into an apprenticeship, applicants must be high school graduates. Further study is often required of sheet metal workers, in order to keep skills up to date and to stay abreast of new technological developments. Workers often complete additional training provided by their union or their employer, to improve existing skills and to acquire new skills.

apprenticeship: A formal, on-the-job training program in which a trainee learns the skills needed for a trade.

A relatively small number of workers pick up the sheet metal trade informally. These people usually start by working as helpers to experienced sheet metal workers. Helpers carry metal and clean up debris, while learning about materials, tools, and their uses. Later, they learn to operate machinery used in their work. In time, helpers learn installation. Most helpers take vocational school courses in mathematics or sheet metal fabrication to supplement their work experience. To be promoted, helpers usually must pass the same written examination as apprentices.

Summary

Sheet metal is made in rolls or flat sheets. Sizes vary, both in length and width, and also in thickness. Generally speaking, sheet metal is any metal with a thickness of less than 1/4". The type of metal used and its size will depend on the product being made. Sheet metal is commonly used in heating, ventilation, and air-conditioning (HVAC) systems, kitchen appliances, and restaurant and hospital equipment.

Sheet metal workers find employment in a wide range of industries, from the building trades to specialty manufacturing. They must be able to envision the way a piece of metal will be transformed into a finished product. An apprenticeship is the best way to learn the trade. Some sheet metal workers do pick up the trade informally, by working for and learning from experienced workers while on the job.

Internet Resources

International Training Institute for the Sheet Metal and Air Conditioning Industry

www.sheetmetal-iti.org

The Sheet Metal and Air Conditioning Contractors'

www.smacna.org

The Sheet Metal Workers International Association

www.smwia.org

Outside Activities

1. If permissible, visit the local sales office of a builder of metal outbuildings. Ask office personnel for copies of brochures that explain and outline the process of constructing metal outbuildings. Answer the following questions. How are the buildings constructed? How do costs of metal outbuildings compare to more traditional outbuildings? Present your findings to your classmates.
2. Visit the local hardware store or national home store to see what type of sheet metal fittings they have available. Prepare a list and drawings of the fittings. Discuss with your classmates the advantages of using these premade fittings. Does it save time? Money? Both?
3. Ask a local restaurant owner if you may look at the ductwork used in the restaurant to remove cooking gases. Is it stainless steel, galvanized steel, or unpainted sheet metal? If someone from the custodial staff is available, ask that person what process is used to clean the ductwork.

Name ______________________ Class ______________ Score __________

Chapter 1
Introduction to Sheet Metal

Review Questions

Write your answers in the spaces provided.

1. Various sheet metal jobs previously done by hand, or using calculators or hand tools, are now done using _____. 1. ______________
2. Sheet metal thickness varies, but generally is less than _____. 2. ______________
   A. 1/8″
   B. 1/4″
   C. 1″
   D. 1 1/4″
3. True or False. Sheet metal is available in a limited number of sizes, dimensions, or gages. 3. ______________
4. Of the following gage numbers, which indicates the thickest metal? 4. ______________
   A. 18 gage.
   B. 20 gage.
   C. 22 gage.
   D. 26 gage.
5. American Standard Wire Gage is the standard measurement used for _____ metals and wire. 5. ______________
6. What is the purpose of coating steel? ______________________________

   ____

   ____
7. A(n) _____ is a combination of two or more metals. 7. ______________

Match each of the following descriptions with the correct term.

8. Has a reddish brown color and is an excellent conductor of electricity 8. ______________
9. Has a bluish black surface. 9. ______________
10. Is most commonly used in HVAC ducts, gutters, and downspouts. 10. ______________
11. Is not processed into its final shape or thickness until is has cooled. 11. ______________

A. Cold-rolled steel.
B. Copper sheet.
C. Sheet steel.
D. Galvanized steel.

12. Which of the following is true of food-safe metal? 12. ______________
    A. It can safely come in contact with food.
    B. It is used for kitchen utensils.
    C. It is used in dairy equipment.
    D. All of the above.

13. Brass is an alloy made of copper and _____. 13. ______________

14. What is the purpose of an HVAC system?______________________________

____

____

15. The best way to learn the sheet metal trade is through a(n) _____. 15. ______________

Chapter 2
Sheet Metal Safety

Objectives

After completing this chapter, you will be able to:

- Understand the importance of working safely.
- List general safety rules.
- Demonstrate safe use of sheet metal machines and tools.
- List safety rules as they apply to clothing, ladders, scaffolding, welding, soldering, and chemicals.
- Identify unsafe practices and work environments.

Technical Terms

Occupational Safety and Health Administration (OSHA)
personal protective equipment (PPE)
scaffolding
solvent
Underwriters Laboratories (UL)

Job Safety

Safety is based on knowledge, skill, and an attitude of care and concern. Workers in all industries must know the correct, safe way to perform their jobs. Working safely is especially important for sheet metal workers because they work with many dangerous materials, tools, and machinery. Sheet metal workers must be familiar with the potential hazards of their craft and understand how to reduce or eliminate them. For example, sheet metal can have extremely sharp edges. In order to reduce the chance of receiving a serious cut, wear gloves for protection when handling sheet steel.

A good attitude toward safety means believing safety is important and working safely at all times. Those who do not believe safety is important will take shortcuts or unnecessary risks in order to finish a job more quickly or less expensively. This is unwise. Shortcuts and speed often contribute to accidents, resulting in injuries, property damage, and sometimes death. Safety is a personal responsibility. Follow safety rules and procedures and refuse to work with those who do not.

Occupational Safety and Health Administration (OSHA): A government agency responsible for setting and enforcing work safety standards.

In 1971, Congress created the ***Occupational Safety and Health Administration*** (***OSHA***). OSHA is responsible for setting and enforcing work safety standards, **Figure 2-1**. In general, employers are required to provide a place of employment free of known hazards likely to cause death or injury. Employers must comply with safety and health standards enacted by OSHA. Employees are also required to comply with these standards.

Figure 2-1. An OSHA employee instructs students on safe ladder use. (Georgia Tech photo by Gary Meek)

General Safety Rules for the Shop and On Site

General safety rules apply to all industrial workers. Both the shop and the job site can be very dangerous. Sheet metal workers, who may work in a shop or on site at a job, have additional rules to follow. Be aware of those safety rules that apply to the shop and to on-site projects.

General Safety Rules

- Pay close attention to the work area. What work is being done? What areas should be avoided?
- Move carefully when walking or climbing.
- Lift with leg muscles, not the back, **Figure 2-2**.
- Move projects with care. Avoid other workers.
- Wear protective gloves, pants, and sleeves to avoid cuts and abrasions.
- Do not wear loose, flowing clothing. It can be caught in moving machinery parts.
- Understand and follow proper machine operation procedures for every machine used.
- When working around machinery, tie back long or loose hair and remove jewelry. These can be caught in moving parts.
- Disconnect power tools before adjusting.
- Keep power tool guards in place.
- Avoid interrupting others when they are using power tools.
- Remove hazards as soon as they are noticed.
- Complete American Red Cross certified first aid training.
- Keep an approved first aid kit on the job.
- Use a fireproof can to store shop rags covered in combustible materials, **Figure 2-3**.

Dressing with Safety in Mind

Personal protective equipment (***PPE***) is clothing, hats, face shields, shoes, and other items that protect a person's body from danger. The entire body must be properly protected in the work environment. PPE must meet OSHA standards.

personal protective equipment (PPE): Clothing, hats, face shields, shoes, and other items that protect a person's body from danger. PPE must meet OSHA standards.

Safety Shoes

Safety shoes are shoes or boots worn to protect the feet. They generally have steel toes, reinforced or steel soles, and slip-resistant soles. Steel toes protect feet from falling or rolling objects. Reinforced or steel soles prevent penetration by sharp objects. Slip-resistant soles protect workers from wet, slippery, or other hazardous surfaces.

Never wear canvas shoes of any type on the job. They are too fragile to protect feet. Avoid shoes with leather soles because they do not provide traction on smooth surfaces.

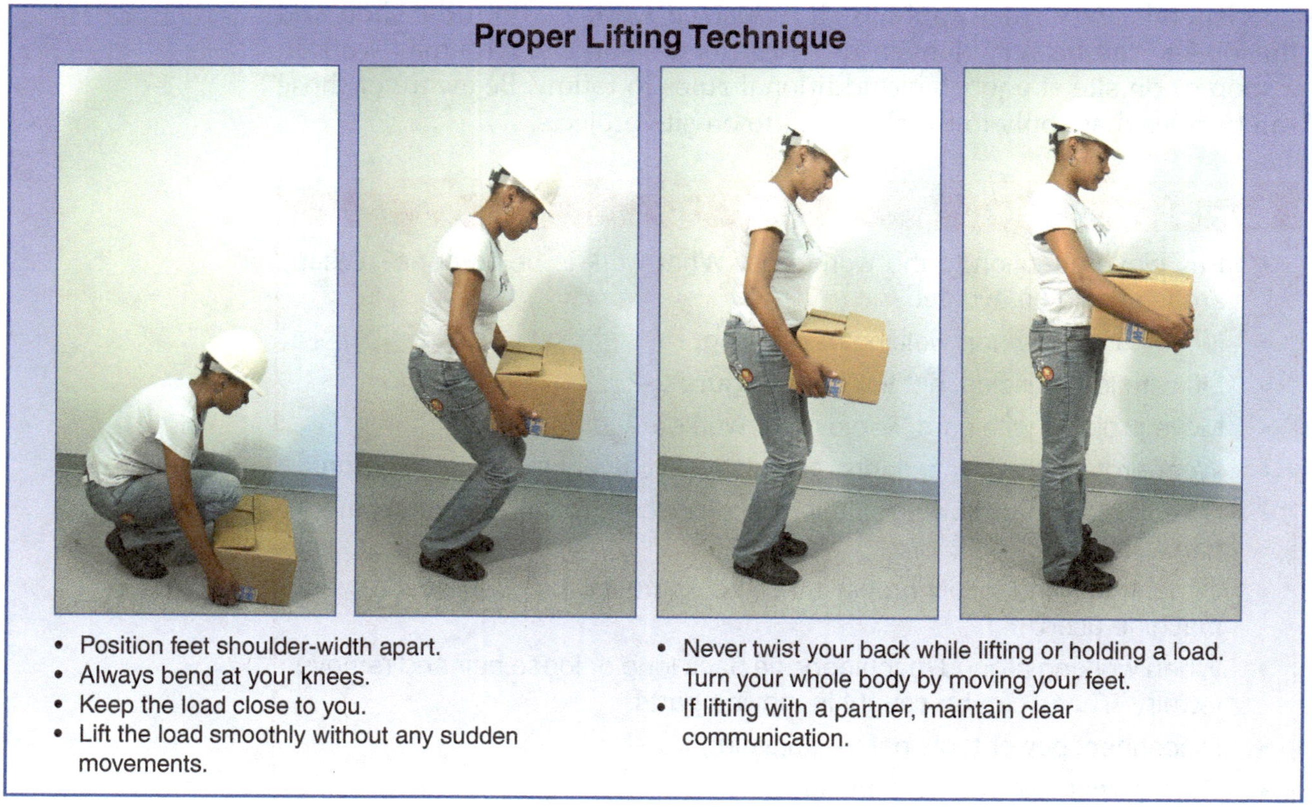

Figure 2-2. Always follow proper lifting procedures to reduce back strain.

Figure 2-3. Soiled shop rags are a fire hazard. Place them in a fireproof container after use. (Justrite Manufacturing Co.)

Gloves

Gloves are worn by sheet metal workers for several reasons. They protect hands from splinters on metal and other rough materials, protruding screws or rivets, and extreme surface temperatures on materials. Work gloves are commonly made of canvas, leather, or rubber.

Canvas gloves and leather gloves are used as general work gloves. They provide protection from dirt, sharp objects, and heat. Welders wear flexible, lightweight gloves to protect their hands from heat. However, the leather gradually hardens and stiffens from exposure to heat and ultraviolet light. To protect the gloves from as much heat as possible, use tongs or pliers to handle hot pieces of metal.

Safety Glasses

Safety glasses include traditional spectacles, goggles, and face masks, **Figure 2-4**. In order to meet OSHA standards, glasses must not shatter or otherwise break during testing. Testing requirements are set by the American National Standards Institute (ANSI). Safety glasses can be designated as either basic impact glasses or high impact glasses.

Wear safety glasses whenever eyes are exposed to a hazard. Much of the work done by sheet metal workers exposes the eyes to danger. Therefore, safety glasses are worn often in the trade.

Hard Hats

All hard hats are designed to protect the head from falling objects. Some hard hats also offer protection from electrical hazards. These hats are divided into classes based on the amount of electrical protection they offer.

Wear a hard hat:

- Whenever there is a danger of falling objects.
- In areas with exposed electrical conductors that might come in contact with the head.

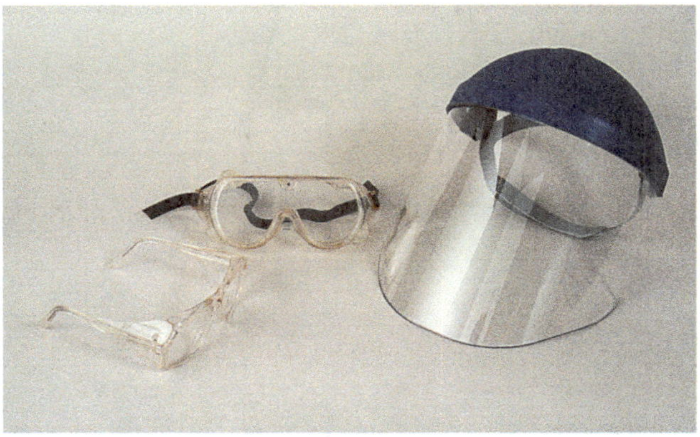

Figure 2-4. Safety glasses come in many styles. Glasses, goggles, and face shields are most often worn in the metal shop.

When hanging ductwork, for example, sheet metal workers stand on scaffolding or ladders. Any objects they hold, such as tools, nails, and ducts, become a hazard to those working below. Exposed electrical wires near the ceiling also are a hazard to those hanging ductwork. Hard hats must be worn in these instances.

Some hard hats are combined with face shields to protect both the eyes and face. The face shield is lifted and lowered, depending on the needs of the wearer.

Working Safely on Ladders, Scaffolding, and Scissor Lifts

Sheet metal workers often install ductwork, gutters, and other items that are located out of reach. This results in a great deal of working on ladders, scaffolding, and scissor lifts. Working above the ground is dangerous. Another danger is that items made of metal can conduct electricity. Therefore, following safety rules is very important when working with these products.

Ladder Safety

There are three types of ladders: stepladders, straight ladders, and extension ladders, **Figure 2-5**. Ladders may be made of wood, metal, or fiberglass. Because metal and water conduct electricity, *never* use a metal, metal-reinforced, or wet ladder where direct contact with an electrical power source is possible.

Straight ladder   Stepladder   Extension ladder

Figure 2-5. Three common types of ladders. (Werner Ladders)

Underwriters Laboratories (*UL*) is an independent, not-for-profit product safety certification organization. Products that pass their testing are said to be UL approved and a seal is affixed to the product. Look for this seal on any ladders used on the job.

Underwriters Laboratories (UL): An independent, not-for-profit product safety certification organization. Products that pass their testing are said to be UL approved and a seal is affixed to the product.

Safety Rules for Ladders

- Check the ladder before using it. Make certain it is in good working condition.
- Do not position a ladder where it can make contact with electrical lines. Check overhead for electrical lines *before* positioning the ladder.
- Properly place an extension ladder or a straight ladder against its support point, **Figure 2-6.**
- If there is any chance the ladder may slide, engage ladder safety shoes, **Figure 2-7**.
- Do not place a ladder in a doorway. If this placement is unavoidable, lock the door and have someone stand at the base of the ladder until the job is complete.
- Never place a ladder on an unstable base, such as boxes or a stack of wood, in an attempt to extend its height.
- Do not join two straight ladders to obtain more height. Find a longer ladder instead.
- Extension ladders must overlap at least three feet between sections.
- Lock spreader assembly on a stepladder and rest legs firmly on a solid footing.
- To raise an extension ladder or straight ladder, walk forward under the ladder as shown in **Figure 2-8**. Grasp other rungs while moving forward.
- Face the ladder when climbing it.
- Stay centered on the ladder. Do not lean past the side rails.
- Do not stand higher than the safe standing level noted on the ladder.
- Use a rope or line to raise tools and equipment. Do not carry tools while climbing the ladder.
- Remove tools from the top of the ladder before moving the ladder.
- When the job is complete, fold the ladder and store horizontally in a cool, dry location, **Figure 2-9**.

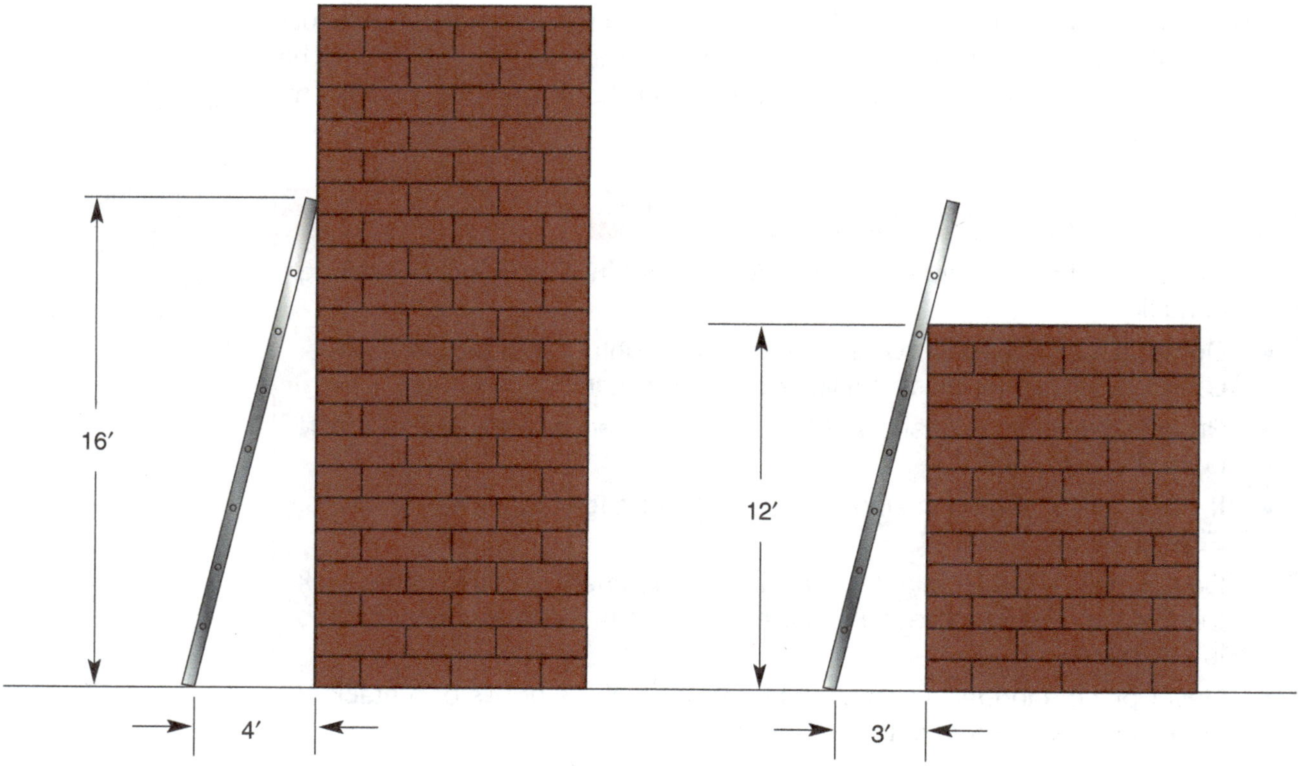

Figure 2-6. Using a ratio of 4:1, place base of ladder one foot out for every four feet of height to the support point.

Figure 2-7. Safety shoes prevent ladders from moving on slippery surfaces. (Werner Ladders)

Figure 2-8. The safe way to lift an extension ladder or a straight ladder.

Figure 2-9. To prevent sagging, store extension ladders and straight ladders horizontally, using supports. (Werner Ladders)

Scaffolding Safety

scaffolding: A temporary platform used to reach elevated work areas.

In commercial and residential construction, it is sometimes necessary to construct a temporary platform to reach elevated work areas. This platform is called ***scaffolding***, **Figure 2-10**. Because it is elevated and people will stand on it, scaffolding must be securely supported and constructed. Many scaffolds have wheels for maneuverability. Lock these during use to provide stability. Materials used to build scaffolding include metal for the frame and wood or metal for the platform. For this reason, use near power lines must be carefully monitored.

Scissor lifts are often used in place of scaffolding, **Figure 2-11**. Sheet metal workers often use them indoors. Basic scaffolding safety rules apply also to scissor lifts.

Safety Rules for Scaffolding

- Construction of the scaffolding should be done by an experienced, skilled scaffolding builder.
- Check the plumb and level as the scaffolding is built or assembled.
- Follow all local, state, and federal codes regarding scaffolding.
- Check the scaffolding closely each day before use.
- Install guardrails, toeboards, and screens on the platform area.
- Maintain the required distance from any power lines to avoid electrocution hazard.
- Do not place makeshift ladders or other devices on the scaffolding to increase its height.
- Use planking that is heavy enough to handle the load to be placed on it.
- Planking must not extend more than 12″ to 18″ (305 mm to 460 mm) beyond supports.
- Remove all items from mobile scaffolding before moving it.
- Keep platform height on rolling scaffolding at or lower than four times the smallest base dimension.
- Store the scaffolding in a safe location when not in use. Do not block any passageways.

Welding Safety

Welding is used in sheet metal work to fasten sheet metal pieces and can be very dangerous work. The light created by the arc is harmful to the eyes. This brilliant light can also cause a severe sunburn on exposed skin. Melting metals create toxic fumes. High temperatures and sparks can cause fires. Dressing safely for the job is important, **Figure 2-12**. Study the following safety guidelines and use them on the job.

Figure 2-10. Properly constructed scaffolding. (Werner Ladders)

Figure 2-11. Follow scaffolding safety rules when working on a scissor lift. (Genie Industries)

Figure 2-12. This welder is dressed properly for the job.

Safety Rules for Welding

- Cover all skin, including neck and forearms. Wear a leather jacket, pants, and gloves when welding. Wear a leather welding cap if welding overhead.
- Do not carry a lighter or matches.
- Never look at an arc with the naked eye. This can damage the eye.
- Wear a helmet with a proper shade to protect the eyes from light and the face from sparks.
- Do not weld metals containing cadmium, zinc, beryllium, or lead without fume extractor or respirator.
- Weld in a well-ventilated area only. If the area lacks proper ventilation, wear a respirator to avoid toxic fumes.
- If using shielding gas in an enclosed area, good ventilation is necessary to replenish air supply.
- Keep the work area clear of combustible materials.
- Properly maintain welding equipment.
- Keep a fire extinguisher handy.
- Post a fire watch if working on one side of a large assembly, and the other side is hidden from view.
- Post a fire watch when work must be done around combustible materials.

Soldering Safety

Soldering is used to join pieces of metal together or to prevent leakage in containers. While welding and brazing are now more widely used in sheet metal work, soldering continues to be a useful process in some applications.

The tip of the soldering iron or copper can reach 500°F (260°C). Burns and fires are always a danger. Learn how to properly operate soldering equipment before working on any job.

Safety Rules for Soldering

- Handle soldering iron carefully. Avoid the tip.
- Use a piece of metal or an insulated board as a rest for the soldering iron.
- Work in a well-ventilated area. Soldering fumes are dangerous.
- Do not handle soldered materials until cool.
- Do not use solder containing lead on any materials that will be used to store or transport food or water.

Working Safely with Chemicals

A variety of chemicals are used in sheet metal work to clean surfaces, remove rust, and to dissolve other substances. These substances include solvents, solutes, and fluxes.

Solvents can be corrosive and can damage the skin on contact. Flux, used in soldering, can create harmful gases that must be properly ventilated. All of these substances must be stored and disposed of properly.

solvent: A liquid that dissolves another substance and forms a solution.

Every community has disposal methods in place for environmentally hazardous materials. Check with the local municipality's sanitation department to learn proper disposal methods.

Safety Rules for Working with Chemicals

- Store all chemicals properly: in proper containers, in a safe location, locked if necessary. Follow all applicable laws.
- Wear a face shield to protect eyes from splashing substances.
- Cover all skin including arms, hands, and neck.
- Work in a well-ventilated area or wear a respirator, if harmful gases will be created.
- Know the location of the nearest eye washing station.

Summary

A good attitude toward safety is important. Safety is based on knowledge, skill, and an attitude of care and concern. Know the correct and proper procedures for performing work assignments. Be familiar with potential hazards and how they can be reduced or eliminated. Understand OSHA standards as they apply to the sheet metal trade.

Work safety begins with protecting the body. Wearing safety shoes, gloves, safety glasses, and hard hats helps reduce the risk of injuries and death.

Follow all general safety rules that apply to industrial work environments. In addition, follow rules that address specific dangers in a metalworking environment. Learn first aid.

Ladders and scaffolding are used often in sheet metal work. They are both sources of many accidents. Understand how to use them properly. The UL label helps identify safe ladders.

Scaffolding is particularly prone to abuse and misuse. Understand the safety rules outlined to work safely on scaffolding.

Welding and soldering are very dangerous due to the high temperatures required for operation. Learn and follow all safety rules. The chemicals used in welding and soldering are environmentally hazardous. They must be stored, handled, and disposed of in a specific manner. Follow these rules to protect yourself and the environment.

Internet Resources

Underwriters Laboratories

www.ul.com

Occupational Health and Safety Administration

www.osha.gov

The Hand Tools Institute

www.safety.hti.org/index.htm

The Power Tool Institute

www.powertoolinstitute.com/safety.html

Outside Activities

1. Arrange to talk with a sheet metal worker about safety. Ask if you can conduct the interview on the job site or in the shop. Have a list of questions prepared before arriving. Include questions about OSHA standards that apply to the shop or job site, what PPE he or she wears, if there has ever been an accident at the shop. Also observe safety precautions being used or in place.
2. Take an inventory of the school shop. Are all safety measures in place? If not, what is missing? As a class, discuss what will need to be done to fix the safety hazards. What will the cost be?

Name ____________________ Class ____________ Score ________

Chapter 2
Sheet Metal Safety

Review Questions

Write your answers in the spaces provided.

1. Which of the following is *not* true regarding shop safety? 1. ________
   A. A worker who is unconcerned with shop safety may take shortcuts or unnecessary risks.
   B. Shop safety is a personal responsibility.
   C. A coworker who is unconcerned about safety poses no danger to you.
   D. Shop safety is based on an attitude of care and concern.
2. What does the acronym OSHA stand for? ____________________
3. What is the purpose of OSHA? ____________________
4. True or False? Safety rules apply to the shop only. There are no safety rules for working on site. 4. ________
5. Name two safety rules to follow regarding first aid. ____________________
6. Personal protective equipment includes ____. 6. ________
   A. shoes
   B. hats
   C. face shields
   D. All of the above.
7. Wear a ____ whenever there is a danger of falling objects. 7. ________
8. Name three types of ladders. ____________________
9. Underwriters Laboratories is a(n) ____. 9. ________
   A. governmental department
   B. safety certification organization
   C. chemical organization
   D. insurance organization.

10. A temporary platform constructed to reach elevated work areas is called _____. 10. ____________

11. Why should you *never* look at a welding arc with the naked eye? ____________

____

____

12. True or False? Good ventilation is not necessary in a shop where welding is done. 12. ____________

13. _____ is used to prevent containers from leaking. 13. ____________

14. _____ can be corrosive and damage the skin or eyes on contact. 14. ____________

15. True or False? The gases created by fluxes are not harmful. 15. ____________

Chapter 3
Measurements and Layouts

Objectives

After studying this chapter, you will be able to:
- Use precision measurement tools.
- Use precision layout tools.
- Read a micrometer.

Technical Terms

caliper
center punch
combination bevel
combination set
combination square
depth gage
dial indicator
divider
firm-joint caliper
graduation
hermaphrodite caliper
hook rule
micrometer
plug gage
prick punch
screw pitch gage
shrink rule
slide caliper rule
snap gage
spring-joint caliper
steel rule
surface gage
swinging-blade protractor
thickness gage
try square
Vernier scale

Using Layout and Measurement Tools

Many sheet metal projects are mass produced using very precise measurements. A slight error in measurement can result in the production of thousands of ill-fitting parts. Understanding how to measure and lay out metal parts is an important skill for the sheet metal worker. A variety of tools are used to make measurements and lay out metal parts.

Steel Rule

steel rule: A metal measuring tool divided into graduations that represent measurement points.

graduation: A line on a steel rule that represents points of measurement.

hook rule: A steel rule with an arm attached to one end that grabs the end of the work piece and allows for hands-free measuring.

The ***steel rule*** is the most common measuring tool used in sheet metal work. Steel rules contain a series of lines, or ***graduations***, that represent the points of measurement, **Figure 3-1.** They use either the U.S. Customary system (inches) or the metric system of measurement. The United States uses the U.S. Customary system, while many other countries use the metric system.

Steel rules vary in length from 1/4″ (7 mm) to more than 12′ (4 m). The 6″ (152 mm) rule is a convenient rule to use in tight spaces. On even smaller work, smaller rules are needed, **Figure 3-2.** A type of steel rule, the ***hook rule***, has an arm attached to one end, **Figure 3-3.** The hook grabs the end of the piece, allowing for hands-free measuring.

Steel rules are normally used to measure flat pieces of metal. But in some instances, rules must be used to measure stock that is not in sheet form. For example, the internal diameter of round stock can be measured with a steel rule, **Figure 3-4.** The external diameter can also be measured using the method shown in **Figure 3-5.**

When using a steel rule, it is a good practice to begin measuring at one inch from the end of the rule. This provides a more accurate measurement because the end of the rule may be worn down from use. Remember to subtract one inch to obtain the correct reading!

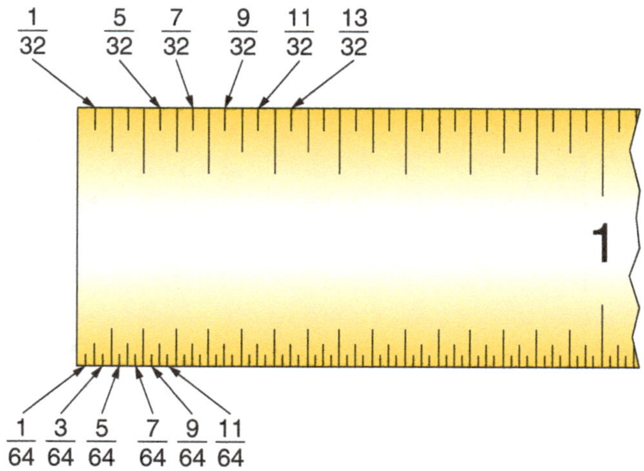

Figure 3-1. Graduation lines on a rule. The divisions are 32nds of an inch on the top of the rule and 64ths of an inch on the bottom of the rule.

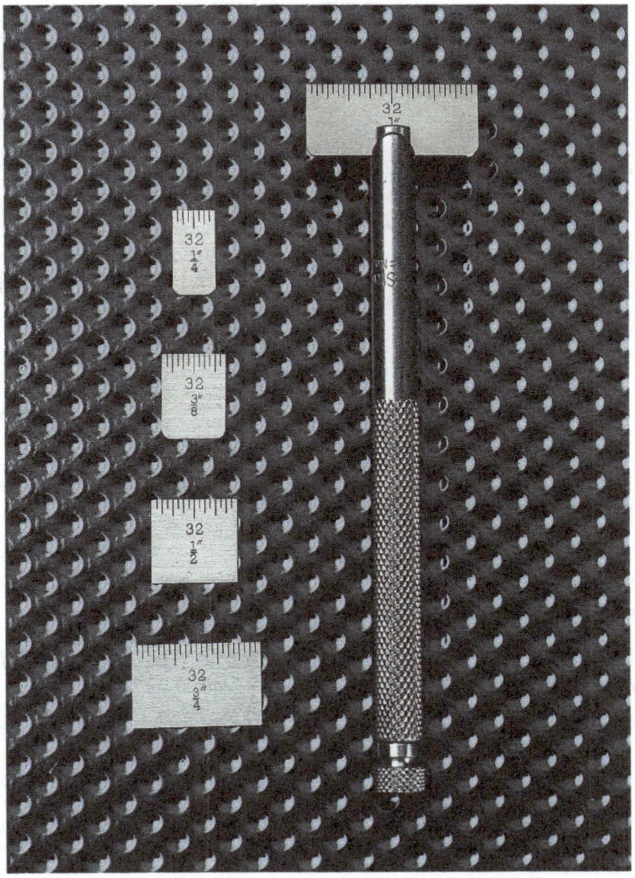

Figure 3-2. Set of small rules with holder. These rules fit where an ordinary rule would not fit. (L.S. Starrett)

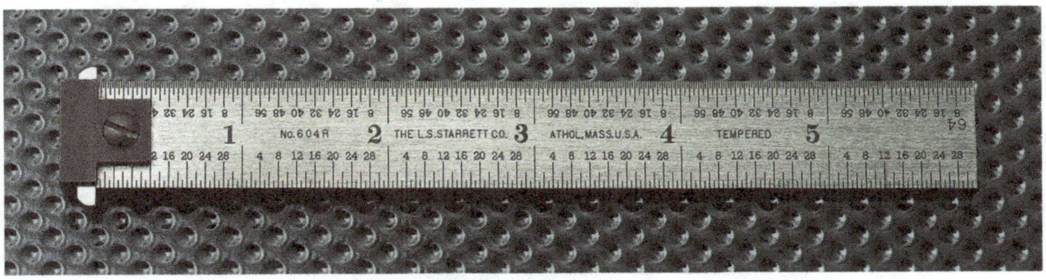

Figure 3-3. A hook rule. (L.S. Starrett)

Steel Rule Measurements

Place the rule correctly in order to avoid errors in measurement.

1. Set the rule to the center of the graduation line, **Figure 3-6**.
2. Measure from that point to the end of the piece being measured.
3. Remember to subtract one inch, if measuring from the one-inch graduation line.

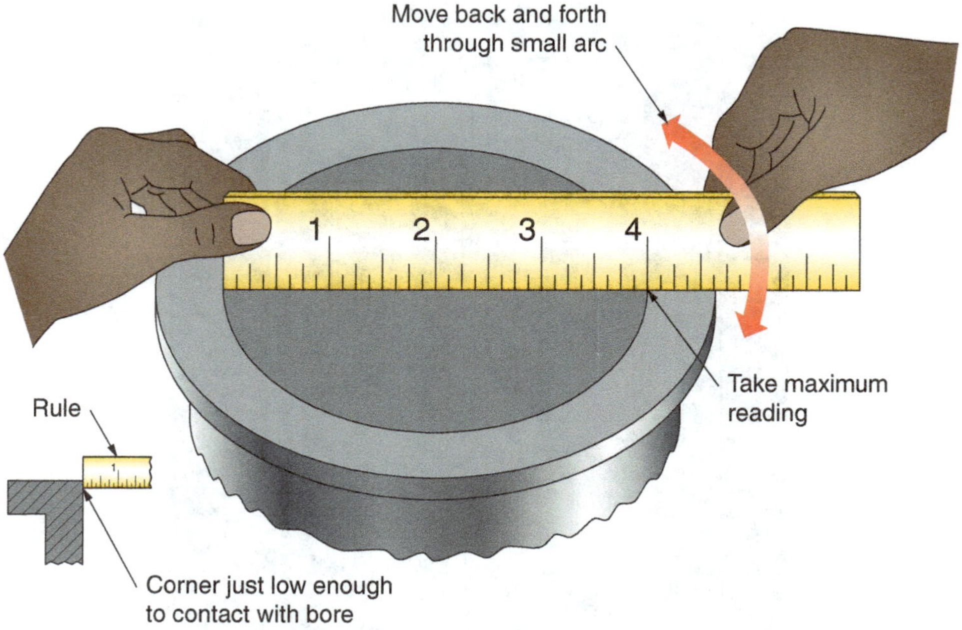

Figure 3-4. Follow this procedure to measure an internal diameter with a steel rule. This method will provide a very accurate measurement.

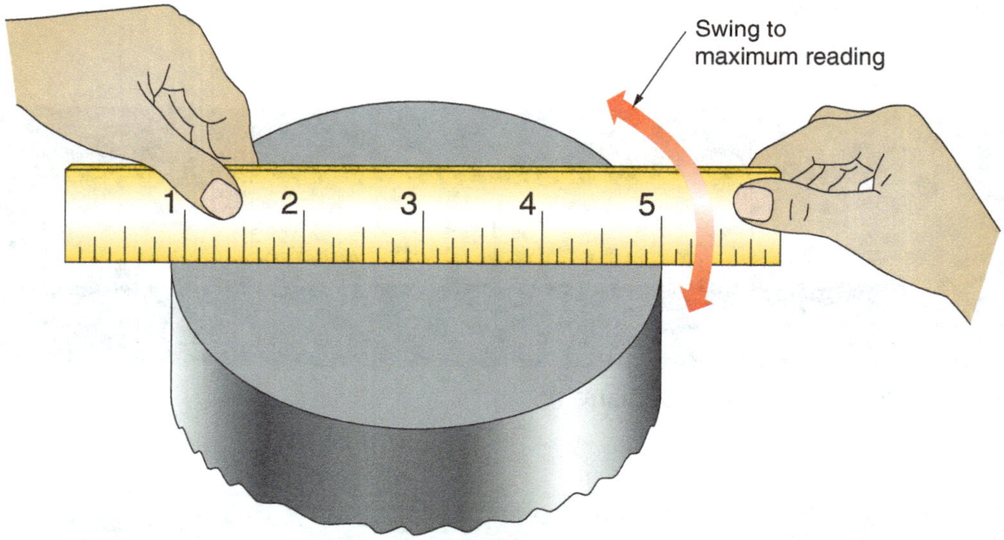

Figure 3-5. Measuring the external diameter of round stock with a steel rule.

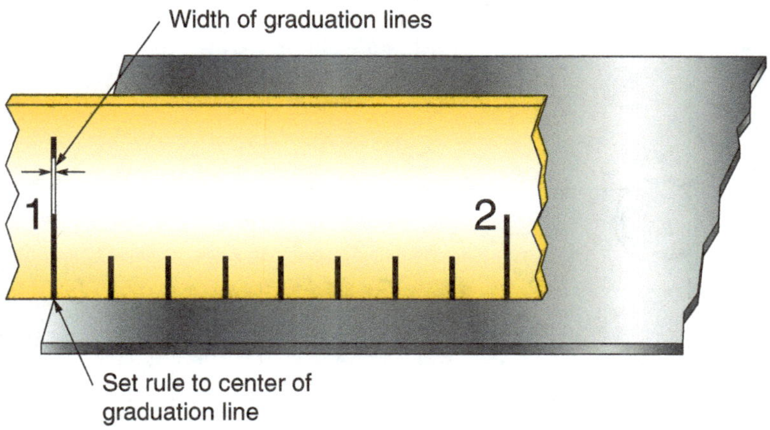

Figure 3-6. This rule is enlarged to illustrate the width of graduation lines. Place the rule correctly to avoid errors in measurement.

Shrink Rule

All molten metals shrink as they cool. This shrinkage has to be accounted for in order to make properly sized finished pieces. Patterns are made oversized. Measurements are made using a ***shrink rule***. A shrink rule appears to be the "normal" length, but is actually slightly longer to allow for shrinkage. For example, a 12″ (305 mm) shrink rule used to measure steel is actually 12 1/4″ (312 mm). As steel cools, it will shrink 1/4″ (7 mm) for every 12″ (305 mm) in length. The amount of shrinkage varies with different types of metal. Shrink rules come in various sizes. Use the shrink rule that corresponds to the metal being used.

shrink rule: A rule made slightly longer than ordinary. Extra length allows for shrinkage of cooled metal.

Slide Caliper Rule

A ***slide caliper rule*** is used to make inside and outside measurements on round or flat stock. It is able to take a more accurate measurement than a steel rule. The slide caliper rule is basically a rule with two jaws attached, **Figure 3-7.** One jaw is fixed and the other jaw moves.

slide caliper rule: A measuring tool used to make inside and outside measurements on round or flat stock.

Making Measurements with a Slide Caliper Rule

1. Place stock between the two jaws.
2. Close the jaws until they lightly touch the sides of the piece.
3. Find the graduation that lines up with the jaw on the right. This is the measurement of the piece.

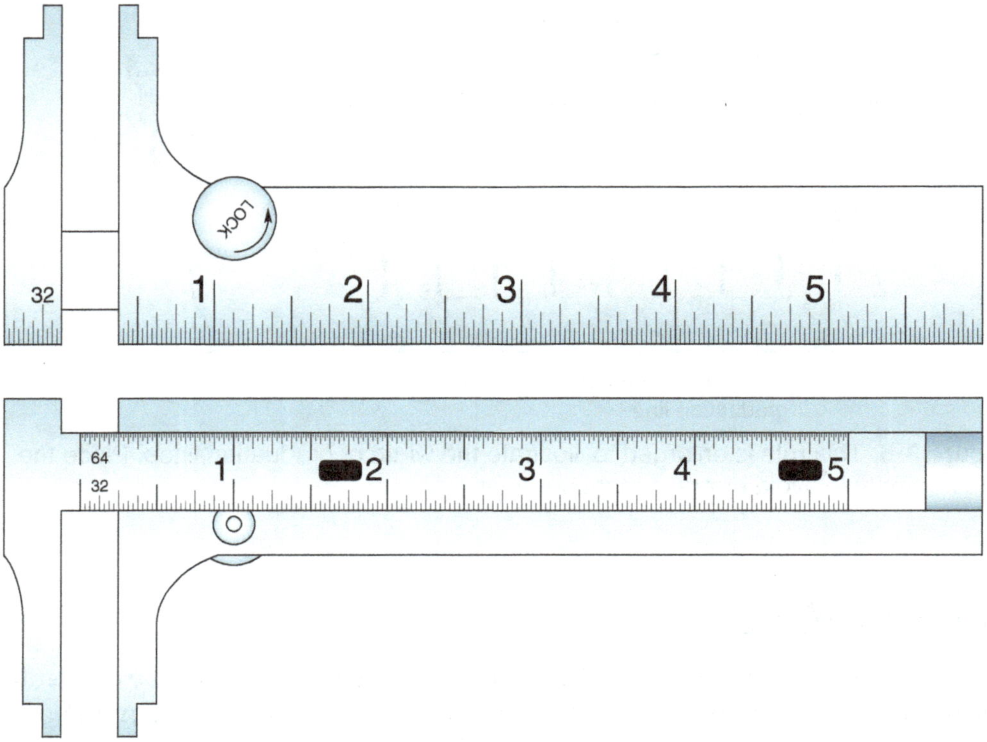

Figure 3-7. These slide caliper rules are small enough to slip in a pocket.

Try Square

try square: A tool used to check the accuracy of 90° angles and to mark 90° angles when laying out patterns.

A ***try square*** is used to check, or *try*, the accuracy of 90° angles, **Figure 3-8.** It is also used to mark 90° angles on work.

The try square blade and stock can wear to the point where they need replacement. Take the try square apart and refit worn parts. After refitting, measure the angle of the try square to ensure it is at 90°. This testing should

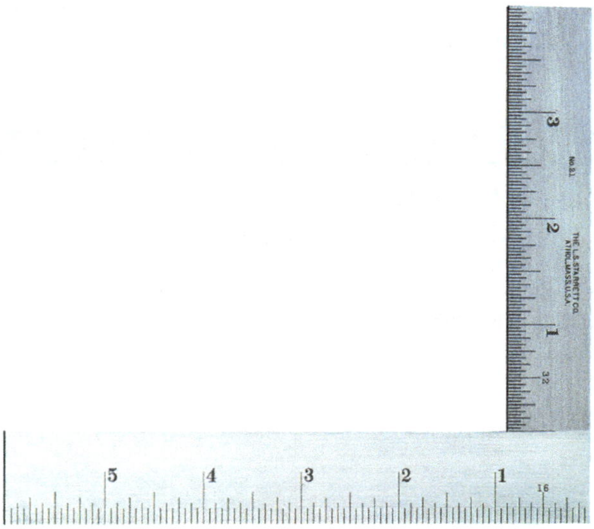

Figure 3-8. The entire outside edge of this try square blade is graduated for use in measuring. Try squares also are made without graduation lines. (L.S. Starrett)

also be done as part of routine maintenance. A try square angle not set to 90° will result in incorrect work. The try square is a precision instrument and should be handled with care.

Testing the Squareness of a Try Square

Place the square in Position A and scribe Line A along the blade.

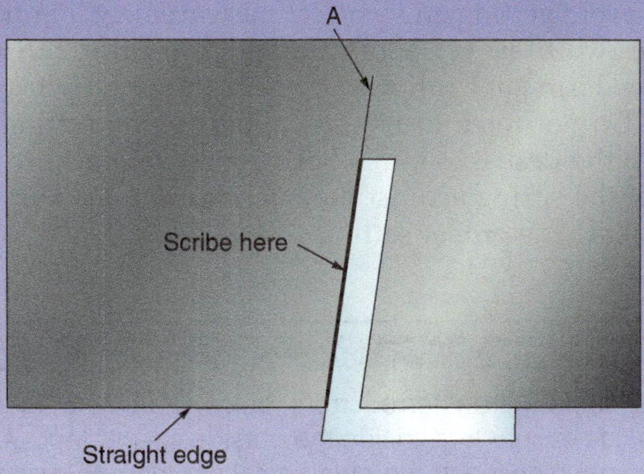

Reverse the square to Position B and scribe Line B.

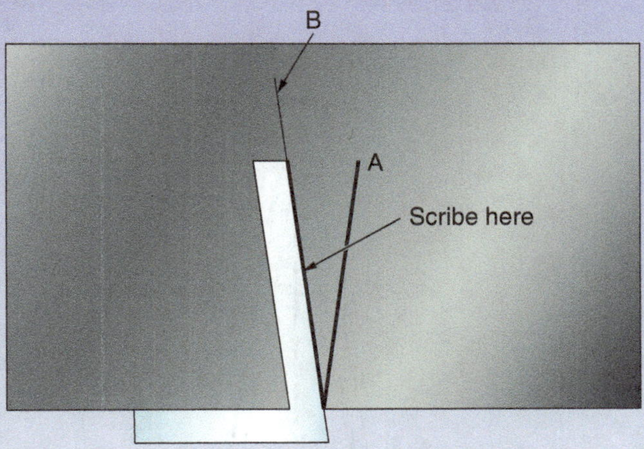

If Lines A and B are not parallel, the try square is not set to 90°. The error is half the angle between the two lines.

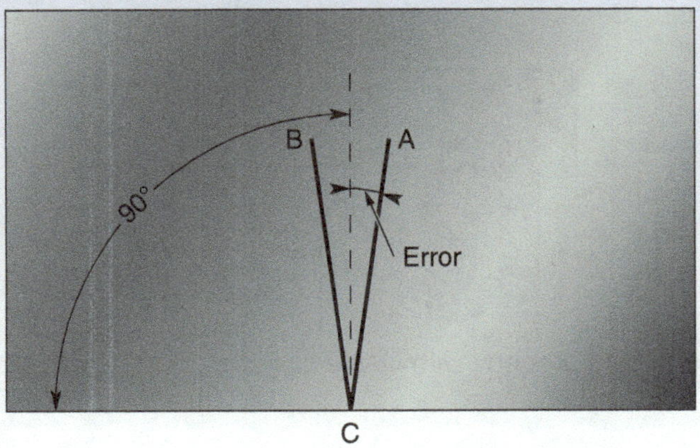

Calipers

caliper: A measuring tool used to take internal and external measurements of diameters, thicknesses, hole depth, and distances between pieces.

firm-joint caliper: A caliper that is held in place by friction at the joint of the two legs.

spring-joint caliper: A caliper that is held in place by the tension of the spring against the nut.

Calipers are used to take internal or external measurements. They can be used to determine a measurement or to set a desired measurement. Diameters, thicknesses, holes, and distances between pieces are all measurements that can be taken using calipers. Calipers do not have graduation lines, so they must be used with a rule. Hold calipers straight across the piece being measured to avoid getting an incorrect measurement, **Figure 3-9.**

There are two basic types of calipers: ***firm-joint calipers*** and ***spring-joint calipers***. Firm-joint calipers are held in place by friction at the joint, **Figure 3-10.** Do not force firm-joint calipers when setting. Instead, gently tap one leg to the desired setting. Spring-joint calipers are held in place by the tension of the spring against the nut, **Figure 3-11.** They are adjusted by turning the nut to the proper setting.

Transferring Settings

Settings can be transferred from outside calipers to inside calipers. Place the inside caliper legs inside the outside caliper legs. Adjust the inside caliper legs until they touch the outside caliper legs, **Figure 3-12.**

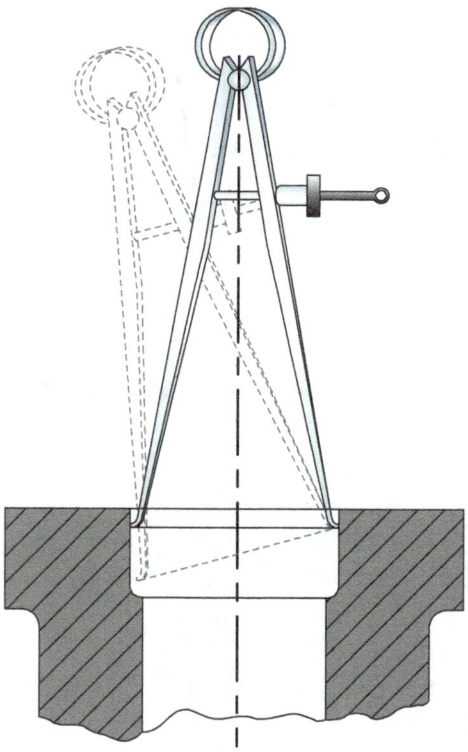

Figure 3-9. Measuring properly with inside calipers.

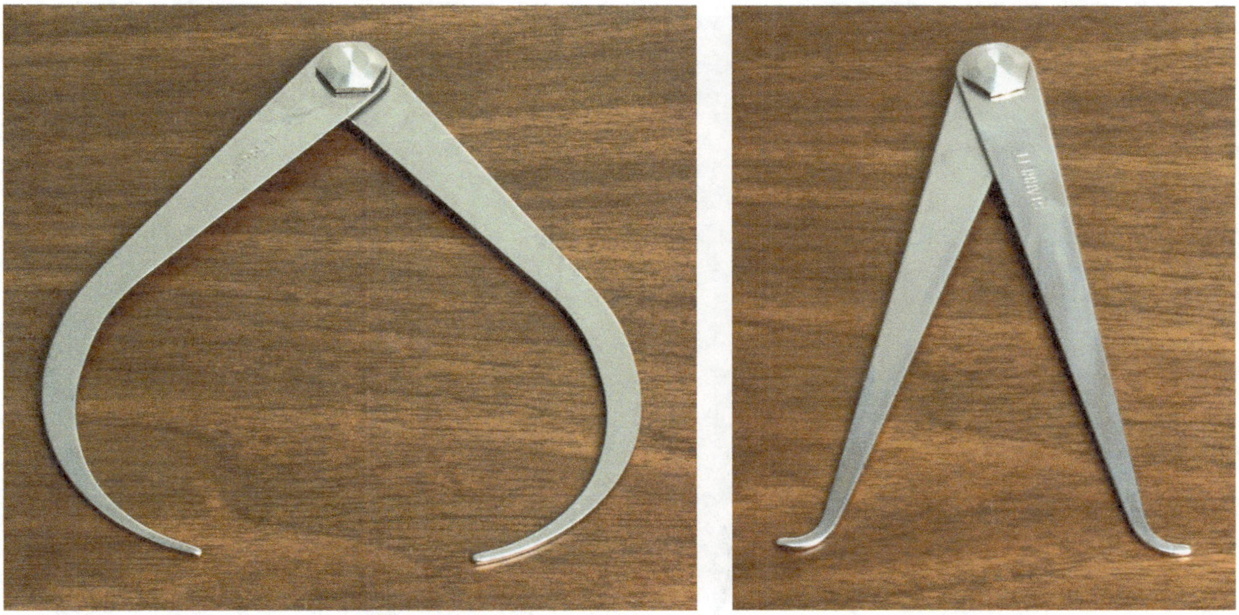

Outside Inside

Figure 3-10. Firm-joint calipers.

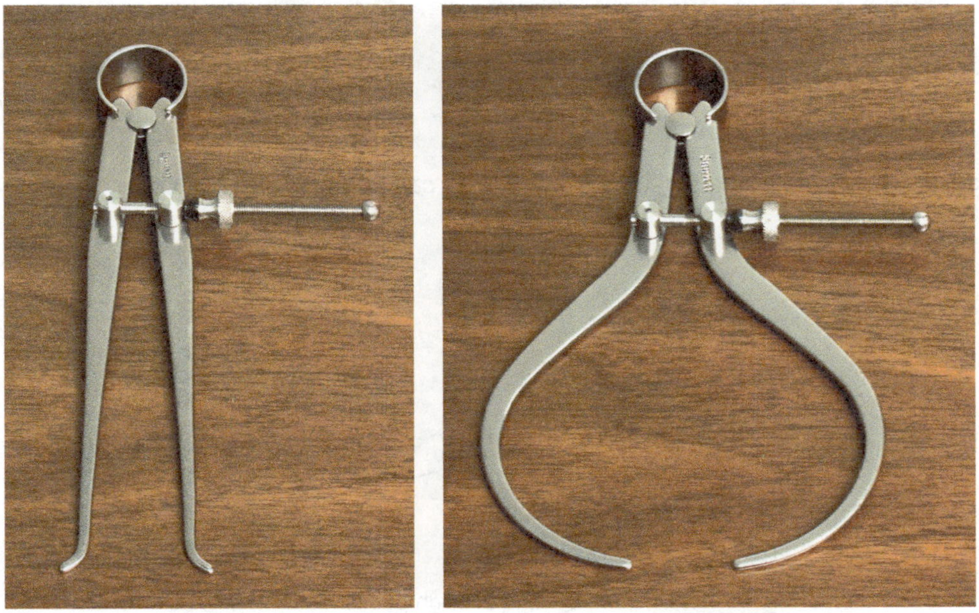

Inside Outside

Figure 3-11. Spring-joint calipers.

Hermaphrodite calipers have a straight, pointed leg and a curved leg. These calipers are used to scribe lines on surfaces parallel to edges or to locate center points on round stock. Set hermaphrodite calipers as shown in **Figure 3-13.**

hermaphrodite caliper: A caliper that has a straight, pointed leg and a curved leg. It is used to scribe lines on surfaces parallel to an edge and to locate center points on round stock.

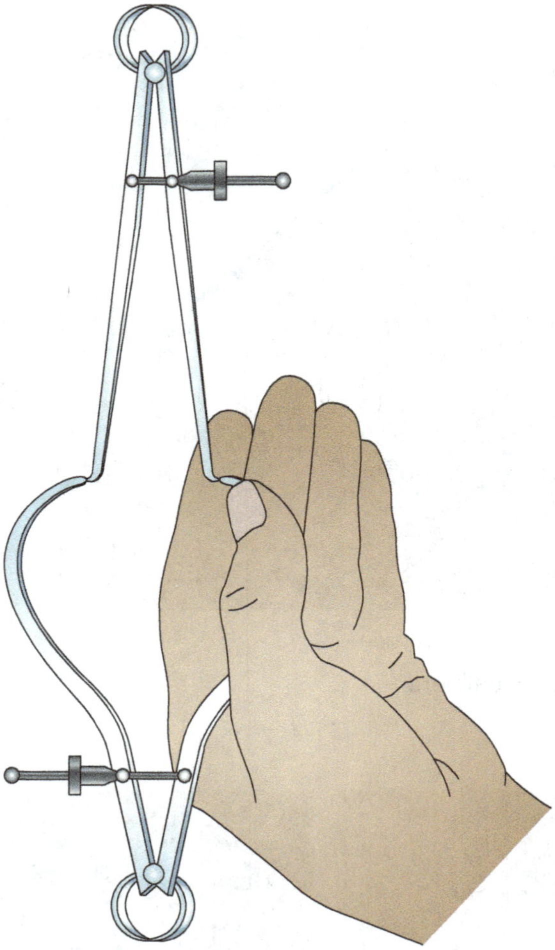

Figure 3-12. Transferring a setting from outside calipers to inside calipers.

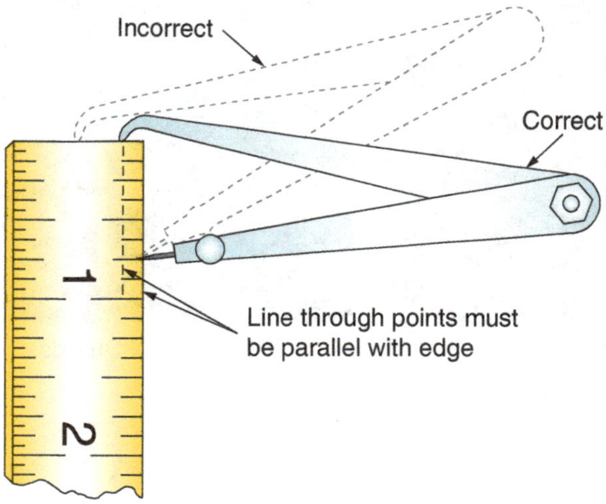

Figure 3-13. Correct and incorrect method for setting hermaphrodite calipers.

Dividers

Dividers are similar in construction to calipers, but both legs end in points, **Figure 3-14.** They are used to transfer measurements from one surface to another. They are also used to scribe arcs and circles in layout work. This is especially useful in sheet metal layouts. Dividers can also be used to locate center points on round stock.

divider: A tool used to transfer measurements from one surface to another and to scribe arcs and circles in layout work.

Check the sharpness of divider points before using. If they are dull, sharpen them on an oilstone. The sharpened point should be straight and in the center of the leg. Move the point back and forth on the stone in a circular motion, presenting a fresh surface to the stone at all times, **Figure 3-15.**

Combination Square

The ***combination square,*** **Figure 3-16,** can be used as a level, a try square, and a miter square. Many combination squares contain scribers that, once removed, are used to scribe lines on sheet metal. The combination square can

combination square: A tool used to check angles, measure lengths, heights, and depths. A scriber contained in the square can be used to scribe lines on work.

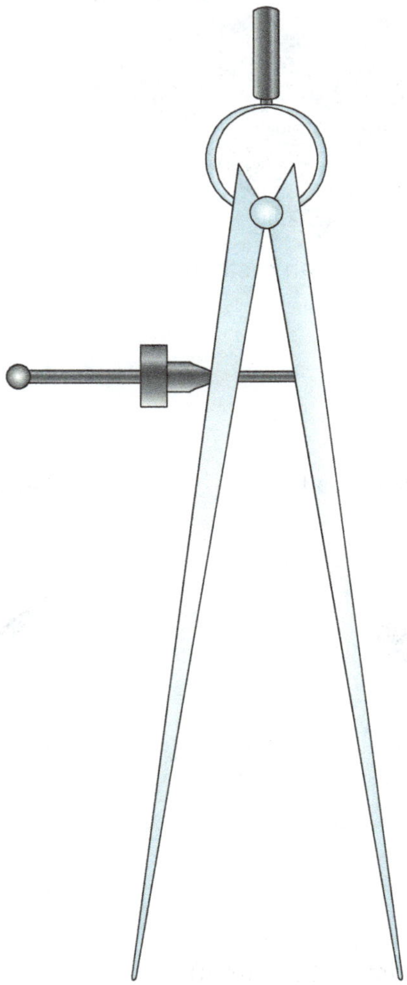

Figure 3-14. A spring divider is made with round, highly polished steel legs. The measuring points come together evenly.

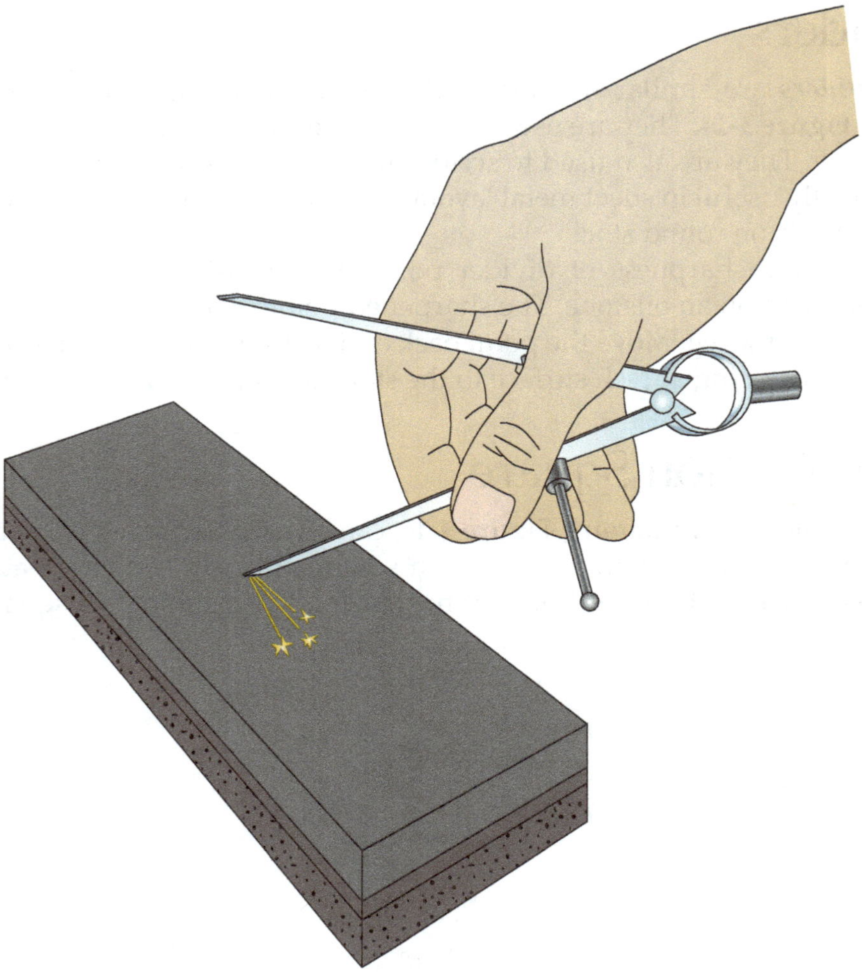

Figure 3-15. Sharpening a divider on an oilstone.

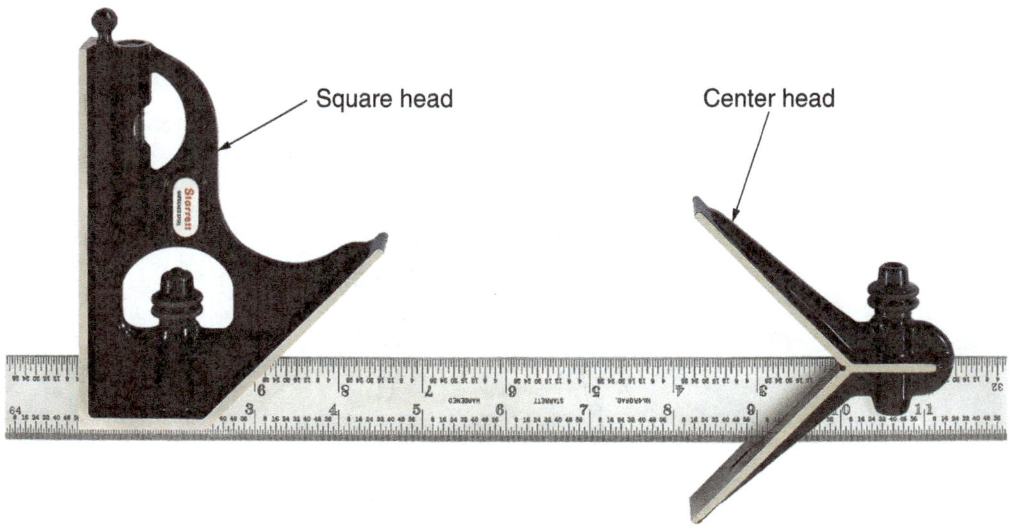

Figure 3-16. Combination square. (L.S. Starrett)

also be used as a height gage or a depth gage, **Figure 3-17.** The head of the combination square slides along a groove in the rule. It can be removed and the rule used separately.

Combination sets, **Figure 3-18,** contain a center head, a square head, and a protractor head. The different heads can be used for different tasks. For example, use the square head to check angle accuracy. Use the center head to locate the center point on round stock.

combination set: A combination square that contains a center head, a square head, and a protractor head. Different heads are used for different tasks.

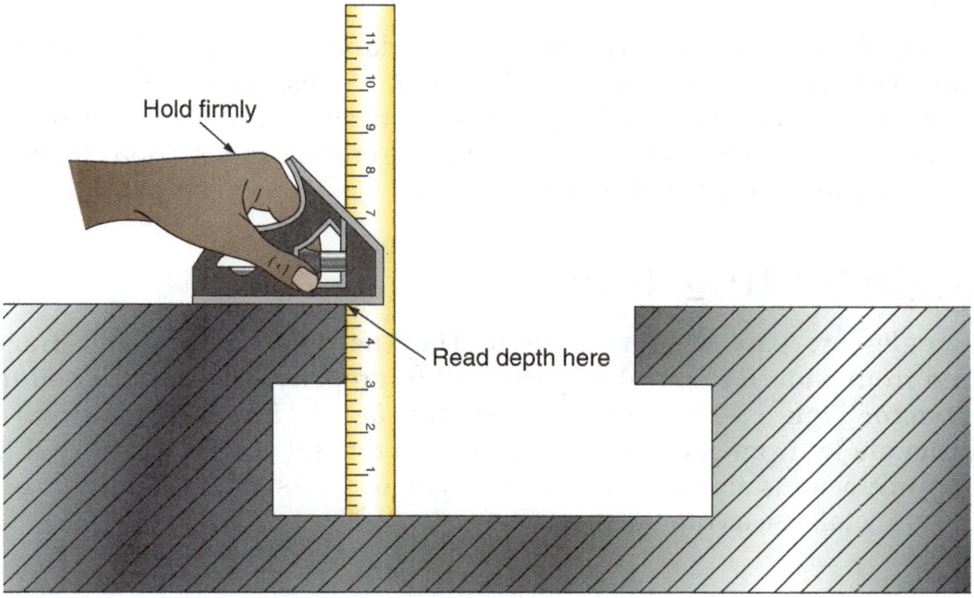

Figure 3-17. Using a combination square as a depth gage.

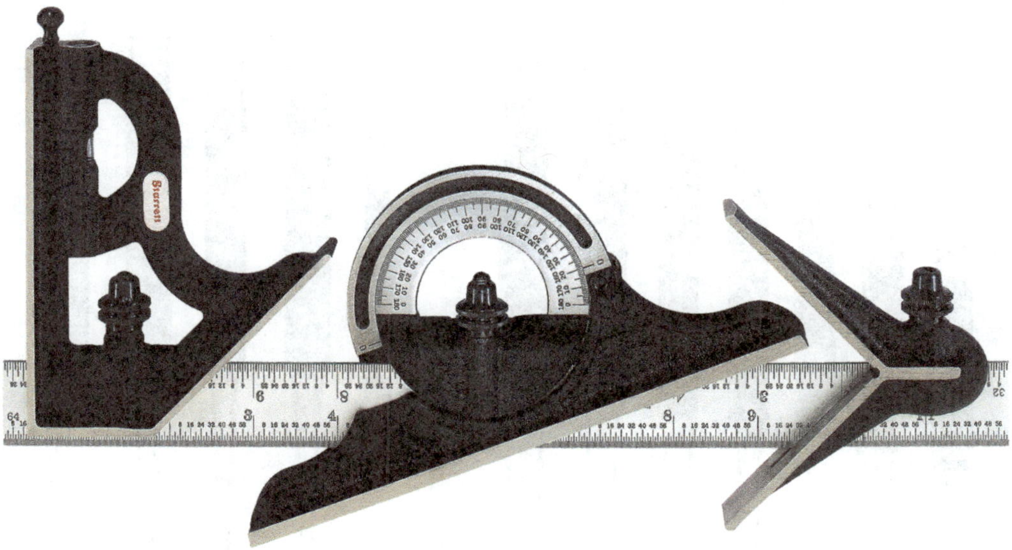

Figure 3-18. Combination set. (L.S. Starrett)

Layout Dye

It is difficult to see scribed lines on sheet metal. Layout dye makes these lines easily visible. Apply the dye with a brush or by spraying onto a clean metal surface. It provides an opaque background that eliminates glare, prevents eye strain, and covers any scratches already on the metal. Once dry, lines can be scribed on the metal and will be much easier to see.

Punches

prick punch: A punch used to make small indentations in metal and to locate points for dividers and fold lines in sheet metal work.

center punch: A punch used to make indentations in metal and on 18-gage and thicker sheet metal.

swinging-blade protractor: A thin plastic or metal tool on which degrees zero through 180 are marked. It is used to draw angles and to check angles.

combination bevel: A measuring device used to transfer and to set angles.

The ***prick punch*** and ***center punch*** are marking tools used in layout work, **Figure 3-19.** The prick punch is used to mark the intersection of centerlines in a layout. The point of a prick punch is more sharply angled than the tip of a center punch. The center punch is used to enlarge indentations in metal and on 18-gage or thicker sheet metal.

Angle-Setting Tools

Angles are part of every pattern. They need to be accurately transferred from pattern to metal so parts will fit properly. The ***swinging-blade protractor,*** **Figure 3-20,** and the ***combination bevel,*** **Figure 3-21,** are used to transfer angles from one surface to another and to set angles. The swinging-blade protractor is also used to square small jobs. To square the corners of a piece of metal, use a common square.

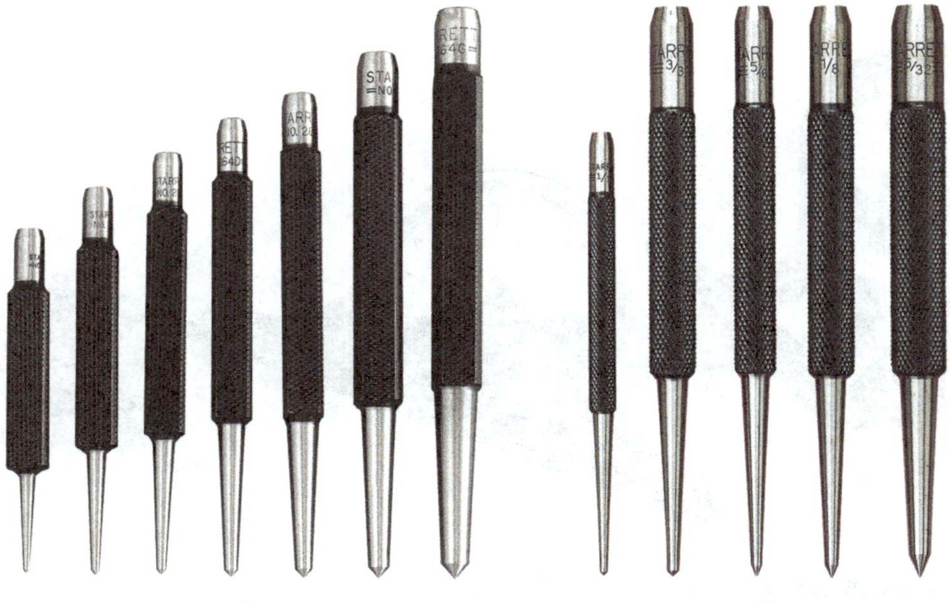

A
Center punches

B
Prick punches

Figure 3-19. Marking tools used for layout work. (L.S. Starrett)

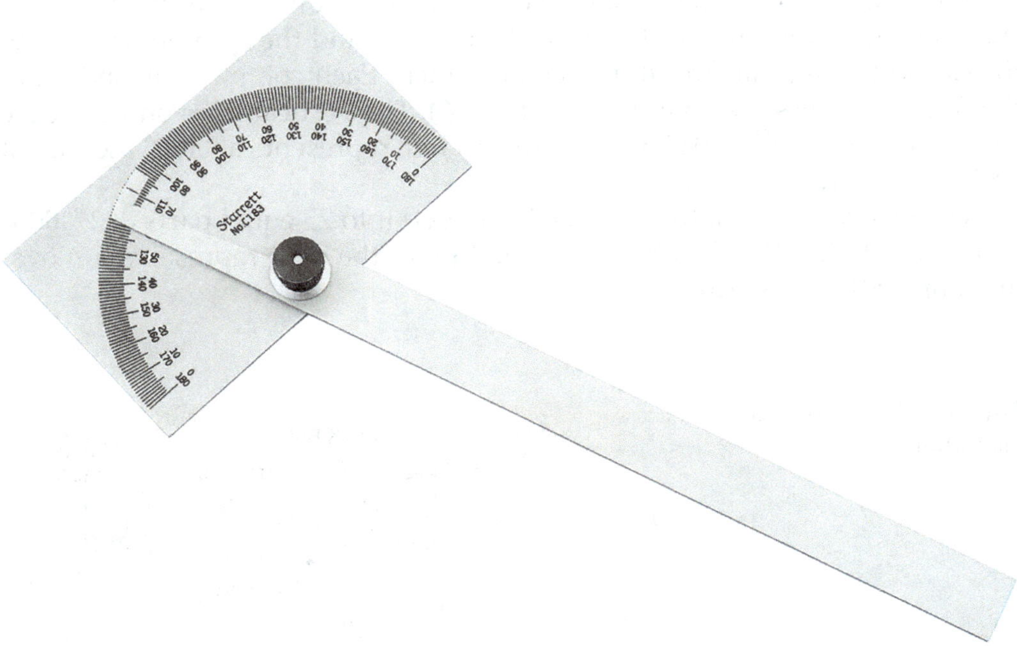

Figure 3-20. Swing-blade protractor used in layout work. (L.S. Starrett)

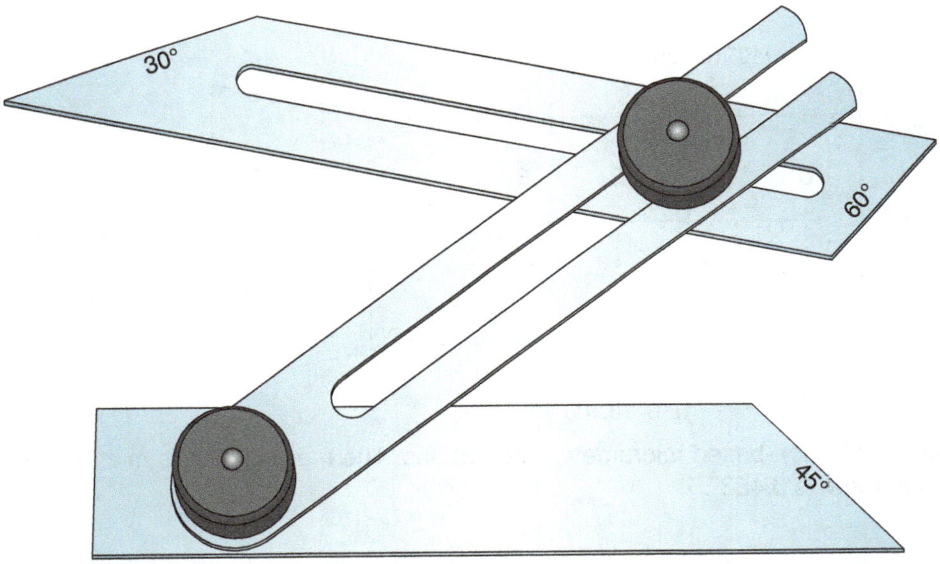

Figure 3-21. The combination bevel is made to swing over the stock and be clamped at any angle.

Micrometers

Micrometers are precision instruments capable of measuring to 1/1000″ (0.001″) and 1/100 mm (0.01 mm). Micrometers that contain a ***Vernier scale*** can measure to 1/10,000″ (0.0001″) and 1/500 mm (0.002 mm). While there are many types of micrometers, the outside micrometer is the type most commonly used in sheet metal work.

micrometer: A precision instrument capable of measuring to 1/1000″ (0.001″) and 1/100 mm (0.01 mm). The outside micrometer is the type most commonly used in sheet metal work.

Vernier scale: A graduated scale used to measure lengths and angles.

The major parts of the outside micrometer are shown in **Figure 3-22.** Not shown is the screw thread that joins the thimble and the spindle. The screw thread on the spindle has 40 threads per inch. Each complete revolution of the thimble moves the spindle 1/40″ (0.025″). The vertical lines on the sleeve each represent 1/40″ (0.025″). Every fourth line represents a 0.100″ increment and is numbered.

The beveled edge of the thimble is divided into 25 equal parts of 1/1000″ (0.001″). Rotating the thimble from one line to the next represents a movement of 0.001″. See **Figure 3-23.**

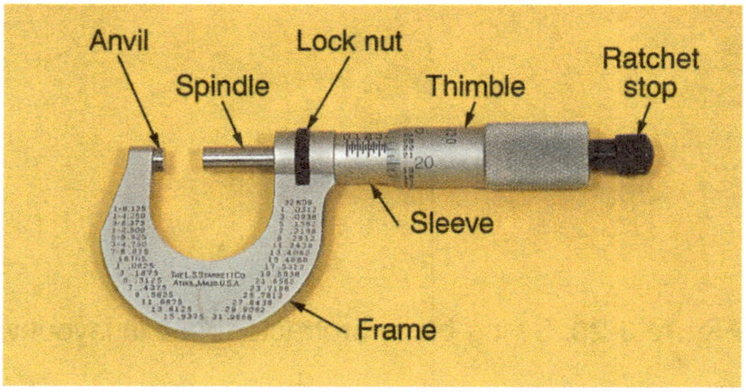

Figure 3-22. Parts of a micrometer.

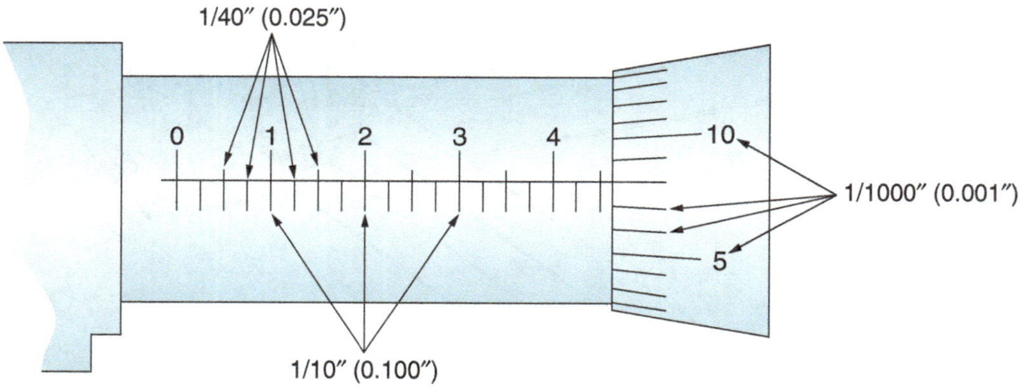

Figure 3-23. Inch-based micrometer graduations. The reading for this micrometer measurement is 0.458″.

Reading a Micrometer

1. The visible "1" on the sleeve represents 0.100″.

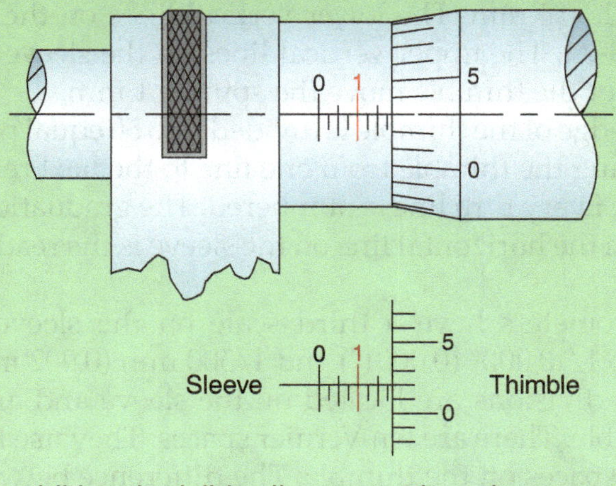

2. The three additional visible lines on the sleeve represent 0.075″. (3 x 0.025″ = 0.075″).

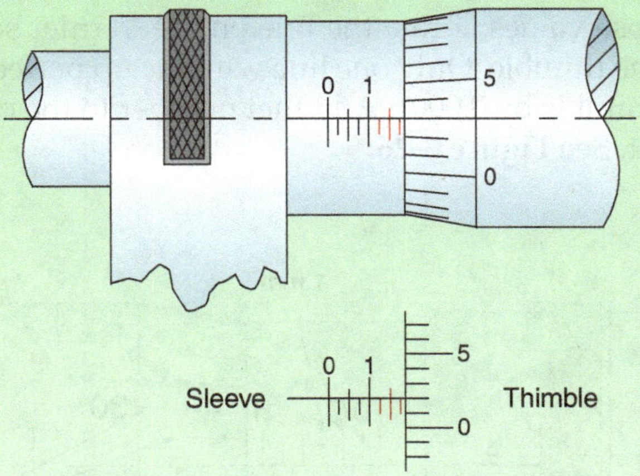

3. The horizontal line on the sleeve coincides with "3" on the thimble and represents 0.003″ (3 x 0.001″ = 0.003″).

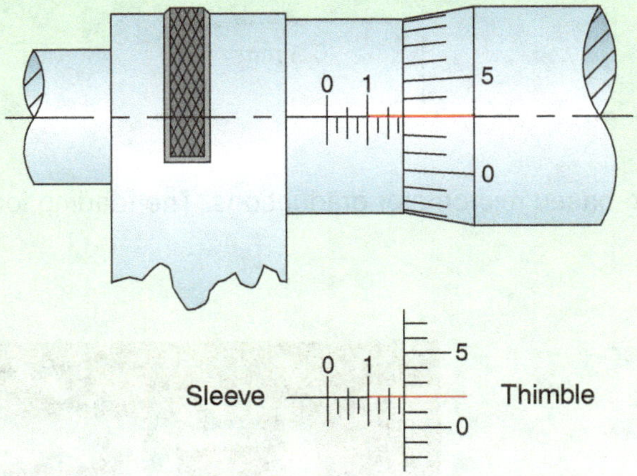

4. The reading on the micrometer is the sum of the three values, or 0.178″.

The measurements will vary from these on metric-based and Vernier micrometers, but they are read in the same way.

Metric-based micrometers can make measurements to 1/100 mm (0.01 mm). They work and are read in the same manner as the inch-based micrometers.

The spindle screw pitch is 0.5 mm. One complete revolution of the thimble moves the spindle 0.5 mm. The lower vertical lines on the sleeve represent 0.5 mm, **Figure 3-24.** The upper vertical lines on the sleeve represent 1 mm. Two revolutions of the thimble move the spindle 1 mm.

The beveled edge of the thimble is divided into 50 equal parts of 1/100 mm (0.01 mm). Rotating the thimble from one line to the next represents a movement of 0.01 mm. Every fifth line is numbered. The graduation on the thimble that lines up with the horizontal line on the sleeve is the reading in 1/100 mm (0.01 mm).

Vernier micrometers have a third scale on the sleeve that can make measurements to 1/10,000″ (0.0001″) and 1/500 mm (0.002 mm), **Figure 3-25.** The Vernier line divisions are located on the sleeve and are parallel to the lines on the thimble. There are ten Vernier spaces. They use the same amount of space as nine spaces on the thimble. The difference between the width of one Vernier space and one thimble space is 1/10,000″ (0.0001″).

To read a Vernier micrometer, follow the directions outlined previously. After finding those values, locate the line on the Vernier scale that lines up with a scale on the thimble. Only one line will line up perfectly. Multiply that number on the thimble by 0.0001. Add that number to the other numbers for the measurement. See **Figure 3-26.**

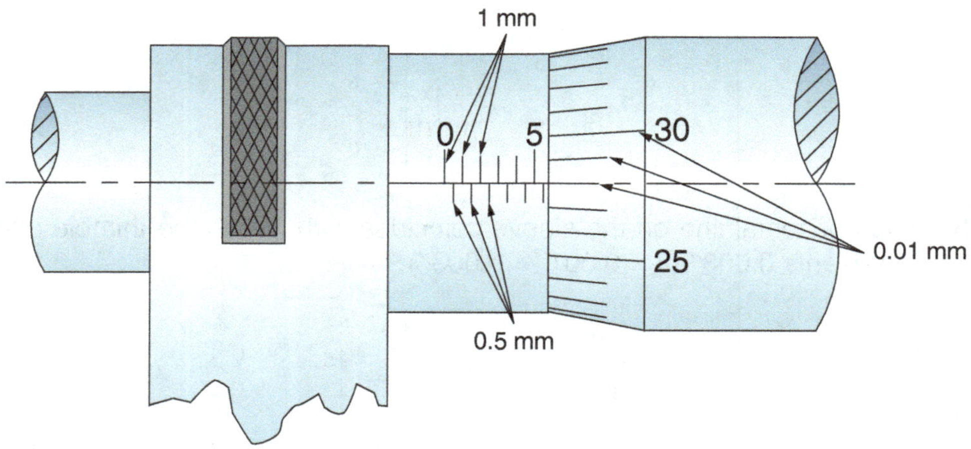

Figure 3-24. Metric-based micrometer graduations. The reading for this micrometer is 5.78 mm.

Figure 3-25. Vernier micrometer.

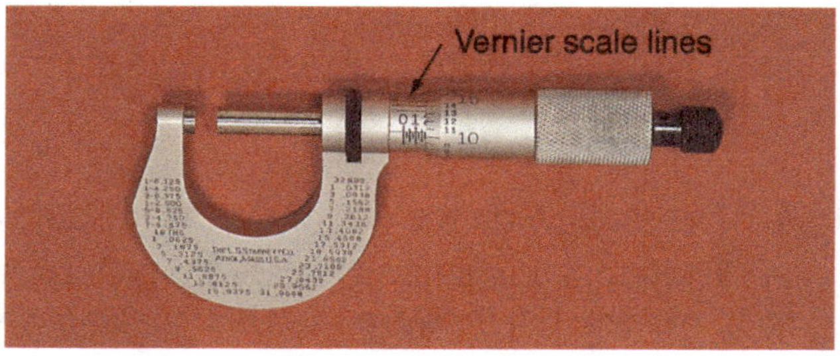

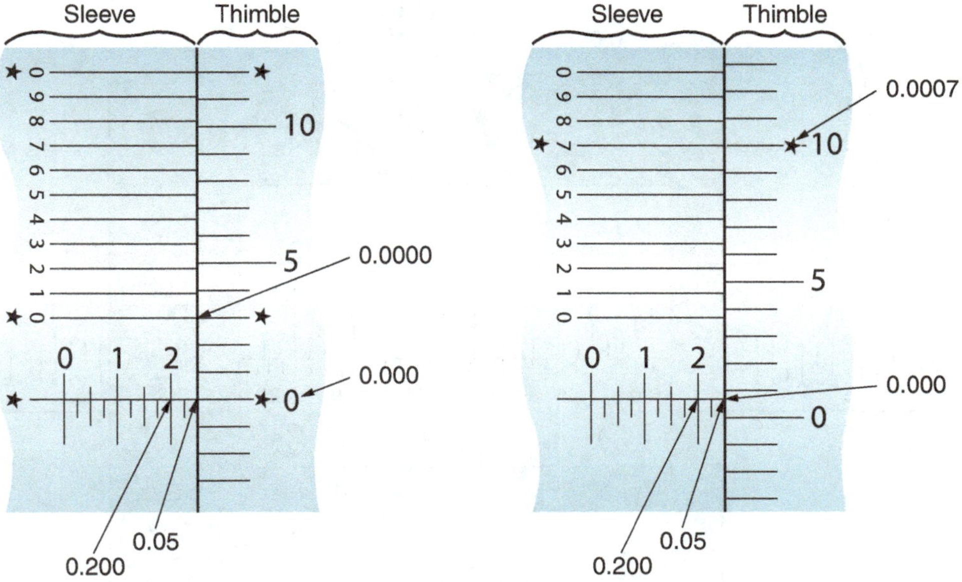

Figure 3-26. The readings for the Vernier micrometers are 0.02500″ on the left and 0.02507″ on the right.

Vernier Calipers

Calipers are used for internal and external measurements. Like other Vernier measuring tools, a Vernier caliper contains a third scale. This scale allows for precise measurements. On both inch-based and metric Vernier calipers, the Vernier scale can be either 25-division or 50-division, **Figure 3-27.** The technique for reading the scale is the same for either division.

The Vernier caliper reading in **Figure 3-28** is 1.436″. Add the inches (1.000), tenths of inches (0.400″), fortieths of inches (0.025″) visible between zero on the bar and zero on the Vernier scale. Then find the graduation on the Vernier scale that lines up with a line on the bar (0.011″). Add the four values to obtain the measurement.

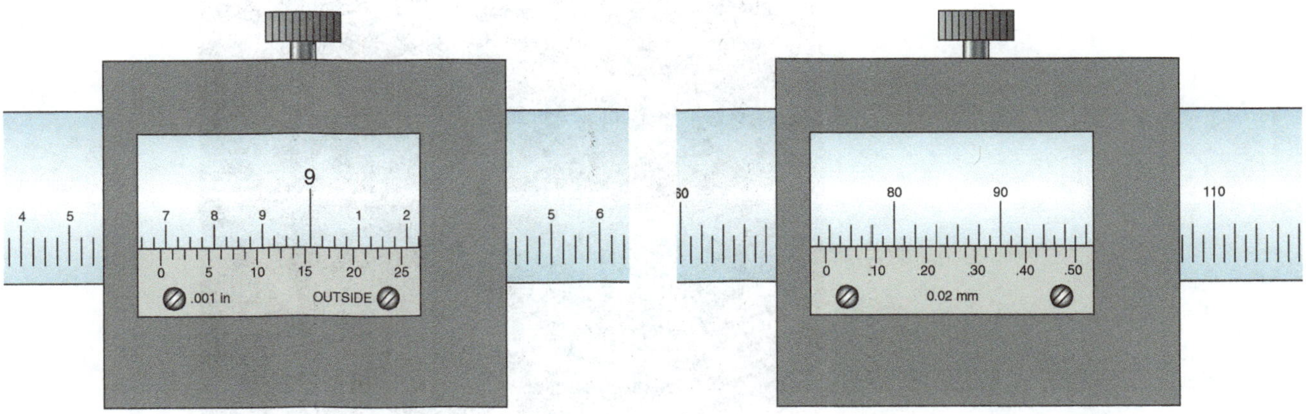

25-division Vernier scale

50-division Vernier scale

Figure 3-27. A Vernier scale can contain 25 divisions or 50 divisions.

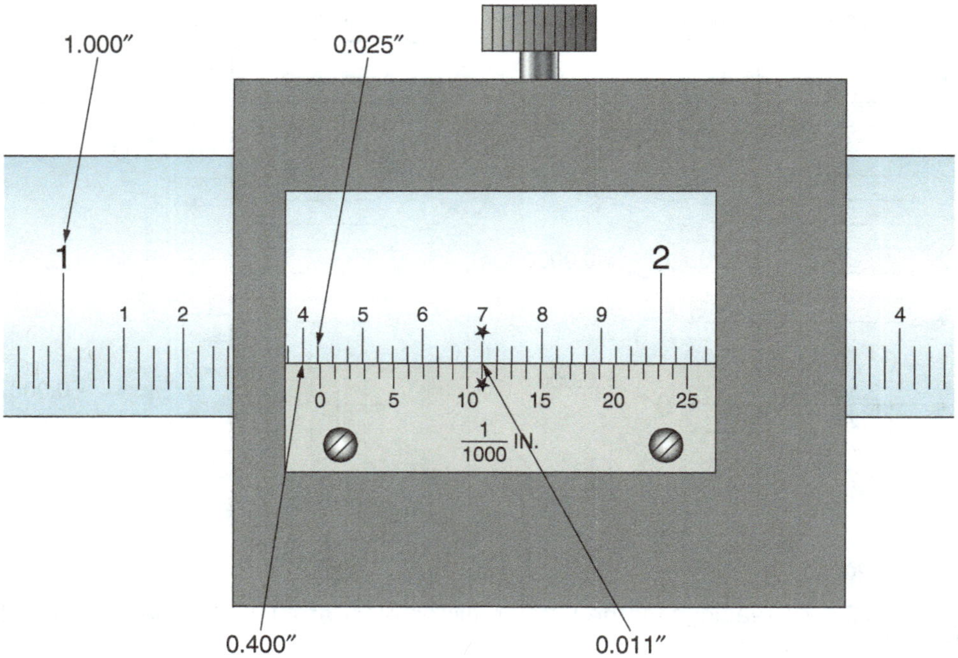

Figure 3-28. The reading on this Vernier caliper is 1.436″.

Gages

surface gage: A gage used to accurately scribe lines and transfer measurements.

The ***surface gage*** is used to scribe lines accurately and transfer measurements, **Figure 3-29.** It is often used with a surface plate for precision work and set with the aid of a combination square. Be certain that the instruments and surface plates are absolutely clean.

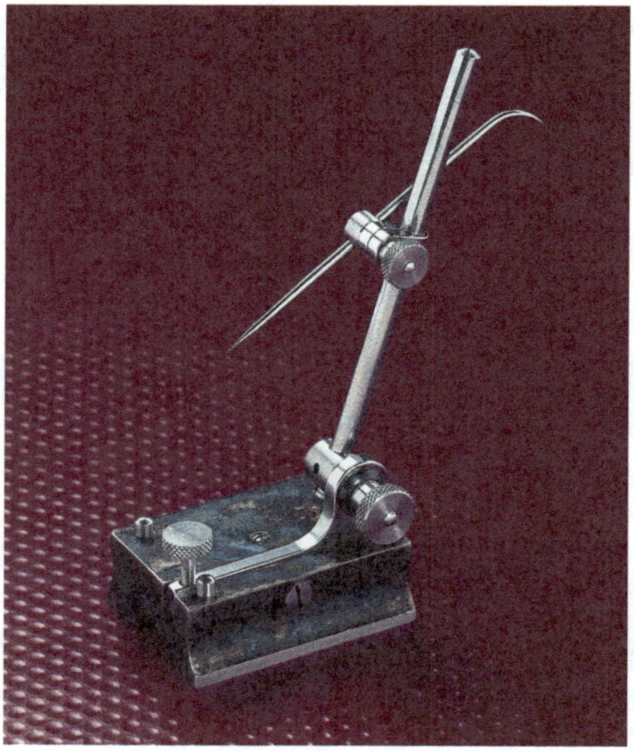

Figure 3-29. Small surface gage. (L.S. Starrett)

The ***depth gage*** is used to measure the depth of holes, slots, and other hard-to-reach places, **Figure 3-30.** Measurements can be made in U.S. Customary or metric. A Vernier depth gage is also available, for more precise measurements.

depth gage: A gage used to measure the depth of holes, slots, and other hard-to-reach places.

Thickness gages are used in the automotive, diesel, aircraft, and service industries to make adjustments and to check work, **Figure 3-31.** Like other gages, measurements can be made in U.S. Customary or metric.

thickness gage: A gage used to make adjustments and to check clearances.

The ***screw pitch gage*** is used to determine the pitch of various screw threads, **Figure 3-32.** Match the teeth of the gage with the thread being measured and read the pitch from the leaf. Screw pitch gages are available in a wide range of sizes.

screw pitch gage: A gage that measures the pitch of screw threads.

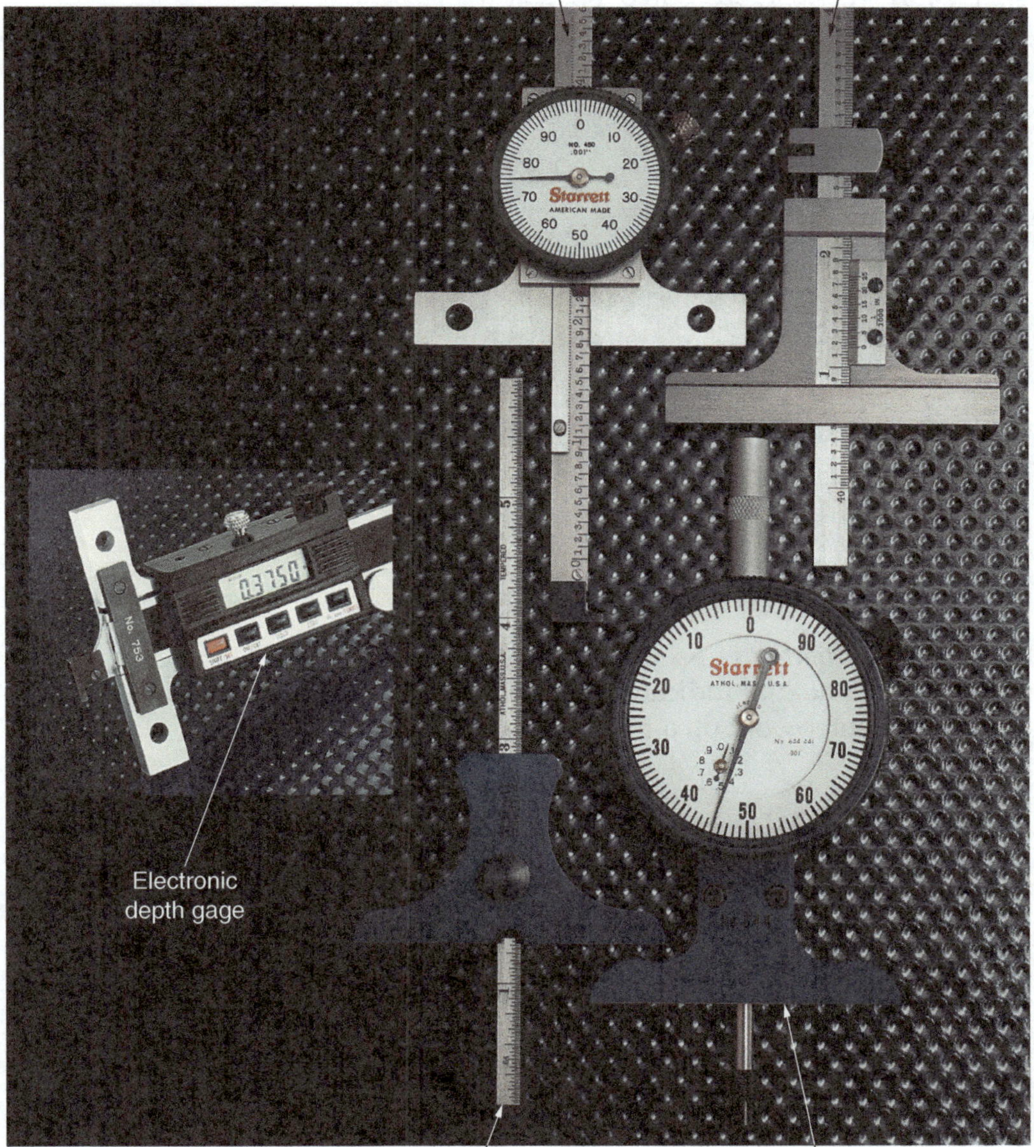

Figure 3-30. Depth gages are made in a variety of styles. (L.S. Starrett)

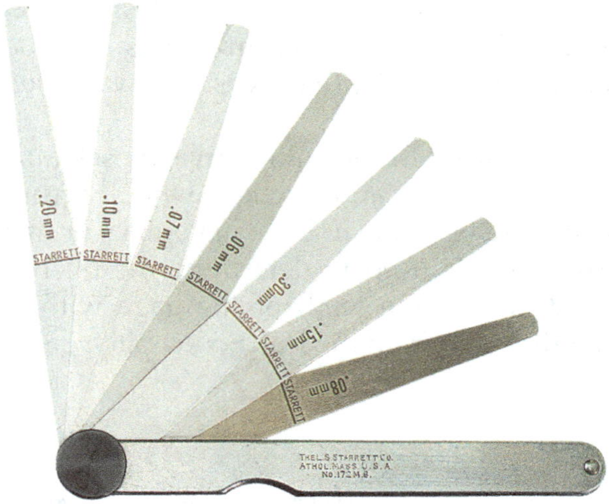

Figure 3-31. Thickness gage. (L.S. Starrett)

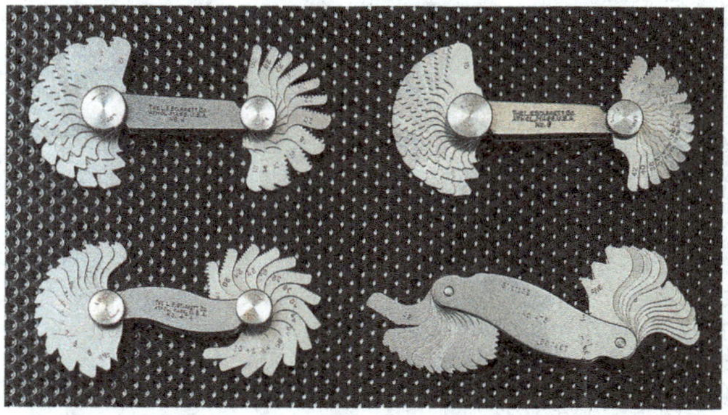

Figure 3-32. Screw pitch gage. (L.S. Starrett)

dial indicator: A precision instrument used to check the roundness of a piece of work or to center work on a metal lathe.

snap gage: A fixed limit gage that has two sets of jaws. The outside jaws test the upper limit tolerance and the inside jaws test the lower limit tolerance.

plug gage: A measuring tool that tests tolerances in two steps. The piece being tested must pass one test (go) and fail the other (no-go). Also called a go no-go gage.

Dial Indicator

A ***dial indicator*** is used for precision checking of manufacturing and repair work. The dial indicator is used to check the roundness of a piece of work or to center work on a metal lathe.

Special Gages

In production work, many duplicate parts are made in large quantities. These mass-produced parts must meet predetermined tolerances. These tolerances are tested using fixed limit gages.

A ***snap gage*,** **Figure 3-33,** has two sets of jaws. The outside jaws test the upper limit tolerance and the inside jaws test the lower limit tolerance. A properly made work piece will pass through the outside jaws and stop at the inside.

A ***plug gage*** tests tolerances using a two-step process, **Figure 3-34.** The work piece being tested must pass one test ("go") and fail the other ("no-go"). These gages are sometimes called "go no-go" gages. When the work is within tolerances, the "go" part of the gage will fit, the "no-go" part will not.

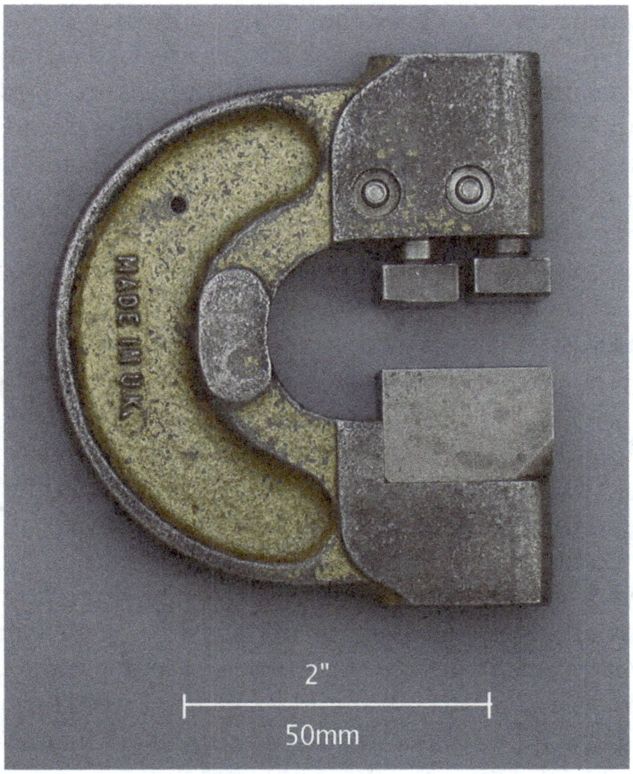

Figure 3-33. Snap gage.

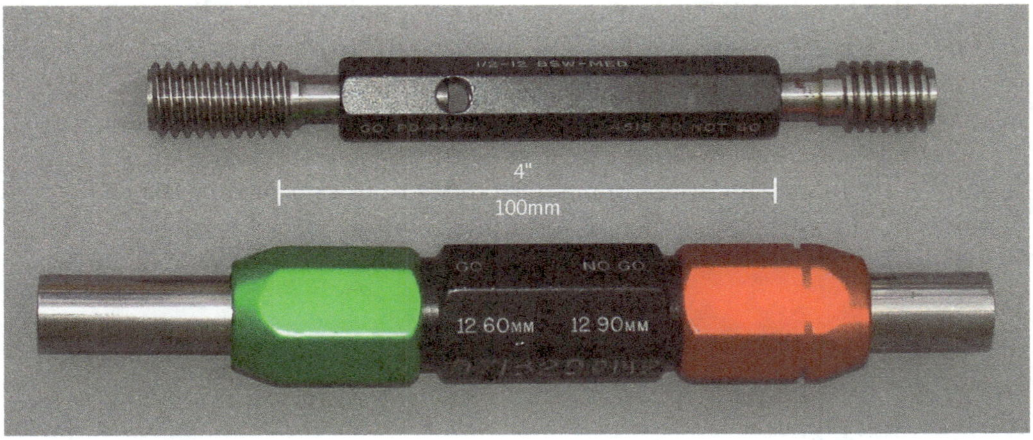

Figure 3-34. Plug gage.

Summary

Many projects in sheet metal are mass-produced and must be made to precise tolerances. Accurate measurement and lay out are required to meet this goal. Measurement and layout tools with very accurate readings are needed.

Rules are a basic measuring tool. Many types are available. Some are calibrated in both metric and U.S. Customary measurement systems. Steel rules are quite basic. Hook rules grab the piece being measured, freeing the hands. The slide caliper rule is used to measure round and flat stock. Shrink rules are used when shrinkage is expected, such as in a foundry casting.

The try square is used to check the accuracy of work. It is available with or without graduation marks on the blade. Calipers are used to make both external and internal measurements. Hermaphrodite calipers are used to scribe lines on a surface parallel to an edge or to locate the center on round stock. Dividers are used in transferring measurements from one place to another as well as for scribing arcs and circles in layout work. The combination square can be used as a level, a try square, and a miter square. A scriber is also located in many of the combination squares. The scriber can be removed and used to scribe lines on sheet metal. It can also be used as a height gage or a depth gage.

Layout dye is used to prepare a ground piece of flat stock for laying out a machine part. It is also used on clean sheet metal surfaces to provide a background on which scribed lines are easily seen. Punches are used in layout work. The prick punch is used to make small indentions in sheet metal to make reference points for additional work. The center punch is used to make indentations in 18-gage or thicker metal and to mark holes to be drilled.

The swinging-blade protractor is used in layout work and to measure angles in sheet metalwork. The combination bevel is used to set and transfer angles.

Micrometers are precision instruments capable of measuring to 1/1000″ (0.001″) and 1/100 mm (0.01 mm). Micrometers that contain a Vernier scale can measure to 1/10,000″ (0.0001″) and 1/500 mm (0.002 mm). The outside micrometer is the type most commonly used in sheet metal work. Learning to adjust and read micrometers takes some practice.

Gages are important in determining the thickness of sheet metal. The American Standard Wire Gage and the United States Standard Gage are used in the United States. Surface gages are used to scribe lines accurately and transfer measurements. Depth gages are used to measure holes and hard-to-reach places. Thickness gages are used widely in the automotive, diesel, aircraft, and service industries in making adjustments and in checking work.

Dial indicators are used for precision checking of manufacturing and in repair work. In production work where more than one part of the same size is being made, various types of fixed limit gages are used to check the work being produced.

Internet Resources

L.S. Starrett Co.

www.starrett.com

Outside Activities

1. Visit a home improvements store to make a list of available tools for use in sheet metal work. Make a note of the prices of each tool. If they have any free literature available about the tools, bring it back to the classroom to use as a resource. Make a list of the tools that are not available.
2. Make a poster that compares metric and U.S. Customary systems of measurement.
3. Research old British units of measurement such as the peck, the rod, and the chain. Explain what items these units were used to measure. Are any of them still in use?

Name ______________________ Class ______________ Score ________

Chapter 3
Measurements and Layouts

Review Questions

Write your answers in the spaces provided.

1. On a rule, the series of lines that represent points of measurement are called _____. 1. ______________
   A. delineations
   B. separations
   C. graduations
   D. calibrations

2. Which of the following statements is true of the steel rule? 2. ______________
   A. It is the most commonly used measuring tool in sheet metal work.
   B. It can be used to measure round stock.
   C. It comes in a variety of lengths.
   D. All of the above.

3. True or False? All metals shrink the same amount as they cool. 3. ______________

4. A(n) _____ is used to make inside and outside measurements on round or flat stock. 4. ______________

5. A(n) _____ is used to check the accuracy of 90° angles. 5. ______________
   A. hook rule
   B. try square
   C. plug gage
   D. combination square

6. Name the two basic types of calipers. ______________________________

   ____

7. Dividers are used to _____. 7. ______________
   A. transfer measurements from one surface to another
   B. scribe arcs and circles in layout work
   C. locate center points on round stock
   D. All of the above.

Match each of the following descriptions with the correct term.

8. Makes scribed lines easily visible. 8. ____________
9. Used to transfer angles from one surface to another. 9. ____________
10. Can be used as a level, a try square, or a miter square. 10. ____________
11. Used to mark the intersection of centerlines in a layout. 11. ____________
12. Contains a center head, a square head, and a protractor head. 12. ____________
13. Used to enlarge indentations in metal. 13. ____________

A. Combination set.
B. Combination square.
C. Center punch.
D. Layout dye.
E. Prick punch.
F. Combination bevel.

14. True or False? A Vernier micrometer has a third scale that can measure to 0.0001″ and 0.002 mm. 14. ____________
15. Name the two gages used to determine the thickness of metal. ____________

____
16. The screw pitch gage is used to determine the pitch of a screw _____. 16. ____________
17. A(n) _____ gage is also called a "go no-go gage". 17. ____________
    A. plug
    B. dial
    C. snap
    D. depth

Chapter 4
Print Reading and Ductwork

Objectives

After studying this chapter, you will be able to:
- Explain the purpose of a print in sheet metal work.
- Read a print.
- Explain the use of notes and dimensions in a print.
- Recognize the symbols used in sheet metal drawings.
- Explain how to determine duct sizes.
- Outline how a duct system is designed.

Technical Terms

angle changer
clinch-tee
computer-aided drafting (CAD)
curved elbow
ductrun
elbow
general note
local note
note
offset
print
square elbow
two-way Y branch
velocity
working drawing

Drawings are a useful tool in manufacturing and construction because most people find it difficult to explain in words how an object is to be built. Drawing, in this instance, is the preferred form of communication. A drawing can illustrate, step by step, the process to follow in the construction of any type of part or item. They are used in many industries, from automobile and aircraft manufacturing to the decorative sheet metal industry.

print: An original, detailed drawing that illustrates how an object is to be built. It includes information on the size, shape, and tolerances of objects, materials to be used, and finishes needed.

working drawing: A copy of a print made for use in the shop or on the job site.

Prints

Prints are original, detailed drawings that illustrate how an object is to be built. These prints include information on the size, shape, and tolerances of objects, as well as the materials to be used and the finish needed, **Figure 4-1**. The object can be as simple as a bolt or as complicated as an office building. Depending on the complexity of the object, prints can contain hundreds of pages. Because prints are the original set of drawings, copies, which are referred to as ***working drawings***, are made for use in the shop or on the job site.

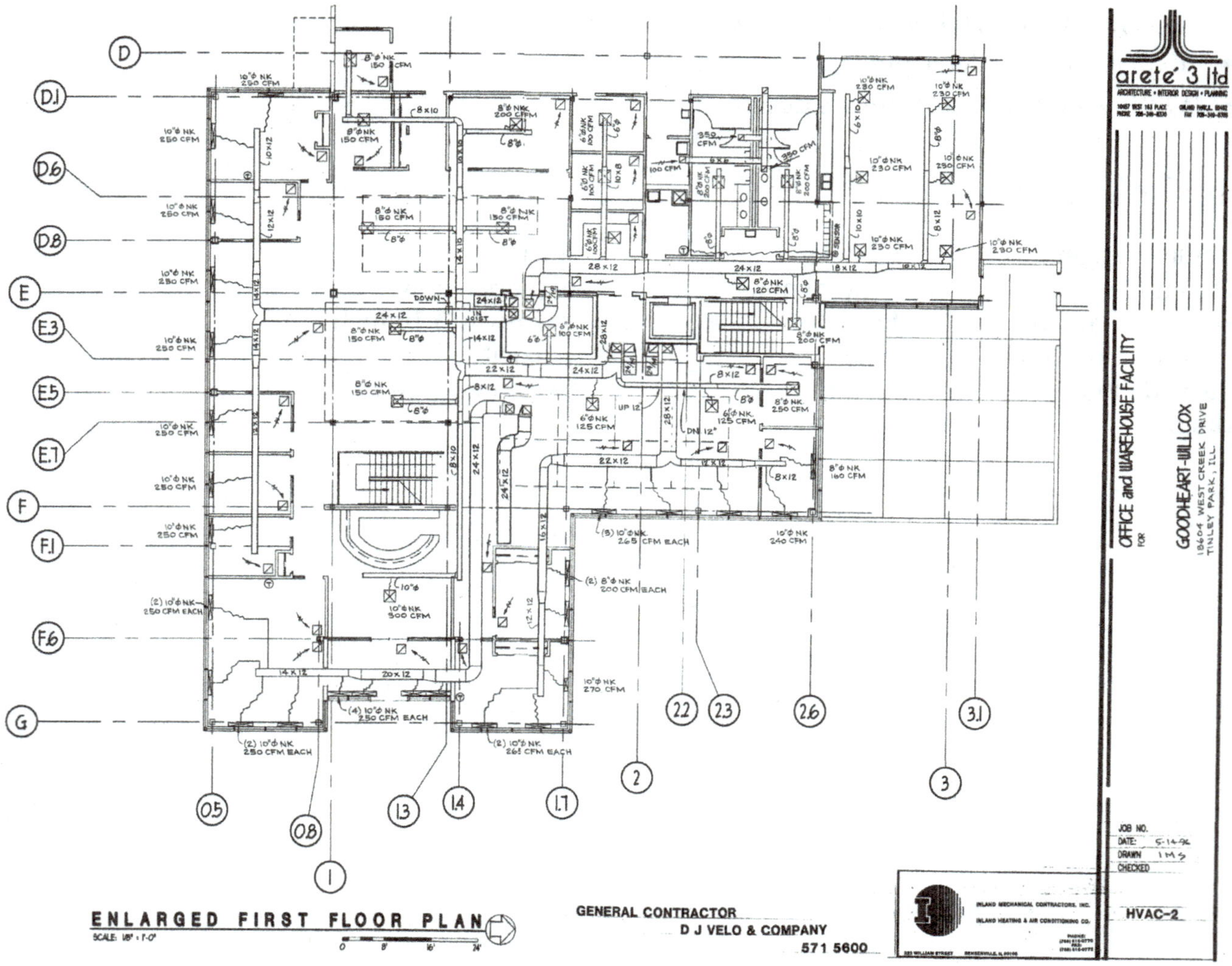

Figure 4-1. This print shows an enlarged first floor plan for an office and warehouse.

A set of working drawings includes all drawings needed to complete a job. For example, a set of drawings for an HVAC system in an office building could contain hundreds of prints. The set of drawings for an industrial kitchen's work surfaces, in contrast, might contain less than one hundred prints.

Views are an important component of working drawings. Each view shows the piece to be built exactly as it appears. The number and type of views used depends on the complexity of the piece being shown, **Figure 4-2**.

A standard set of working drawings for a commercial construction project includes a title sheet, along with sets of civil drawings, architectural drawings, structural drawings, plumbing drawings, mechanical drawings, and electrical drawings. Special drawings are included as needed. They are combinations of several sets, such as a mechanical-electrical drawing. Each set of drawings is designated by a letter. It is the first letter of the set in which the drawing is found. All drawings designated by an *A*, for instance, come from the architectural drawing set. The letter designation is found in the title block. See **Figure 4-3**.

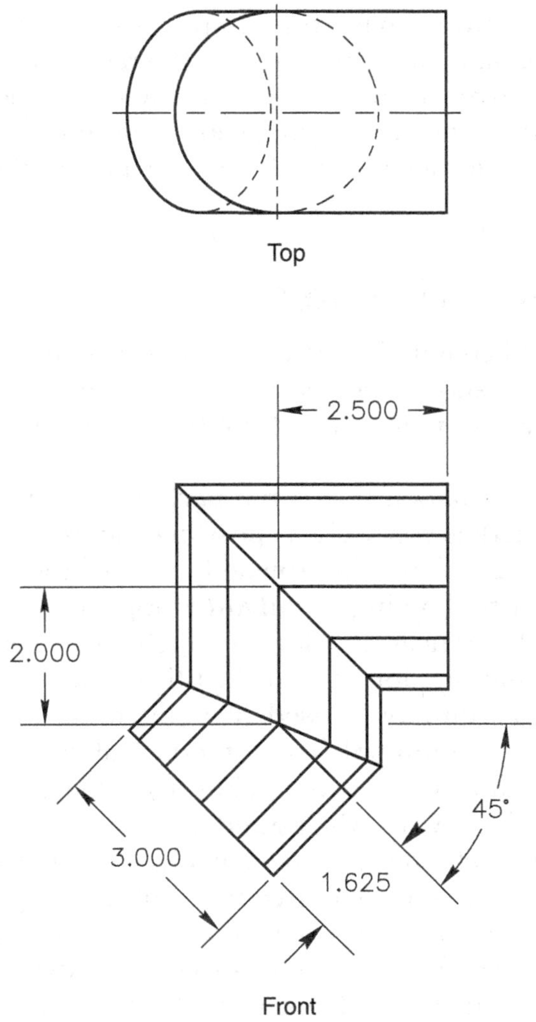

Figure 4-2. One or both of these views may be included in a working drawing.

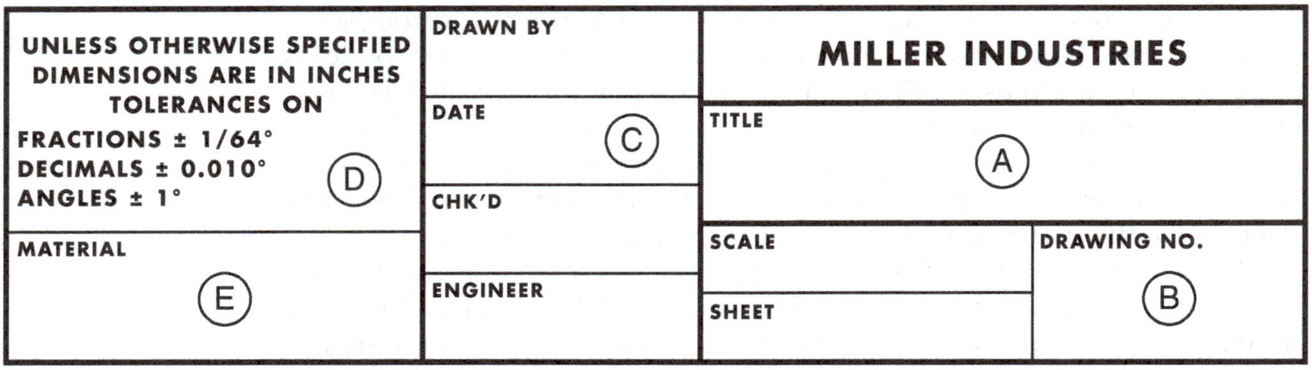

Figure 4-3. Familiarize yourself with the parts of a title block. A—Title name. B—Drawing number. C—Names of drafter, checker, and engineer. D—General tolerances. E—Specifications.

computer-aided drafting (CAD): Software used to create prints and all other types of drawings used in construction.

Working drawings are created either manually or by using ***computer-aided drafting*** (***CAD***) software. Regardless of how the drawings are made, the lines, symbols, notes, dimensions, and shapes used all have standard meanings.

Architects do not always have time to prepare a complete set of drawings for a room or area. Instead, they sketch the requirements for the space. This sketch is then given to the sheet metal worker who determines the types of joints and ducts needed to do the job. Working from a well-thought out set of plans makes for fewer mistakes on the part of the worker and the contractor.

note: An instruction or comment that provides information that is not part of the normal dimensioning system. It is stated as briefly and concisely as possible and is lettered horizontally.

local note: An instruction or comment that provides information about a specific process or feature; placed near the view to which it refers.

general note: An instruction or comment that provides information that applies to the entire drawing.

Dimensions and Notes

A drawing should contain all the dimensions needed to make an object. However, dimensions should not be repeated anywhere on the drawing. Information contained in notes should not be shown again in dimensional form.

Notes provide information that is not part of the normal dimensioning system. They are stated as concisely as possible. Notes are lettered horizontally and should not be placed in crowded areas of the drawing. ***General notes*** apply to the entire drawing. ***Local notes*** apply to a specific view of the drawing. They are placed near the view to which they refer. Materials, tolerances, and specifications are found near the title block.

Abbreviations are commonly used in notes to save space. **Figure 4-4** contains a list of several commonly used standard abbreviations. The American Society of Mechanical Engineers publishes a complete list of accepted abbreviations that comply with ANSI standards.

The work of sheet metal professionals involves construction and installation of sheet metal components. These components are designed by someone other than the sheet metal professional. Therefore, it is important for sheet metal workers to understand the use of mechanical drawing and drafting techniques, whether done manually or using CAD. There must be someone building and assembling the system who understands and can read prints and drawings.

Standard Abbreviations Used on Drawings			
Actual	ACT	Fillet	FIL
Advance	ADV	Gage	GA
Assemble	ASSEM	Inside diameter	ID
Babbitt	BAB	Metal	MET
Bend radius	BR	Not to scale	NTS
Circumference	CIRC	Outside diameter	OD
Clockwise	CW	Stainless steel	SST
Equivalent	EQUIV	Wrought iron	WI

Figure 4-4. The standard abbreviation list is quite long. This short list includes several frequently used abbreviations.

Air Duct Systems

Many sheet metal workers are employed in the HVAC industry. Heating, ventilation, and cooling systems are composed of an arrangement of pipes, ducts, and other special devices. The system is used to circulate conditioned air throughout a building. A properly designed system will operate economically, efficiently, safely, and quietly. The correct amount of air will be delivered to the correct areas. Rooms will be comfortable and relatively draft free. A poorly designed system increases costs and wastes resources.

Ductwork is a combination of ducts (pipes) and fittings used to circulate air. Ducts can be square, rectangular, or round, **Figure 4-5**. Ducts are open at both ends, a uniform size, and, usually, straight. Fittings are irregularly shaped pieces made from an individually designed pattern. Their function

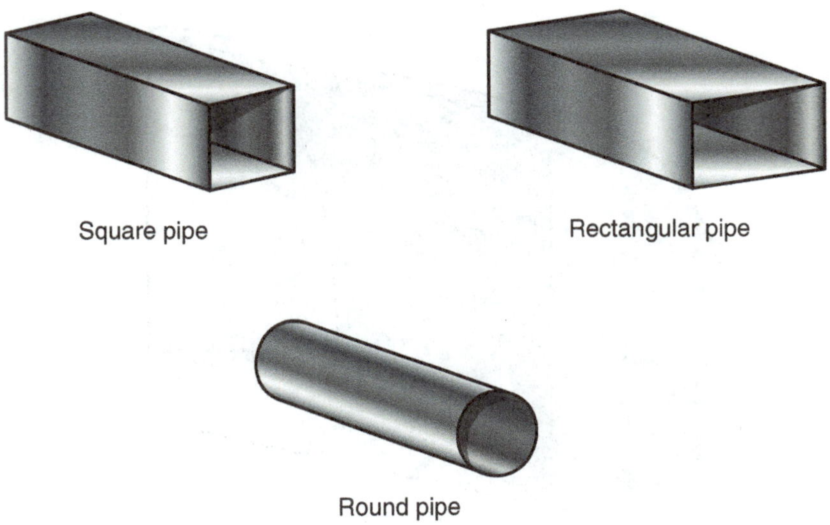

Figure 4-5. Ductwork, or pipes, are made in square, rectangular, or round shapes.

is to redirect airflow, act as a transition between differently shaped ducts, or both, **Figure 4-6**. Fittings include elbows, offsets, and transitions.

Elbows are used to make turns in the ductwork, **Figure 4-7**. An ***angle changer*** makes 45° turns and a ***square elbow*** makes 90° turns. But an abrupt 90° turn can increase resistance to the airflow pattern. To avoid this, a ***curved elbow*** is sometimes used, which turns the ductrun pattern gradually. A ***two-way Y branch*** splits the main air stream into two separate streams, **Figure 4-8**. A ***clinch-tee*** may be straight or have a curved throat. It provides a separate path for airflow anywhere along the pipe. ***Offsets*** are used to move ductwork around an obstacle.

When joined together, all the ducts and fittings in a complete system are called a ***ductrun***.

Air duct systems are designed by engineers. While sheet metal workers are not involved in the design of air duct systems, understanding design concepts is helpful when building the ductwork in the system.

elbow: A duct used to make a turn in a ductrun.

angle changer: An elbow that makes a 45° turn.

square elbow: An elbow that makes a 90° turn.

curved elbow: An elbow that changes ductrun direction gradually.

two-way Y branch: A duct that splits the main air stream into two separate streams.

clinch-tee: A duct with a straight or curved throat. It provides a separate path for airflow anywhere along the pipe.

offset: A flexible duct used to move ductwork around an obstacle.

ductrun: A complete system of ducts and fittings.

Duct System Design

A duct system is designed after calculating heating and cooling loads and choosing the properly sized supply unit. Unit blower size determines how the ducts will be sized. Engineers sketch several designs, based on supply unit placement, main duct and branch lengths and locations, and air register placement and size. The arrangement that best serves the building mechanically and architecturally is chosen, **Figure 4-9**.

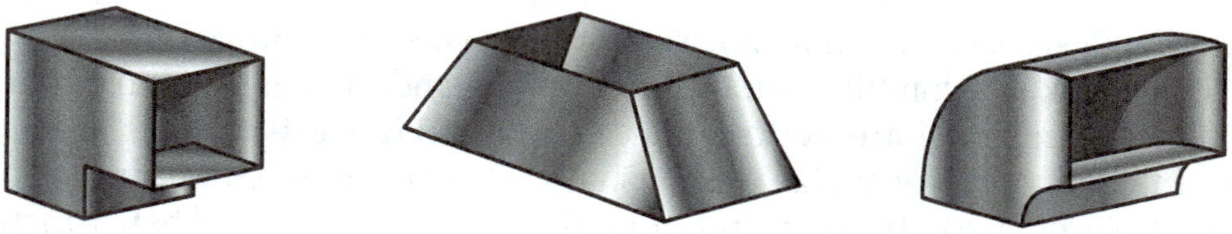

Figure 4-6. Fittings are made in a variety of shapes, based on their function.

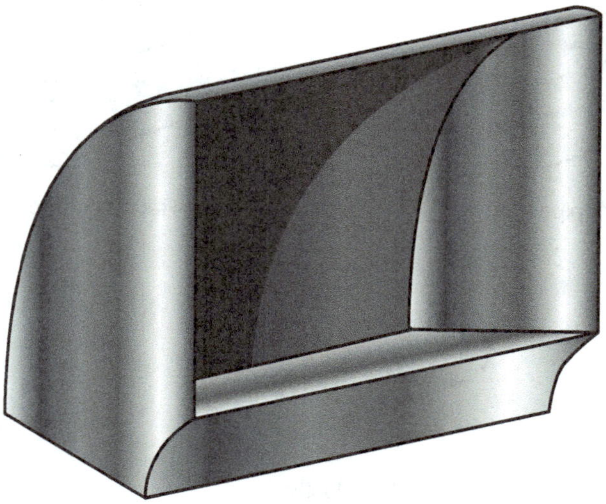

Figure 4-7. A 90° curved elbow.

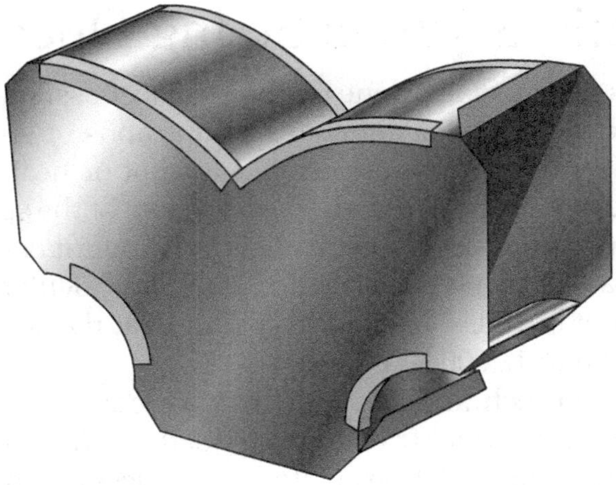

Figure 4-8. A two-way Y branch.

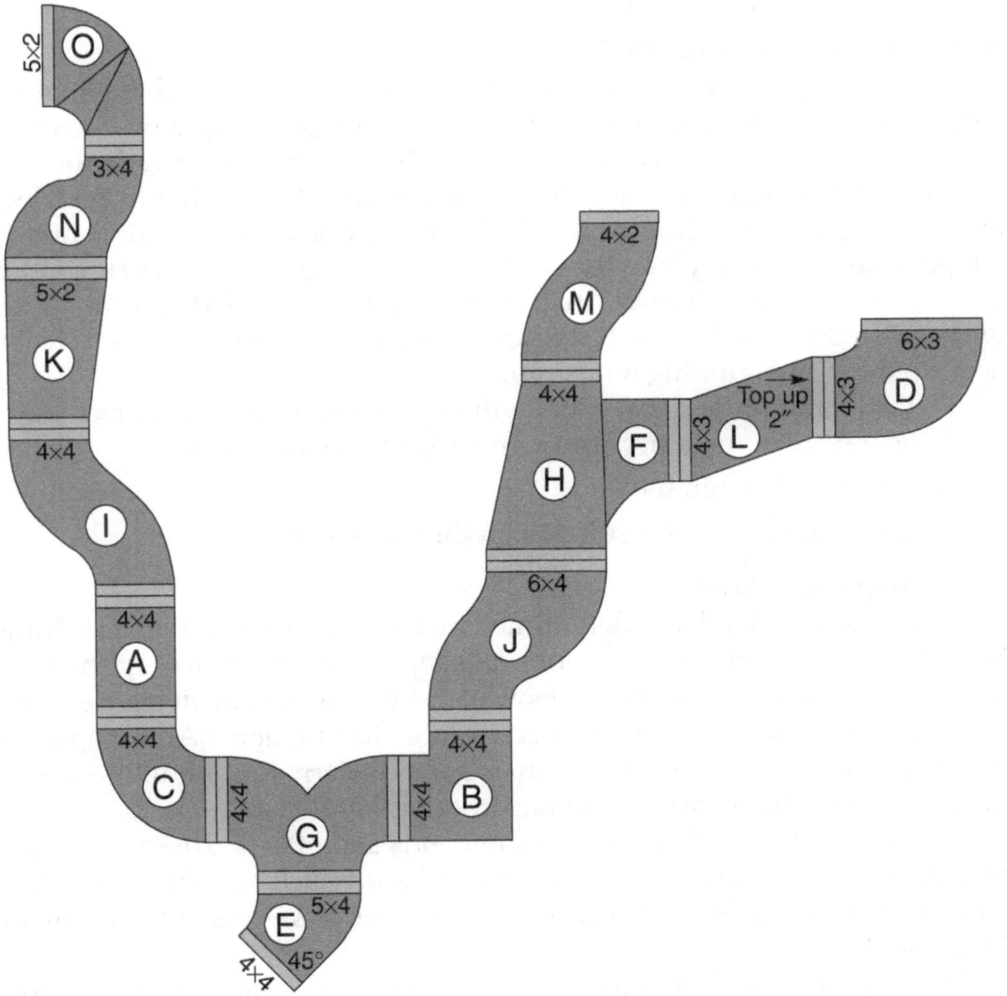

Figure 4-9. A working drawing of a ductrun layout.

Considerations When Sizing Ducts

Duct sizing is a job usually done by the engineer assigned to a particular building being constructed. Using charts and graphs or software, the engineer determines the amount of air needed to cool or heat a space. In most cases, this is done by inputting the area and volume of the space, and number of air changes per hour desired. The result is the amount of ductwork needed to handle that load. If using software to make the calculations, it may also suggest the number of BTUs needed to cool or heat the space. A one-ton air conditioner is equal to 12,000 BTUs.

velocity: Speed of air through ductwork.

Properly sized branch and main ducts have equal frictional losses per foot of length. To create equal frictional losses, the ***velocity*** in smaller ducts must be lower. The highest friction created, regardless of which duct it occurs in, is the resistance offered by the duct system as a whole. This must also be considered when sizing the ducts.

Other factors to consider when sizing ducts include heat gain and loss in ducts and duct shape and area.

Heat Gain and Loss in Ducts

Air moving through a duct system can gain or lose heat. These losses or gains are very important considerations in the design of a system. Causes for losses or gains can be internal or external. Internal causes include the type of lights and heat-generating machines used, and the number of people in the building. External factors include the orientation of the building, outside temperatures, humidity, and the spaces through which the ducts run. Often, ducts will run through areas that are not heated or cooled. The air in the ducts will either lose heat to the cooler surrounding temperature or will gain heat from the surrounding temperature.

The amount of heat gain or loss will depend on the following factors:

- Temperature of the spaces through which the duct passes.
- Air velocity through the duct.
- Type and thickness of insulation on the ductwork.

Duct Shape and Area

Duct size and depth are determined by the available space in a building. Round ducts have the most economical shape. They create less friction and require less metal to construct. However, because round ducts use more overhead space, they are rarely used in most installations. An exception is in warehouse areas, where headroom is not a concern. Rectangular ducts are the preferred shape in areas with headroom limitations.

Ductruns often contain a number of bends and offsets. These are considered obstructions because they disrupt or disturb airflow. When conditions require sharp or right angle bends, use vane elbows across the air stream, **Figure 4-10**.

While a great deal of computing and thought go into sizing ductwork, many of the more difficult computations are done using software or calculators. In the past, calculations were made using charts. This method is still used in some instances.

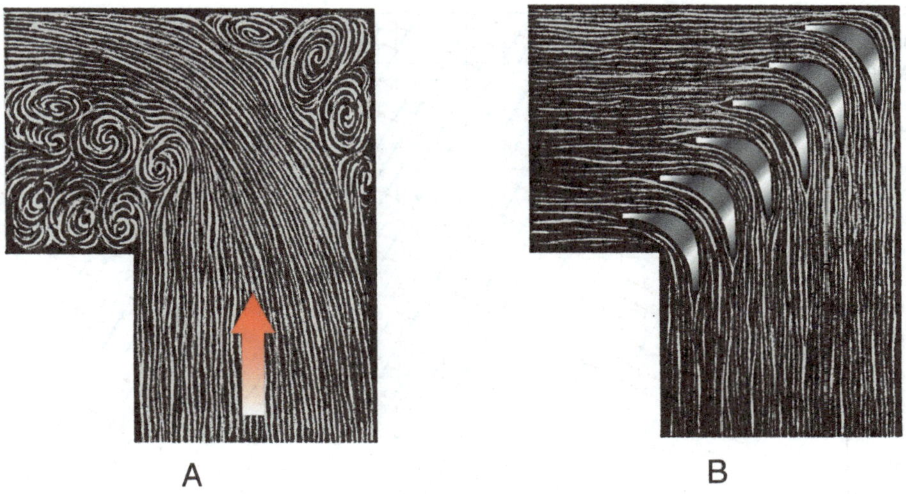

Figure 4-10. Bends in ductwork create their own type of problems. A—Turbulence causes the air to pile up. B—Vane elbows help reduce pressure loss in bends.

Duct Sizing Using Charts

1. A system requires delivery of 5000 cubic feet per minute (cfm) for a distance of approximately 80′, with the longest branch beyond that point conveying 1000 cfm for an additional 70′.
2. Allow a total supply duct resistance of 0.10 inch water-gage resistance pressure.
3. The supply duct is 12″ deep.
4. The total length of the longest run is 150′ (80+70).
5. The resistance for 100 feet of duct is 0.067″ (100÷150 × 0.10 = 0.067″).
6. Find 0.06 resistance along the bottom of the chart in **Figure 4-11**. Follow that line up to its intersection with the horizontal line representing 5000 cfm. Read the diagonal line. This is the equivalent size of the round duct required: approximately 28″ in diameter.
7. Using **Figure 4-12**, find the intersection of the 12″ vertical line with the horizontal 28″ line. Follow the curved line up and to the right and find the value attached to that line. In this case, the width of the rectangular duct will be 60″. The main ductrun dimensions will be 60″ × 12″.
8. Repeat these steps for the 1000 cfm branch.
9. Using **Figure 4-11**, find the intersection of the 0.06″ resistance line and the 1000 cfm line. Follow the diagonal line to the value attached to it. The equivalent round duct required would be 16″. Follow through on **Figure 4-12** to find the width of the branch duct. It will be 12″ × 18″.

Making or Buying Ductwork

It is the job of the sheet metal worker to make the ducts, install any fittings, join and hang all of it together into an entire system. Working drawings provide the dimensions for the work. Ducts and fittings can either be custom-made or purchased ready-made.

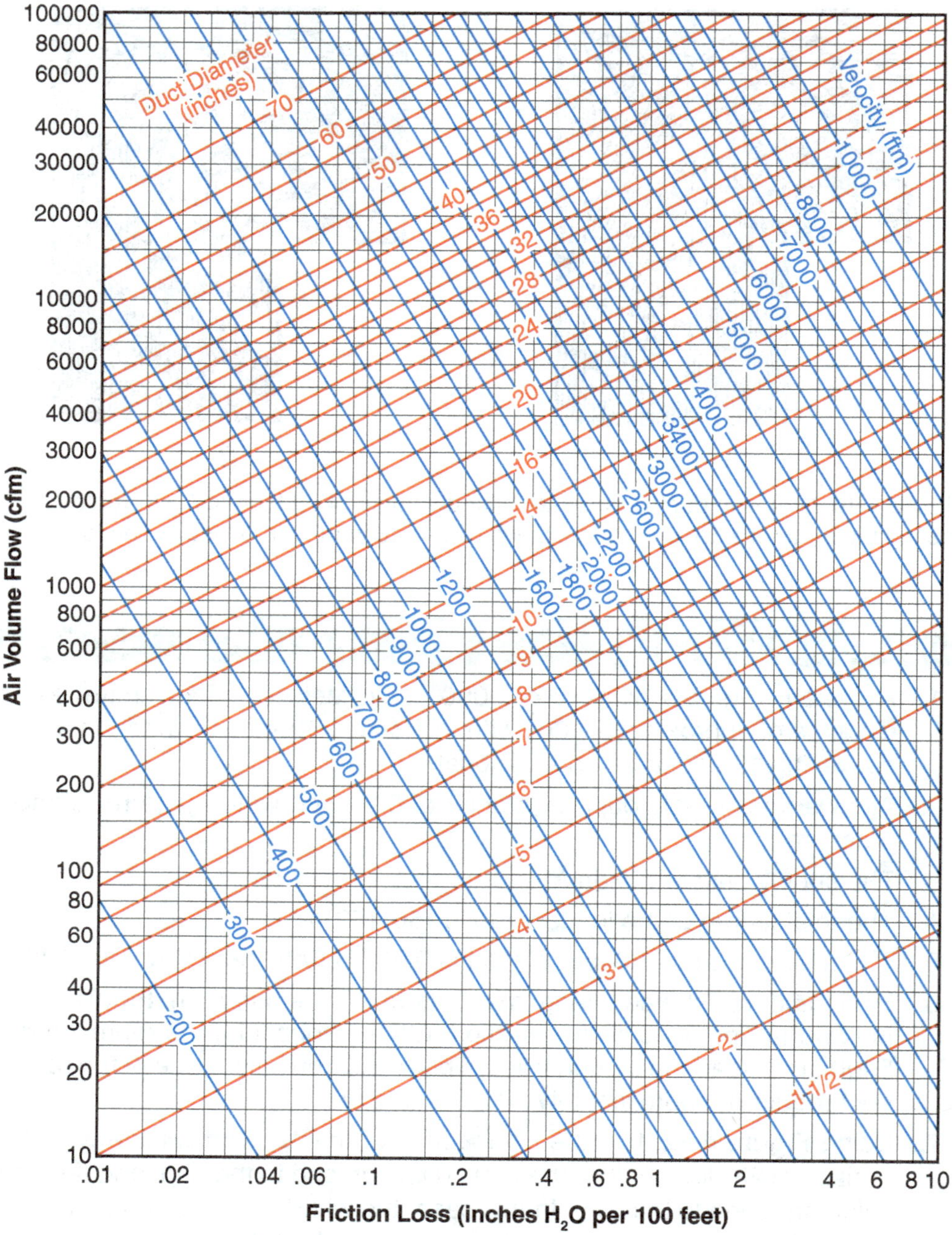

Figure 4-11. Air friction chart. (The Engineering Toolbox)

Residential installations are normally fitted with ready-made ducts and fittings. When repairing or remodeling older homes, custom-made ductwork is necessary because such houses are often one-of-a-kind and have older systems that will not accommodate ready-made parts.

Commercial and industrial installations use both ready-made and custom-made ductwork. The working drawings indicate what type of piece should be used.

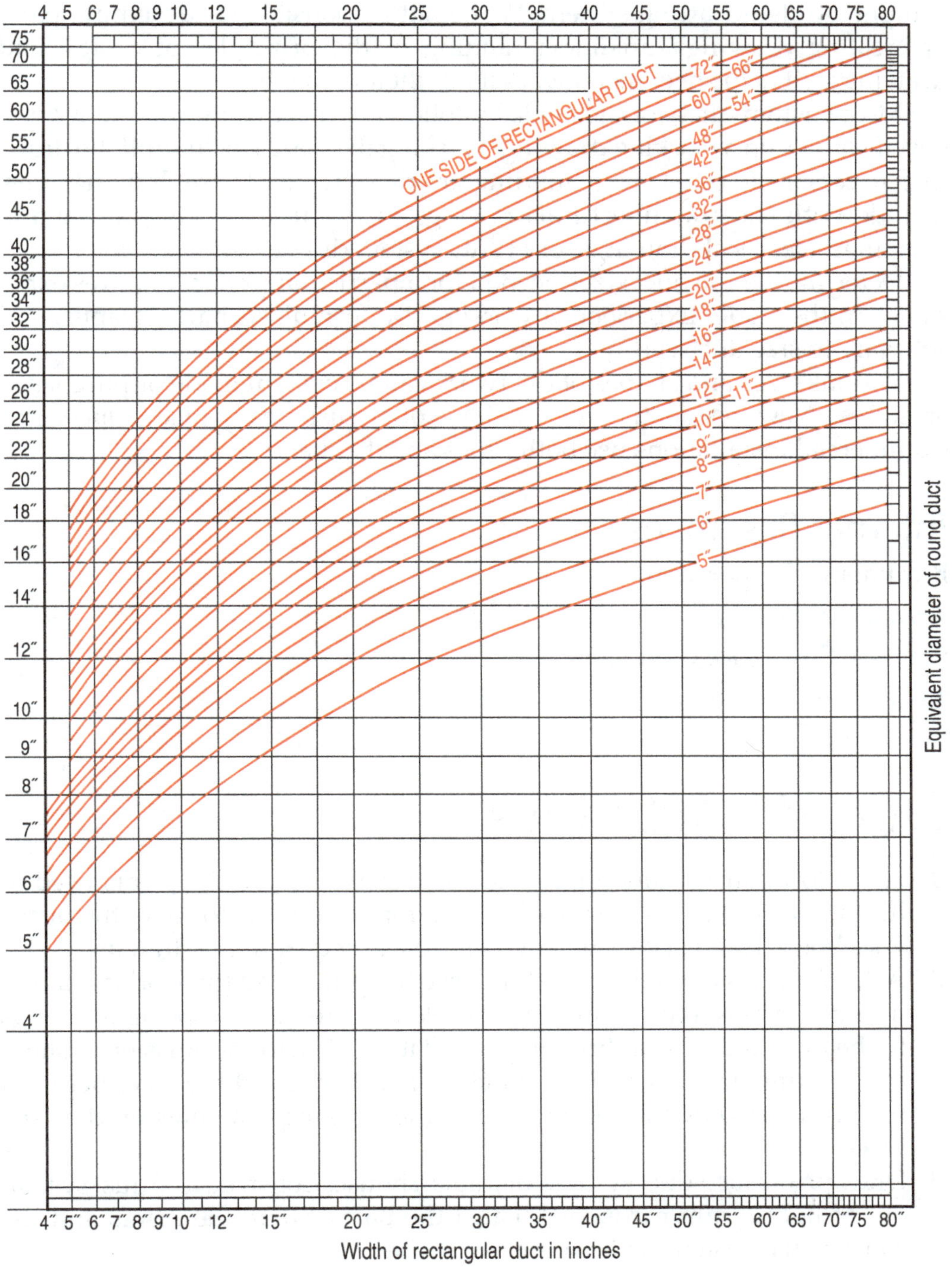

Figure 4-12. Equivalent diameter duct sizing chart.

Summary

In order to build, assemble, and service sheet metal parts, sheet metal workers must understand how to read working drawings. The design of parts and systems is done by an architect or designer. The actual drawings are made by a mechanical engineer, using manual drafting techniques or a computer-aided drafting system. While not responsible for designing parts or creating drawings, an understanding of the principles of design and drawing will help sheet metal workers in their own jobs.

Most custom ductwork is made by following the layouts and the written specifications on a working drawing. Working drawings include all the drawings necessary to construct a building or other project. These drawings may include a title sheet, architectural drawings, mechanical drawings, structural drawings, electrical drawings, or HVAC system drawings.

Determining the duct sizes needed for a particular project requires experience in the trade. Experienced workers use formulas, charts, graphs, and software to size ductwork.

Various sizes and shapes of ductwork can be custom made or purchased at stores. New residential installations often use ready-made ductwork. Commercial installations use both types about equally.

Internet Resources

Brinkmann Corporation

www.brinkmann.net

Simpson Strong Tie Company

www.strongtie.com

Outside Activities

1. Ask a local air-conditioning installer to see a copy of the prints used in his or her most recent installation. Ask for permission to share the prints with your class. Discuss what you have learned from the drawings.
2. Ask a local residential contractor to see the prints and the working drawings of the most recent building completed. Are there working drawings for heating and air-conditioning installation? Ask for permission to share the working drawings with your class. Point out the differences between the prints used by the contractor and the working drawings used by the workers.
3. Draw prints for a duct system for a small house. Construct the duct system out of lightweight cardboard. Label the parts. Share the prints and the constructed system with your class.

Name ______________________ Class ______________ Score ________

Chapter 4
Print Reading and Ductwork

Review Questions

Write your answers in the spaces provided.

1. Which of the following statements about prints is *not* true? 1. ____________
   A. Prints can contain hundreds of pages.
   B. Prints are copies of the original work.
   C. Prints include information on the size, shape, and tolerances of objects.
   D. Prints illustrate how an object is to be built.
2. Mechanical-electrical drawings can be found in a standard set of _____ drawings. 2. ____________
3. True or False? A drawing should contain all the dimensions needed to make an object. 3. ____________
4. Explain the difference between general notes and local notes. ____________
5. _____ is a combination of pipes and fittings used to circulate air. 5. ____________
6. What is the function of a fitting? ____________
7. What type of duct is shown in the following drawing? 7. ____________

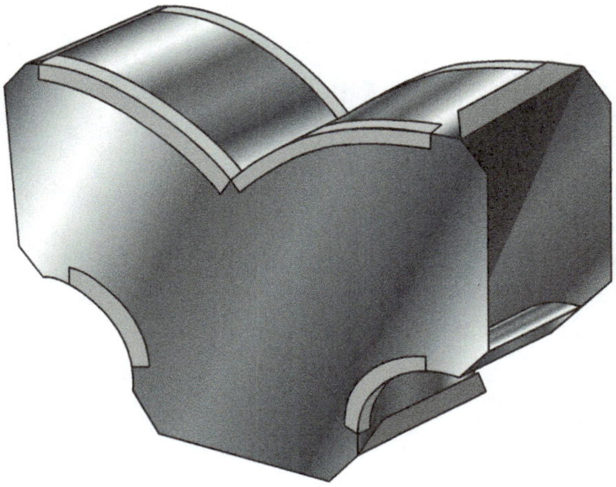

8. True or False? Air duct systems are designed by sheet metal professionals. 8. ____________________

9. Heat gain or loss depends on which of the following? 9. ____________________
   A. Air velocity through the duct.
   B. Temperature of the spaces through which the duct passes.
   C. Type and thickness of insulation on the ductwork.
   D. All of the above.

10. _____ ducts have the most economical shape. 10. ____________________

11. What is the purpose of the vane elbows, shown in the following drawing?____________________
____

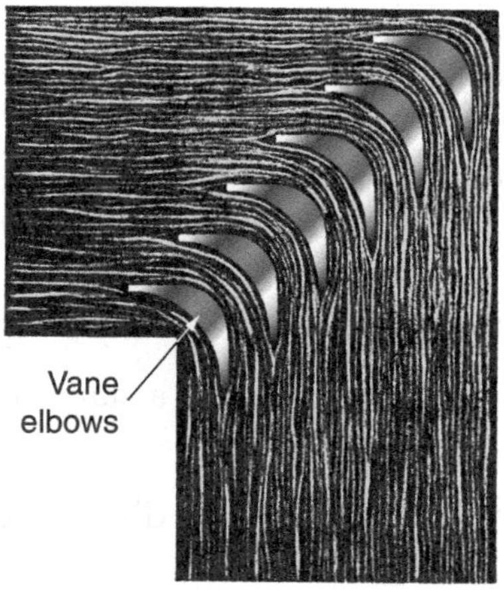

12. The sheet metal worker's job includes _____. 12. ____________________
    A. making the ducts
    B. installing fittings
    C. joining and hanging the ducts and fittings into a finished system
    D. All of the above.

Chapter 5
Sheet Metal Hand Tools and Machinery

Objectives

After studying this chapter, you will be able to:

- Identify various hand tools and machines.
- Explain the purpose of each tool and machine.
- Demonstrate the proper way to use these hand tools and machines.
- Discuss safety rules associated with different tools and machines.

Technical Terms

Allen wrench
bar folder
bead
beading machine
box and pan brake
brake
burring machine
cape chisel
cornice brake
crimper
diagonal pliers
diamond-point chisel
double-seaming machine
draw filing
edge
face
flat chisel
hand punch
hand snips
heel
hem
kerf
needle-nose pliers
pin punch
point
press brake
rotary machine
round-nose chisel
seam
set
setting-down machine
shears
slip-rolls forming machine
slip-joint pliers
solid punch
squaring shears
straight filing
tang
temper
tensioning
throatless shears
turning machine
utility pliers
wiring machine

Tools and machines are an integral component of sheet metal work. They make it possible to lay out, cut, form, bend, and join metal. The tools and machinery discussed in this chapter are used to create all types of sheet metal products, from a simple downspout to the housing for a jet engine. Knowing when and how to use each item is an important responsibility of a sheet metal worker.

Hand Tools

There are a variety of hand tools that are still important to the sheet metal worker. Even with the use of machinery, some tasks cannot be done without the use of a hand tool. Hand tools tend to be used in specialized circumstances, to do custom and small-run jobs. Important hand tools for the sheet metal worker include hacksaws, files, chisels, hammers, mallets, screwdrivers, pliers, wrenches, and hand punches.

Hacksaw

A hacksaw is used to saw metal, **Figure 5-1.** Hacksaws have either a fixed frame or an adjustable frame. Fixed-frame hacksaw blades have a fixed-length blade. Most hacksaws used now have adjustable frames. Adjustable-frame hacksaws have several different blade lengths, including 8″, 10″, and 12″ (203 mm, 254 mm, and 305 mm). The number of teeth per inch on the blade also varies. Common numbers per inch include 14, 18, 24, and 32. The type of metal being cut will dictate the number of teeth needed, **Figure 5-2.**

The ***set*** in the blade teeth determines the width of the cut and prevents the blade from binding in the cut. Teeth are arranged in one of three ways: undulating, raker, or alternate, **Figure 5-3.** Each arrangement produces a different width. This slot produced by the blade is called the ***kerf.***

set: The tooth pattern on a saw. Determines the width of the cut and prevents blades from binding in the cut.

kerf: The slot produced by a saw blade.

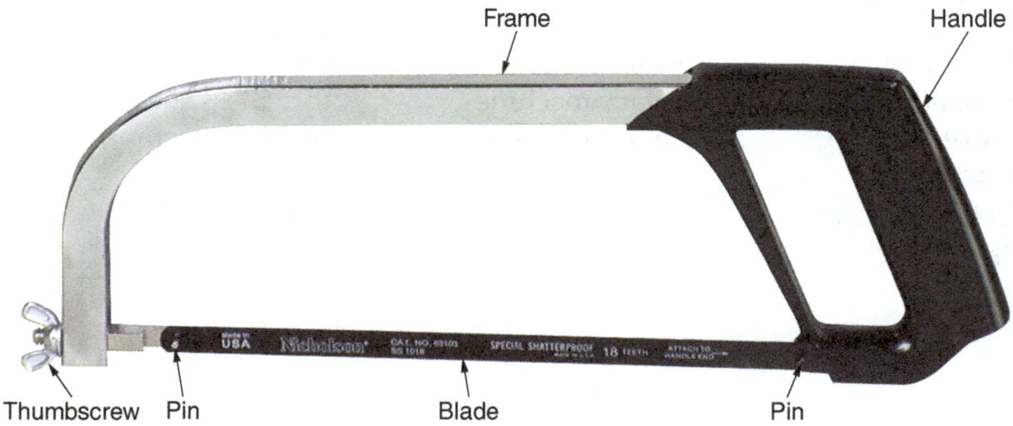

Figure 5-1. A hacksaw has three parts: frame, handle, and blade. This is a fixed-blade hacksaw. (Nicholson)

Hacksaw Blade Usage	
Number of teeth per inch	**Use to Cut These Materials**
14	Material 1″ or thicker Cast iron, machine steel, brass, copper, aluminum, bronze, or slate
18	Material 1/4″ to 1″ thick Annealed tool steel, high speed steel, copper, bronze, aluminum Light structural shapes
24	Material 1/8″ to 1/4″ thick Iron, steel, brass and copper tubing, wrought iron pipe Drill rod, conduit, light structural shapes, metal trim
32	Material 1/8″ and thinner Tubing and sheet metal

Figure 5-2. The type of metal being cut will determine what blade should be used.

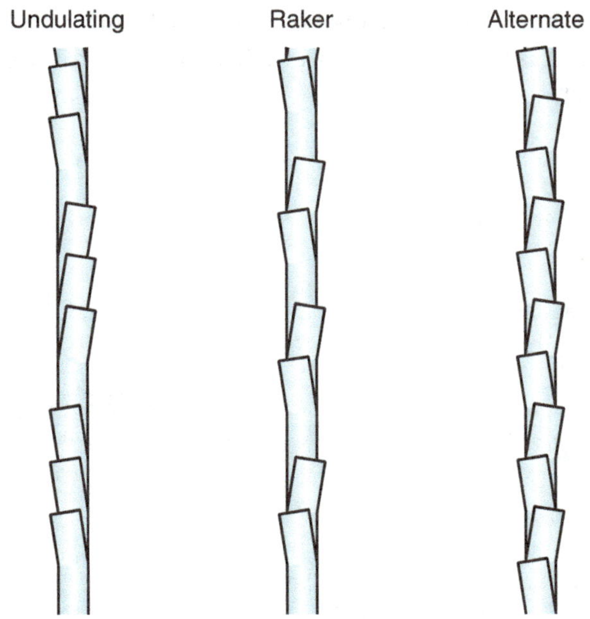

Figure 5-3. The set of the teeth determines the width of the cut.

Preparing to Cut

Adjust the frame to accommodate the length of the blade. Pins located on the hacksaw frame hold the blade in place. With the blade teeth pointing away from the handle, place the hole in the blade over the pin on the front of the frame. Then place the hole at the back of the blade over the pin at the back, **Figure 5-4.**

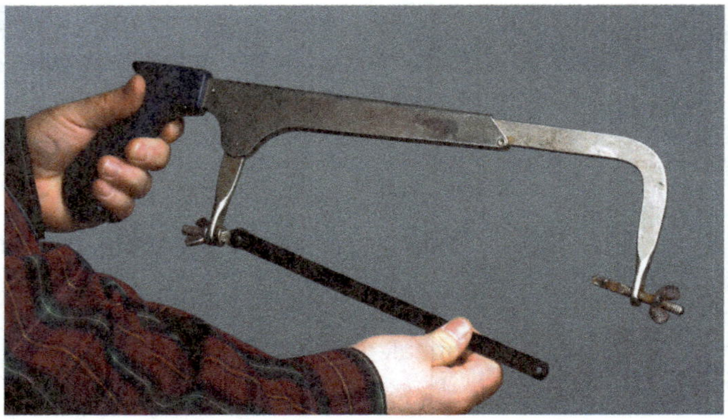

Figure 5-4. Inserting the blade in the frame.

After the blade is placed on the pins, it will need to be adjusted. Adjust the blade by turning the thumbscrew. Keep turning until the blade makes a pinging sound when hit with the thumb. This adjustment, or ***tensioning***, ensures that the blade will cut properly, cleanly, and safely. Poorly tensioned blades tend to break or cut poorly.

tensioning: Proper adjustment of a saw blade. Assures the blade will cut properly, cleanly, safely.

Fasten the metal securely in a vise before cutting. There are a number of ways to position the material in the vise, based on its shape, **Figure 5-5.** Position the work so it is as secure as possible in the vise. Oddly shaped pieces are harder to secure than rectangular or square pieces. Hold oval or circular pieces in wood, leather, or copper vise jaw covers. Sandwich thin sheet metal between pieces of scrap wood. Then cut through both wood and sheet to make cutting easier.

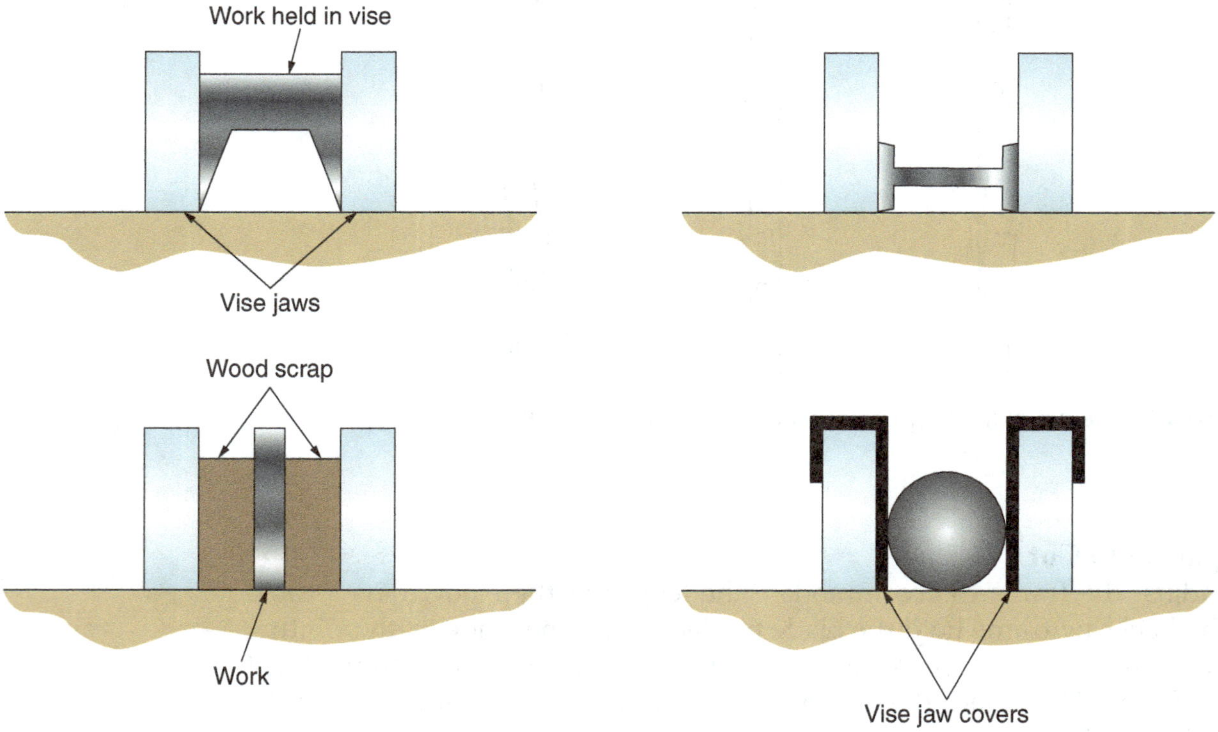

Figure 5-5. Secure material in vise according to its shape.

Cutting with a Hacksaw

Cutting into hard material is made easier when a guiding notch is made with a saw or file. Using your thumb as a guide, saw slowly using short strokes, **Figure 5-6.** As the cut deepens, take your thumb off the material, grip the front of the frame firmly, and take one full length stroke. Keep at least three teeth in the work at all times.

The metal is now ready for sawing. Hold the handle and the front of the frame firmly. Stand at ease, feet at least 12" (305 mm) apart, facing the work. In addition to the motion of the arms, sway the body by flexing the knees in line with the stroke. Move the blade in a straight line and maintain even pressure to avoid binding or twisting the blade, causing it to break. Use greater pressure on heavy work. Cut using the full blade length.

Apply enough pressure on the forward stroke to engage the teeth into the cut. Do not permit the teeth to slip over the material. This dulls the teeth and can break the blade. Lift the blade slightly on each return stroke to avoid dulling the teeth, but do not lift the blade out of the cut.

Work the hacksaw at a rate of 40–60 strokes per minute. If the saw is worked much faster than this, excess heat will be generated by the teeth. Excess heat ruins the ***temper*** of the blade, making it useless. Operate the hacksaw at a slower rate when sawing very hard metals.

temper: Hardness and strength of metal.

Blades will break occasionally and must be replaced. When using a new blade, start a new cut in the material. Do not use the old cut. The kerf made by a dull blade is more narrow than the kerf made by a new blade. Sawing in the same cut will cause the new blade to bind, stick, and sometimes break.

The service life of a hacksaw blade depends upon choosing the correct blade for the job. To select the correct hacksaw blade, consider the shape and kind of metal to be cut. When the hacksaw is not in use, hang it up. Coat the blade with a film of oil to prevent rusting.

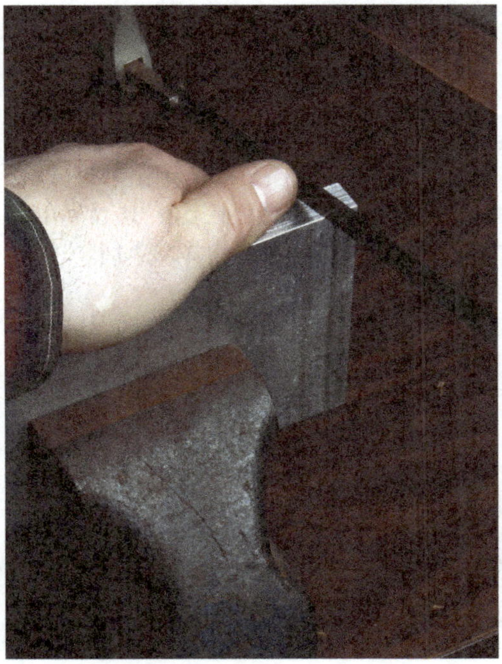

Figure 5-6. Start a cut using your thumb as a guide.

File

In sheet metal work, files are used to smooth, shape, and finish metal. File shapes, lengths, and abrasiveness vary, but they all contain a tang, heel, face, edge, and point, **Figure 5-7.** The ***tang*** is tapered and designed to fit into a handle for safe use. The tang begins at the ***heel.*** The ***face*** of the file contains the teeth. The shapes of the ***edge*** and ***point*** are determined by the shape of the file.

tang: Tapered portion of a file designed to fit into a handle.

heel: Bottom portion of a file, nearest the tang.

face: The surface of a file that contains the teeth.

edge: Surface of a file that is adjacent to the face. Shape is determined by shape of the file.

point: Tip of the file, at the opposite end of the tang.

File Categories

Files are categorized according to their shape, size, teeth, and coarseness. Files come in two basic outline shapes: blunt and tapered. Shape is also determined by the view of the file from its end, or the cross section. Size is measured in inches, from the heel to the point. See **Figure 5-8.**

File teeth are single cut, double cut, rasp cut, or curved tooth, **Figure 5-9.** Single-cut files have a single row of teeth and are used to produce a smooth finish. They are generally used with light pressure. Double-cut files have two rows of teeth that cross each other. They remove material quickly, create a rough finish, and require heavy pressure. Rasp-cut files have individually formed teeth. They are used for wood working and on soft metals. The teeth on a curved-tooth file are undercut. Curved-tooth files remove material quickly, can be used on all types of metals, provide a smooth finish, and tend not to clog.

Files are also categorized by the coarseness of the file teeth. Levels of coarseness, from greatest to least, are coarse, bastard, second cut, and smooth.

File Selection

With all the possible combinations of shapes, sizes, and teeth, thousands of files are available. Selecting the correct file for a job is important. Factors to consider during the selection process include:

- Type of metal being filed.
- Surface finish needed.
- Shape, location, and size of the part to be filed.
- Amount of material to be removed.

After selecting the correct file for the job, clamp the part in a bench vise at elbow height. This height is the most comfortable position for filing. Clamp the piece tightly enough so that it does not vibrate during filing. If the vise has rough jaws, cover them with soft aluminum or copper caps to prevent damage to the part.

Before using the file, attach it to a tightly fitting handle. Select a handle with a hole large enough for the tang. Insert the tang in the hole and tap the end of the handle on a hard, flat surface to drive the file into place.

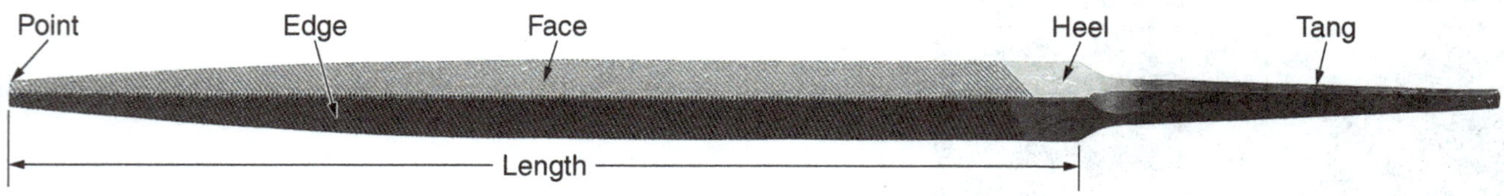

Figure 5-7. All files contain these parts. The length is measured as shown. (Nicholson)

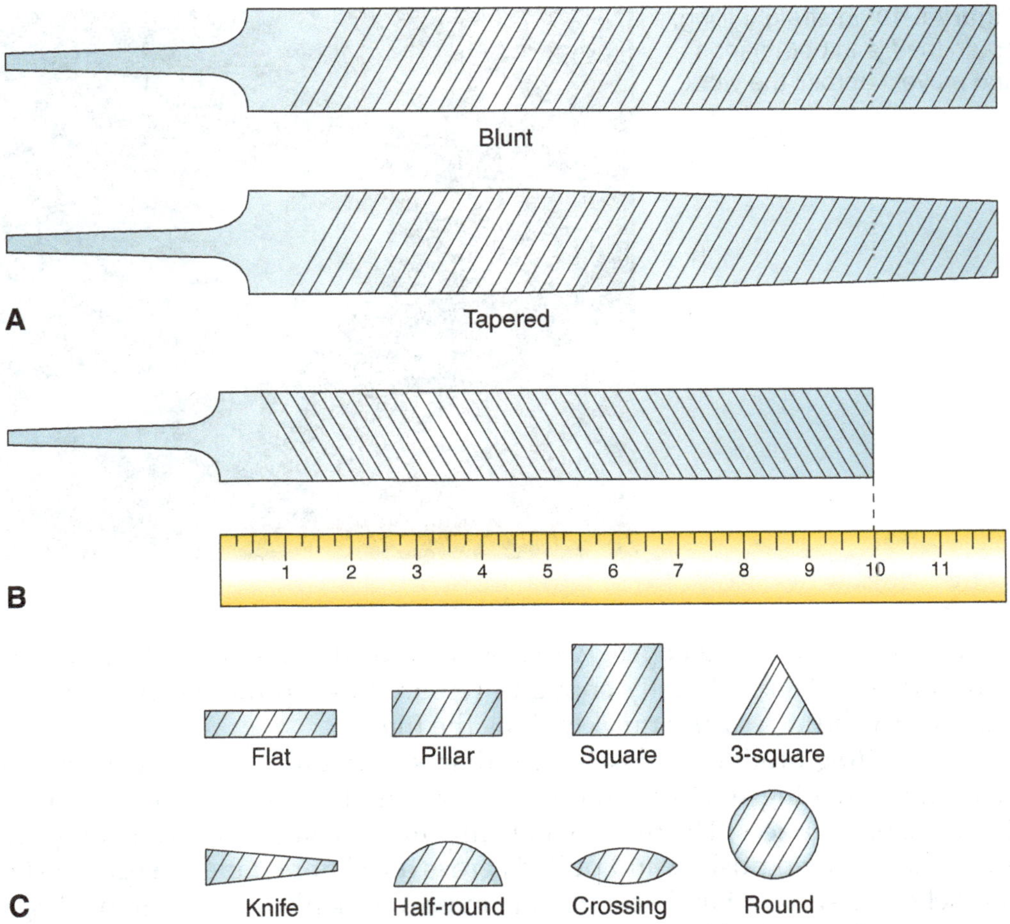

Figure 5-8. File terminology. A—Blunt files have parallel edges. Tapered files decrease in width from heel to point. B—File length is measured in inches, from heel to point. C—File cross section shapes are either circular, three sided, or four sided.

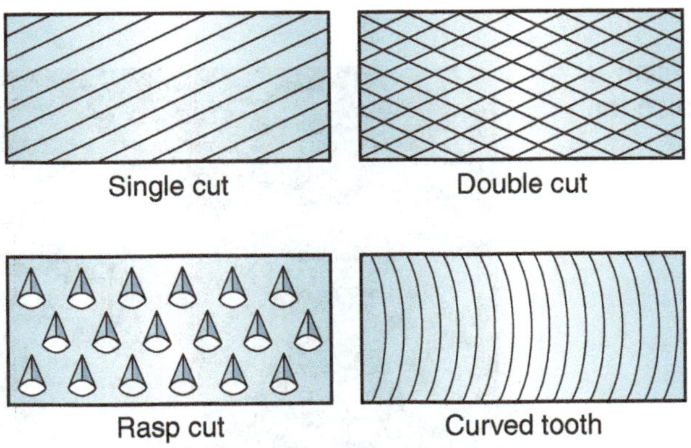

Figure 5-9. Files can be classified according to the types of teeth they contain.

Filing Methods

There are two basic filing methods: straight filing and draw filing. In ***straight filing,*** the file is pushed lengthwise straight ahead or at a slight diagonal across the work, **Figure 5-10.** The file cuts on the forward stroke and is raised off the work on the return stroke. Dragging the file on the backstroke causes the teeth to dull. Use full strokes and keep the file horizontal to

straight filing: Filing method in which the file is pushed lengthwise straight ahead or at a slight diagonal to the work.

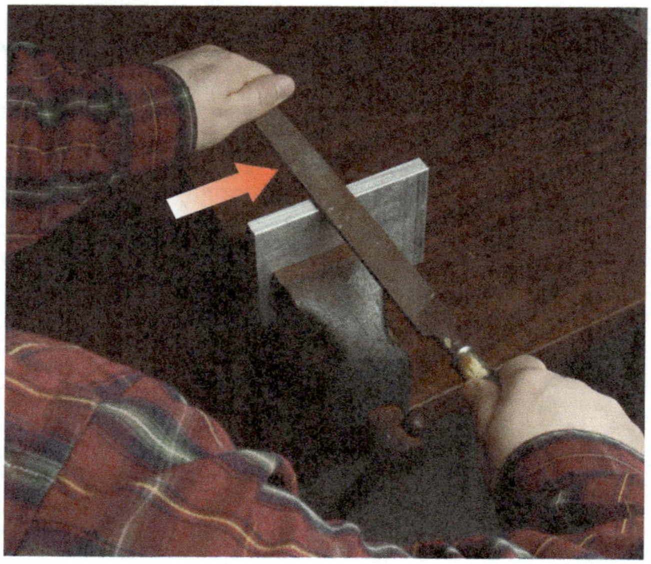

Figure 5-10. In straight filing, grasp the file in both hands and move it across the work.

ensure square sides and edges. Avoid using a rocking motion, which results in rounded ends. Apply only enough pressure to keep the file cutting. Too much or too little pressure quickly dulls the file.

draw filing: Filing method in which the file is pushed and pulled across the work and at a right angle to the work. Produces a finer finish than straight filing.

Draw filing produces a finer finish than straight filing. It is a convenient method for making smooth square edges on small surfaces or for accurately finishing flat surfaces. The file is held firmly in both hands, at a right angle to the work. It is then pushed and pulled across the clamped piece, **Figure 5-11,** parallel to the surface. Single-cut files are most commonly used for draw filing.

During filing, the file teeth clog with metal chips and filings. These chips and filings must be removed to avoid scratching the filed surface. Chips and filings are removed with a file card and a file brush, **Figure 5-12.** The card loosens the debris and the brush removes it from the file.

Figure 5-11. Draw filing produces a finer finish than straight filing.

Store files separately from other tools. Hang them by their handles in a rack or store them in a soft file caddy. Keep them away from moisture, to prevent rusting.

Chisel

Chisels are used to cut, shear, or chip cold metal. Chisels are used with a hammer in a two-handed operation. Hold the chisel in one hand against the cutting surface and the hammer in the other hand. Tap the hammer against the chisel to make the cut.

Four commonly used chisels are the flat chisel, the cape chisel, the round-nose chisel, and the diamond-point chisel, **Figure 5-13.** Each chisel has a slightly different head shape, designed to perform certain tasks.

The ***flat chisel*** is used for general cutting work. It can be used to chip metal, cut thin metal sheets, and cut rivets and bolts. The cutting face angle on a flat chisel is 35°. The ***cape chisel*** tapers to a narrow cutting edge. It is designed to cut grooves and keyways in corners. The ***round-nose chisel*** is designed to cut grooves and to chip inside corners. The ***diamond-point chisel*** is used to cut grooves and to square corners. It is square at the point and ground at an angle across the diagonal corners, making the cutting face diamond shaped.

flat chisel: Chisel used for general cutting work.

cape chisel: Tapered chisel used to cut grooves and keyways in corners.

round-nose chisel: Chisel used to cut grooves and to chip inside corners.

diamond-point chisel: Chisel used to cut grooves and to square corners. Cutting face is diamond shaped.

Figure 5-12. This combination file card and brush will keep files clean. (Nicholson)

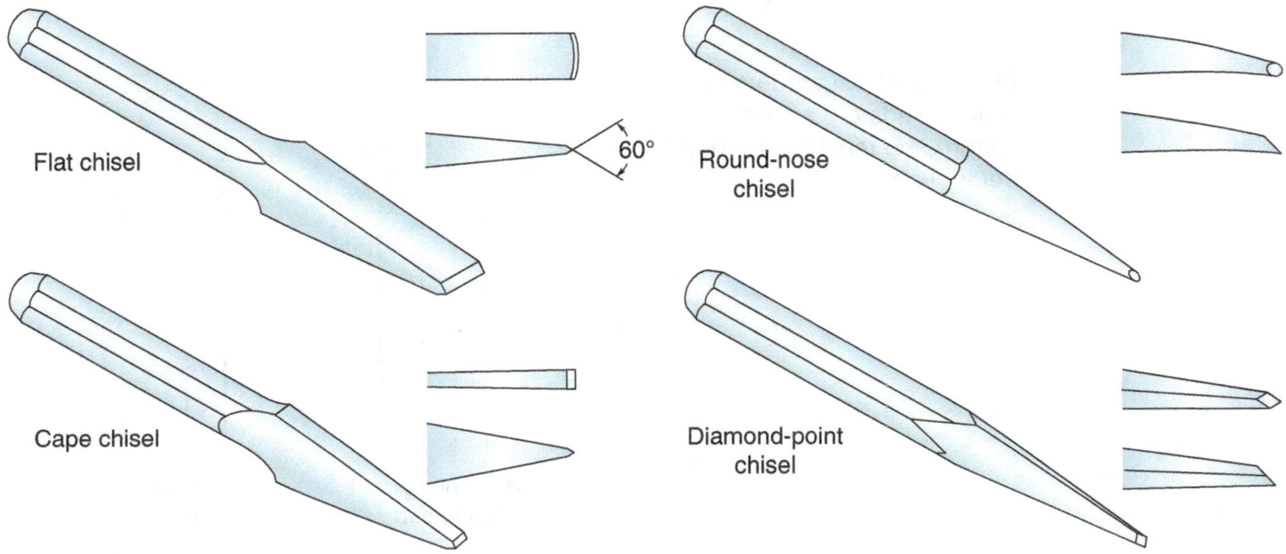

Figure 5-13. Each of these chisels has a uniquely shaped tip, designed to do a specific job.

Chisel Safety

- Wear safety glasses when using a chisel to avoid flying metal chips.
- Secure work in a vise.
- Do not use a chisel with a mushroomed head, **Figure 5-14**. Regrind the chisel before using.
- Work away from your body. Hold the equipment in a way that reduces the possibility of striking your hand with the hammer.
- Remove sharp edges that result from the cutting action. These sharp edges can be removed by grinding or filing the work.

Using a Chisel

Select the chisel based on the type of cut to be made. Choose a hammer that is heavy enough to move the chisel along the work. Hold the chisel loosely in one hand and the hammer in the other. Place the chisel against the work at an angle that will cause the chisel to enter the metal. Once started, continue cutting to the proper depth. Changes in the angle or depth of the cut can be made by changing the angle of the chisel.

Hammer and Mallet

Hammers and mallets are striking tools. They both consist of a head and a handle. Hammers are used with chisels and punches and for other general striking tasks. Hammers are classified by the weight of the head, without the handle. Hammer heads are commonly made of steel or alloys, while mallet heads are normally made of leather, plastic, rubber, or wood.

Commonly used metalworking hammers are the machinist's hammer, the riveting hammer, the setting hammer, and the mallet. The machinist's hammer comes with either a ball peen, straight peen, or cross peen. The ball peen is the most commonly used, **Figure 5-15.** The riveting hammer is used to flatten seams and rivets. It has a smaller head to eliminate damage to the

Figure 5-14. Mushroomed chisel heads are a safety hazard and must be reground before use.

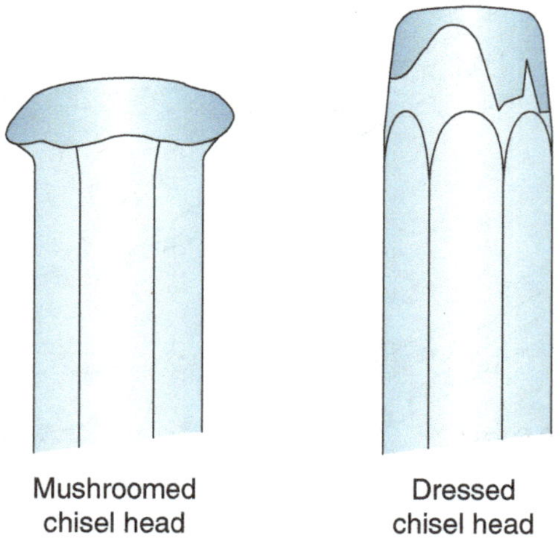

surrounding metal. The setting hammer is also used to flatten seams and to flatten or form metal. The mallet is used to apply light blows to materials that might be damaged by a heavier blow from a hammer.

Using a Hammer

Do not choke up on the handle. Instead, grip the handle as near the end as possible. Each blow to the work surface will be more efficient than when holding the handle near the head. Strike the work surface with the full face of the hammer head.

Metal Stakes

Metal stakes are used with hammers and mallets to bend and fold sheet metal, **Figure 5-16.** The metal is placed on the stake and is then struck with either the hammer or the mallet. The stakes are designed to fit into a bench plate attached to a table. This holds them securely during use.

Figure 5-15. Ball peen machinist's hammer. (Klein Tools)

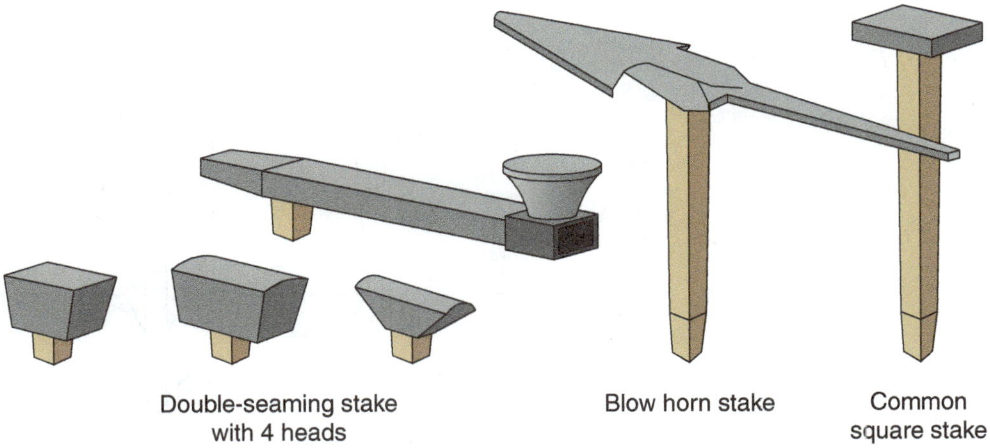

Figure 5-16. Metal forming stakes comes in a wide variety of styles.

Screwdriver

A screwdriver is used to tighten or loosen screws. There are many types of screwdrivers, but the most common are the standard and the Phillips, **Figure 5-17.** Standard screwdrivers have a straight, flat tip that fits into the screw slot. Phillips screwdrivers are used on Phillips-head screws. On these screws, two slots intersect each other at the center of the screw head. These slots prevent the screwdriver from sliding out of the screw slot and marring the work.

Choose a screwdriver whose blade fits snugly in the screw slot. A poorly fitting screwdriver can damage the screw and screwdriver blade. Check the screwdriver blade before using. The sides of a properly ground blade are nearly parallel and the blade is slightly concave. Damaged blades must be reground on a grinding wheel.

When regrinding *any* hand tools, do not hold the tool against the grinding wheel for longer than is recommended or the tool blade may overheat. An overheated blade will draw the temper, after which the tool is soft and useless.

Pliers

The primary purpose of pliers is to grip, but they are also used to cut and form metal. There are many different shapes and sizes to allow for a variety of situations, **Figure 5-18.** ***Slip-joint pliers*** have an adjustable hinge, useful for holding different sizes of work. ***Needle-nose pliers*** are used to hold work in tight spaces and to shape light metal and wire. They have a long nose and flat gripping surface. ***Diagonal pliers*** are used to cut wires, nails, pins, and screws. ***Utility pliers*** can hold large work in their parallel jaws. They are ideally suited for heavy-duty work.

slip-joint pliers: Pliers with an adjustable hinge, useful for holding different sizes of work.

needle-nose pliers: Pliers with a long nose and flat gripping surface, used to hold work in tight spaces and to shape light metal and wire.

diagonal pliers: Pliers used to cut wires, nails, pins, and screws.

utility pliers: Pliers with parallel jaws which can hold large work. Useful for heavy-duty work.

Figure 5-17. Two types of screwdrivers.

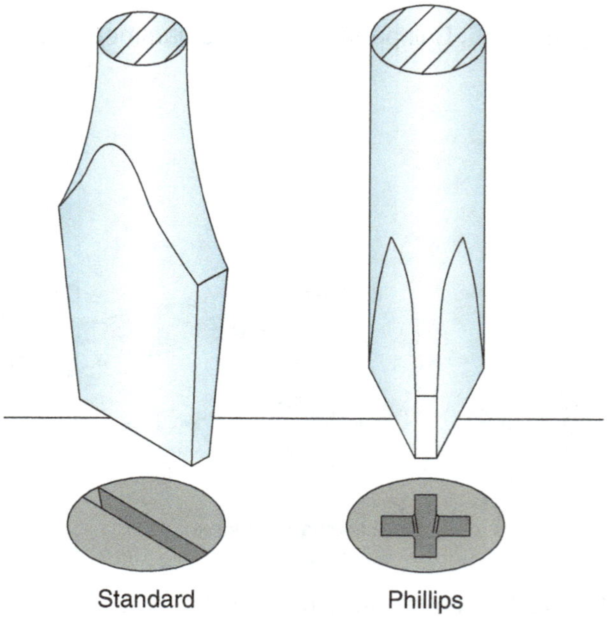

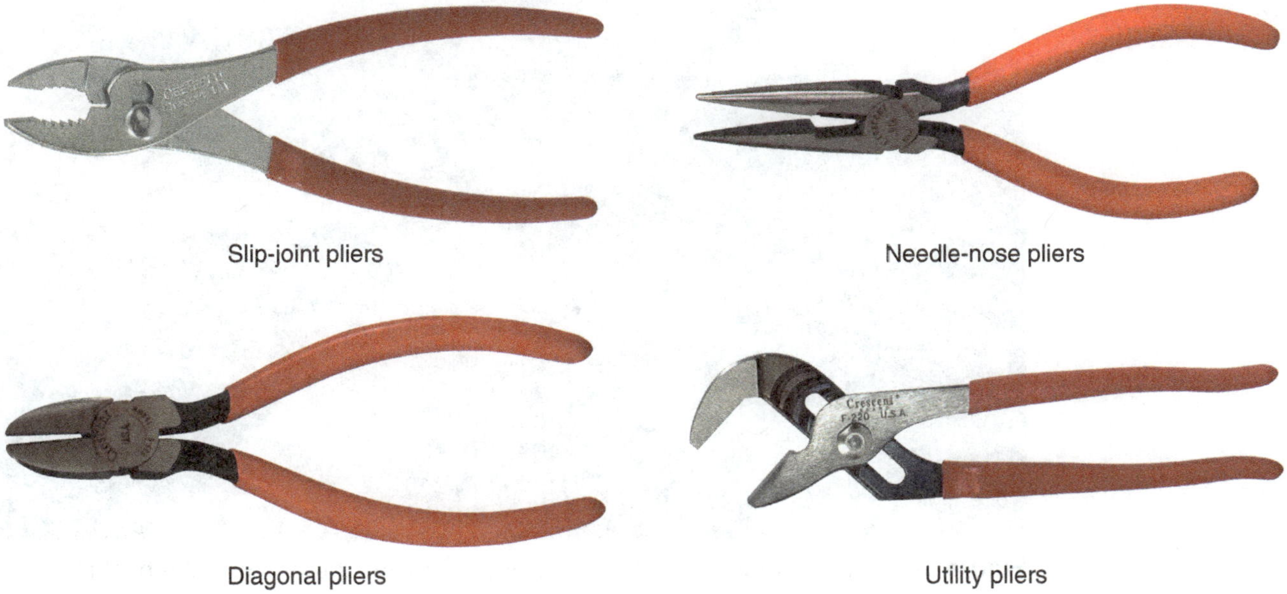

Figure 5-18. Types of pliers. (Crescent)

Wrench

Wrenches, both adjustable and nonadjustable, are used to tighten and loosen nuts and bolts. Adjustable wrenches are available in different sizes and have an adjustable jaw to accommodate different size nuts and bolts. When using an adjustable wrench, place the wrench on the nut or bolt so the force is on the fixed jaw. Then tighten the adjustable jaw until the wrench fits the nut or bolt snugly, **Figure 5-19.** These wrenches are not designed for heavy-duty work.

Nonadjustable wrenches have fixed openings to fit nuts and bolt heads. The distance between the jaws determines the wrench size. Choose a wrench that fits squarely on the sides of the nut or bolt head, **Figure 5-20.**

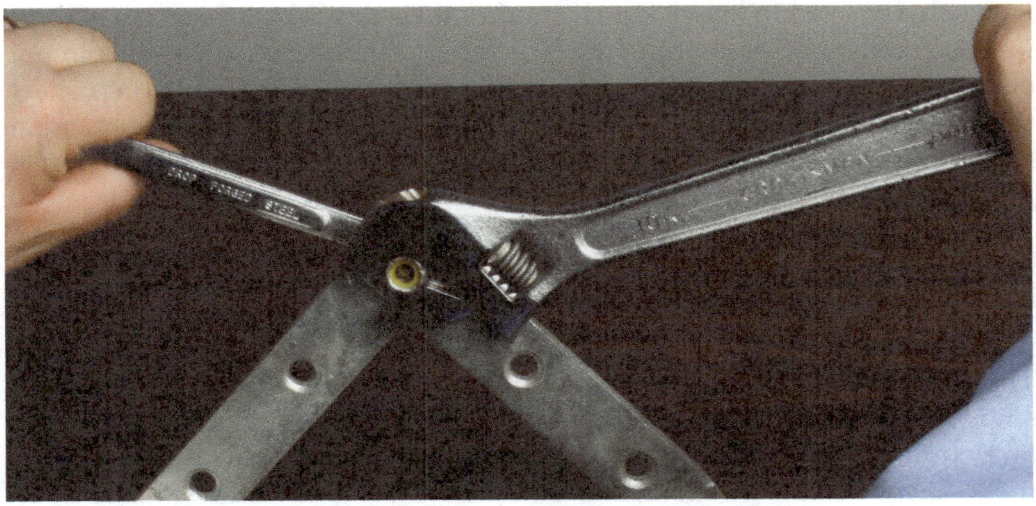

Figure 5-19. This adjustable wrench is properly tightened onto the nut.

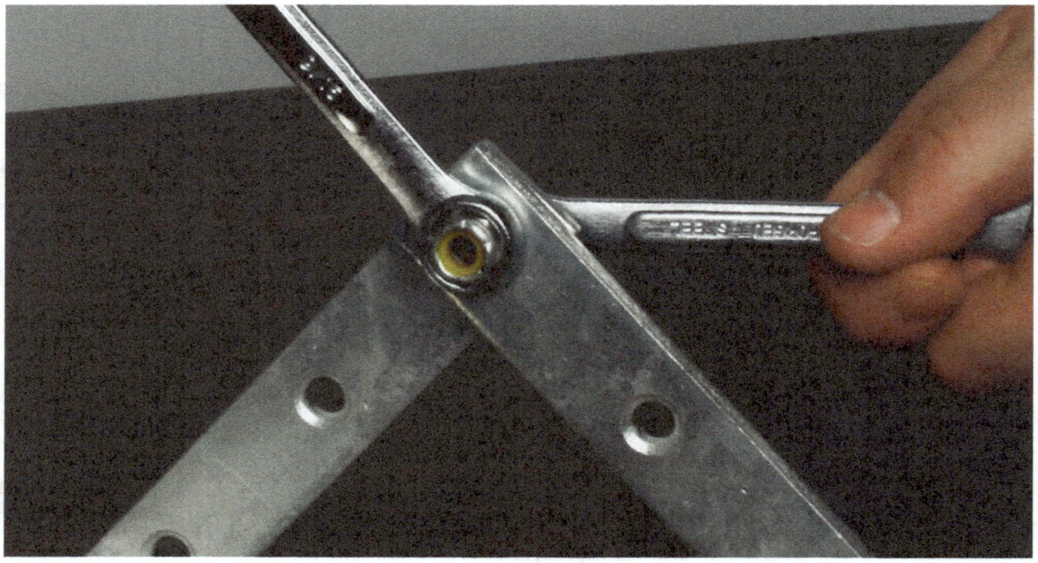

Figure 5-20. Try several nonadjustable wrenches to find the one with the best fit.

It is much safer to pull the wrench rather than push it. A pushed wrench can break loose suddenly, resulting in an injury to the hand. If a wrench must be pushed, use the palm of the hand so that the knuckles will not be injured if there is a slip.

Allen wrench: A hexagonal wrench with a 90° bend. It comes in a number of sizes and fits screw heads that have a similarly shaped hole.

Allen wrenches are designed to be used with headless setscrews. They are available singly or in sets, **Figure 5-21.** The hexagonally shaped wrench is designed to fit the hexagonally shaped socket in the setscrew.

Hand Punch

hand punch: A blunt tool used to pierce a solid material.

solid punch: A hand punch used to make small diameter holes in light-gage metal.

pin punch: A hand punch used to drive pins out of holes.

A ***hand punch*** has either a pointed or blunt tip. It is used to make indentations, mark locations for holes, knock out rivets, loosen pins, and drive objects through metal. Punches used often in metalworking are the ***solid punch, pin punch,*** prick punch, and center punch, **Figure 5-22.**

The solid punch is used to make small diameter holes in light-gage metal. It can also be used to knock out rivets after their heads have been cut off and to loosen pins. After loosening the pins, use a pin punch to drive the pin out of the hole. The prick punch is used to transfer measurements onto sheet metal. The center punch is used to start a hole to be drilled. The indentation made by the punch prevents the drill from wandering. The center punch point has an angle of 90°.

Hand Snips

hand snip: A cutting tool used to cut sheet metal. Another name for *shears.*

shears: A cutting tool or machine used to cut out sheet metal patterns.

Hand snips, or ***shears,*** work much like scissors but are used to cut metal. The blades on the snips are either straight or combination. Straight-blade snips are used to cut straight lines. The blade on combination snips bends away from the cut, causing the metal to slide over the top of blade. This design makes combination-blade snips ideal for cutting curves. Hand snips used by the metalworker include general-use snips and aviation snips.

Figure 5-21. Allen wrenches are also known as hex wrenches. (Klein Tools)

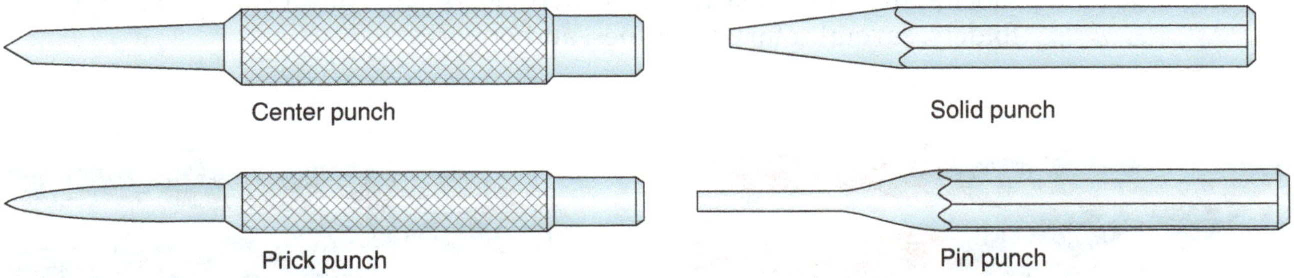

Figure 5-22. Punches are used to perform a variety of striking jobs. Note the different ends.

General-purpose snips are used to cut metal that is 26 gage or lighter. They have either a straight blade or a combination blade and are used for most general cutting jobs. Aviation snips are used to cut both straight and curved cuts in thin-gage sheet metal. They are made in three styles, **Figure 5-23.** The serrated blade edge and compound-levered cutting action provide easy one-hand operation.

Using Hand Snips

1. Choose snips based on the gage of the material and on the desired cut (straight cut, right curve, left curve).
2. Open the snips as far as possible. Insert the metal. Hold the cutting blade at a 90° angle to the work.
3. Close the snips slowly. Follow the layout line. Stop cutting when the metal is 1/4″–1/2″ (7 mm–13 mm) from the end of the blade. Open the snips again and repeat the process.
4. When the cut is finished, deburr the metal to prevent cuts to the hands. Deburring can be done with a file, but using a hand- or machine-operated deburring tool is more efficient.

Do not use hand snips to cut wires and rods. Instead, use tools designed for that job. Use a rod parter to cut small rods to length.

Sheet Metal Machinery

Sheet metal machinery is available to do all sheet metal jobs that can be done by hand. Because the machines do the work more quickly, production increases. More work is completed in less time. Larger pieces of metal are cut and formed. Multiple pieces are made more quickly. Seams can be created more easily. These machines include shears, slip-rolls forming machines, bar folders, several rotary machines, and brakes.

Left cutting

Right cutting

Straight cutting

Figure 5-23. Offset aviation snips. (Klein Tools)

Shears

After a layout is complete, the material is cut out of the metal sheet. In addition to hand snips, several types of hand-operated machines are available for cutting sheet metal.

Throatless shears are used extensively in sheet-metal shops, **Figure 5-24.** As stock is passed through the shears, the sheared sections separate, going to the right and left, under and over the spiral-shaped head. Because the shears have no handles to impede the passage of the sheet metal, there is no limit on the width of stock that can be sheared. The machine can also be used to cut irregular and circular patterns. The upper cutter is serrated to feed the stock, while the lower cutter revolves freely in a sleeve.

throatless shears: A cutting machine used to make straight, curved, and irregular cuts in sheet metal. It is designed so that the piece of metal being cut can be moved into different positions for difficult cuts.

Foot-operated ***squaring shears,*** **Figure 5-25,** are a necessity in most sheet metal shops. Most squaring shears can cut up to 15-gage mild steel sheet metal. Squaring shear size is based on the maximum width of sheet metal the machine can handle. Sizes are usually between 30″ and 36″ (762 mm and 914 mm).

squaring shears: A foot-operated machine used to trim or square large metal sheets. It can cut metal up to a maximum gage.

Long sheets of metal are inserted into the back of the squaring shears. Most machines have a graduated scale for setting the front gage to the desired length. The front gage can also be used when cutting several pieces of identical length. Smaller lengths of metal may be cut by inserting from the front, against the back gage.

Do not cut band iron, wire, or heavy gage metal on the squaring shears. These can nick the blade, causing irregularities in the cut edge.

Figure 5-24. Throatless shear.

Figure 5-25. Foot-operated squaring shear. (Tennsmith, Inc.)

Squaring Shear Operation

To safely use foot-operated squaring shears, review these steps.

1. Make certain that the machine is in workable order before proceeding. The machine should be cleaned, oiled, and adjusted.
2. Inspect the bed and blade for any wires, tools, or rules left on the machine. Remove them as they can get under the blade and cause damage.
3. Set the back gage to the dimension required for the length of the material. Check both ends of the gage with a flexible rule.
4. Do not exceed the capacity of the machine. In most instances, 16-gage mild steel is the limit.
5. Place the metal against the left-hand side gage. Slip it under the hold-down until the metal contacts the back gage.
6. Firmly press the side of the sheet in place, using both hands. Keep fingers away from the cutting blade and foot behind the treadle.
7. Press down on the foot treadle with the other foot.
8. Once the metal is cut, release the foot treadle slowly back to its original position.
9. Remove the metal from the shelf behind the shears.
10. Deburr the sharp edges with a file.
11. Remove and store leftover material.

Ring and Circle Shears

Ring and circle shears are used for cutting rings and circles inside pieces of sheet metal. This machine is used when the cut cannot begin on an edge of the metal sheet. With the clamping head positioned for the desired diameter, the metal is inserted, the cutting disk is lowered, and the cut is made.

Motor-Driven Shears

Many of the hand-operated shears are also available with motors. Motors allow an increase in productivity over hand-operated machinery. Motor-driven machines are found mostly in large sheet metal shops, where a great amount of work needs to be completed in a minimum amount of time.

Large special-purpose shears can be found in sheet metal industries. These large shears that can handle sheets or plates of metal up to 1/2″ (13 mm) thick and 12′ (4 m) wide. Hydraulic shears are available in standard shearing capacities for 1 1/2″ (38 mm) plate.

Forming Machines

In addition to the cutting and punching operations performed with snips, shears, and punches, metal can also be formed into arcs, spirals, circles, cylinders and cones by means of forming equipment. These machines vary from small hand-operated forming rolls to large motor-driven bending rolls and forming presses.

Slip-Rolls Forming Machine

The ***slip-rolls forming machine*** is used to bend sheet metal into a curved form, **Figure 5-26.** The machine has two geared rollers in front and one loose roller in back. The sheet metal is formed by all three rollers. The two front rollers grip the metal sheet and force it against the rear roller. This action bends the metal around the front upper roller, forming a cylinder, **Figure 5-27.** After the cylinder is formed, the upper roller is slipped out of place to remove the cylinder.

slip-rolls forming machine: A forming machine used on light gage sheet metal to bend stock into cylindrical shapes.

The distance between the geared rollers depends on the thickness of the metal. This distance can be changed using an adjusting screw on the lower front roller. Raising the rear roller will create a smaller radius and lowering it will form a larger radius. The lower front roller and the rear roller have grooves cut along their right-hand edges. These grooves are used to either form wire or to form cylinders with the wire edge already installed. The standard forming rolls are 18″ (457 mm) wide.

The powered slip-rolls forming machine is used in large production shops to perform the same function as the hand-operated machine. It is used on 16 gage and lighter sheet metal. The roll drive is controlled by a toggle control. The covered foot pedal provides instant forward and reverse action. All three rolls are gear-driven to ensure even starting and feeding of materials. The machine has large roll adjusting screws, scales to speed repeat setups, a wide clearance top-roll latch, and an end-mounted handle for easy lifting of the upper roll.

Figure 5-26. Hand-operated slip-rolls forming machine. (Tennsmith, Inc.)

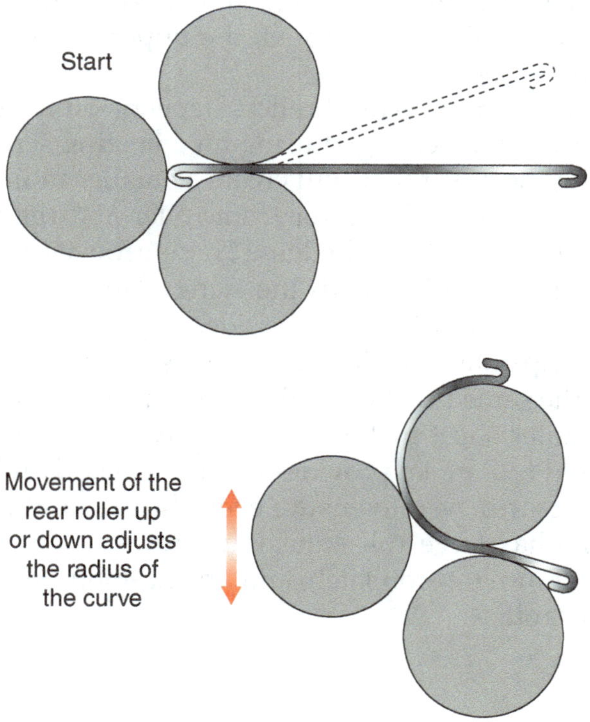

Figure 5-27. Follow the path of the metal through the rollers.

Bar Folder

The ***bar folder*** bends sheet metal to make formed edges, ***seams,*** and ***hems,*** **Figure 5-28.** It clamps and folds in one motion. The bar folder is also used to form edges to hold wire. The width of the folded edge is determined by the setting of the depth gage. The angle of the fold is determined by the wing. Both 45° and 90° bends can be made on bar folders.

bar folder: A forming machine used to bend sheet metal for formed edges. It is also used to form seams.

seam: A method of joining sheet metal pieces together. Ends of sheet metal are folded together and then fastened, usually with rivets or solder.

hem: An edge used to strengthen sheet metal objects. The edge has a rolled-over or rounded appearance.

Grooving Machine

Seams can be closed using a grooving machine or a hand groover. A grooving machine can tighten or close the seam on cylindrical shaped piece of metal. To close a seam by hand, follow the steps outlined here.

Making a Hand-Grooved Seam

1. Hook together the pieces to be joined, **Figure 5-29A.**
2. Position the pieces on a flat stake and slightly flatten the seam with a mallet.
3. Choose a hand groover with a slit approximately 1/16″ (2 mm) wider than the seam, **Figure 5-29B.**
4. Position the groover over the seam at one end and strike it lightly but firmly with a hammer. Repeat the process at the other end.
5. Groove the rest of the seam by moving the groover along the seam and striking the hammer.
6. Make two or three passes over the seam. This will make the seam neat.
7. Use a mallet to flatten and smooth the seam. Do not use a hammer as it will mar the finish of the metal.

Rotary Machine

Rotary machines are used to make edges on round and cylindrical objects, **Figure 5-30.** The machines contain rollers of different shapes. The rollers pull the metal piece through them, forming the edge. This action is similar to, but not the same as, the action of a hand-operated can opener. Combination machines come with a variety of interchangeable rollers.

rotary machine: Forming machine used to form edges on round and cylindrical object. Comes with interchangeable rollers.

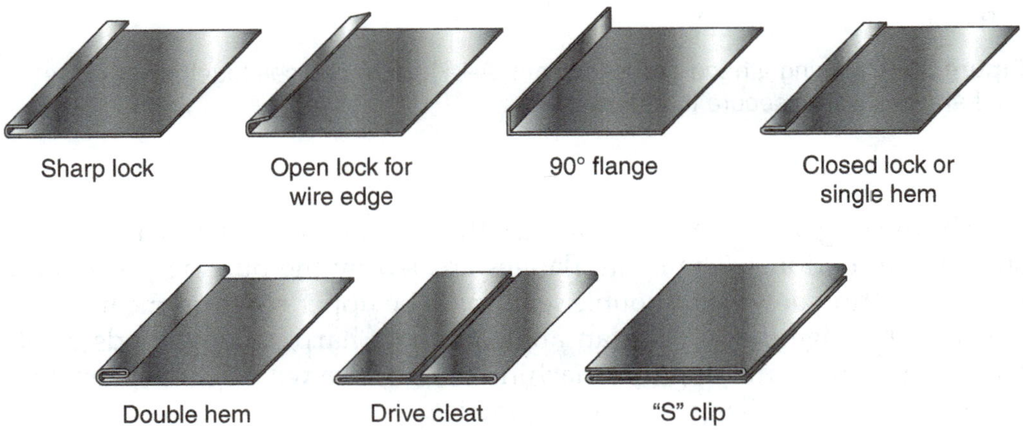

Figure 5-28. Some of the forming work that can be done on a bar folder.

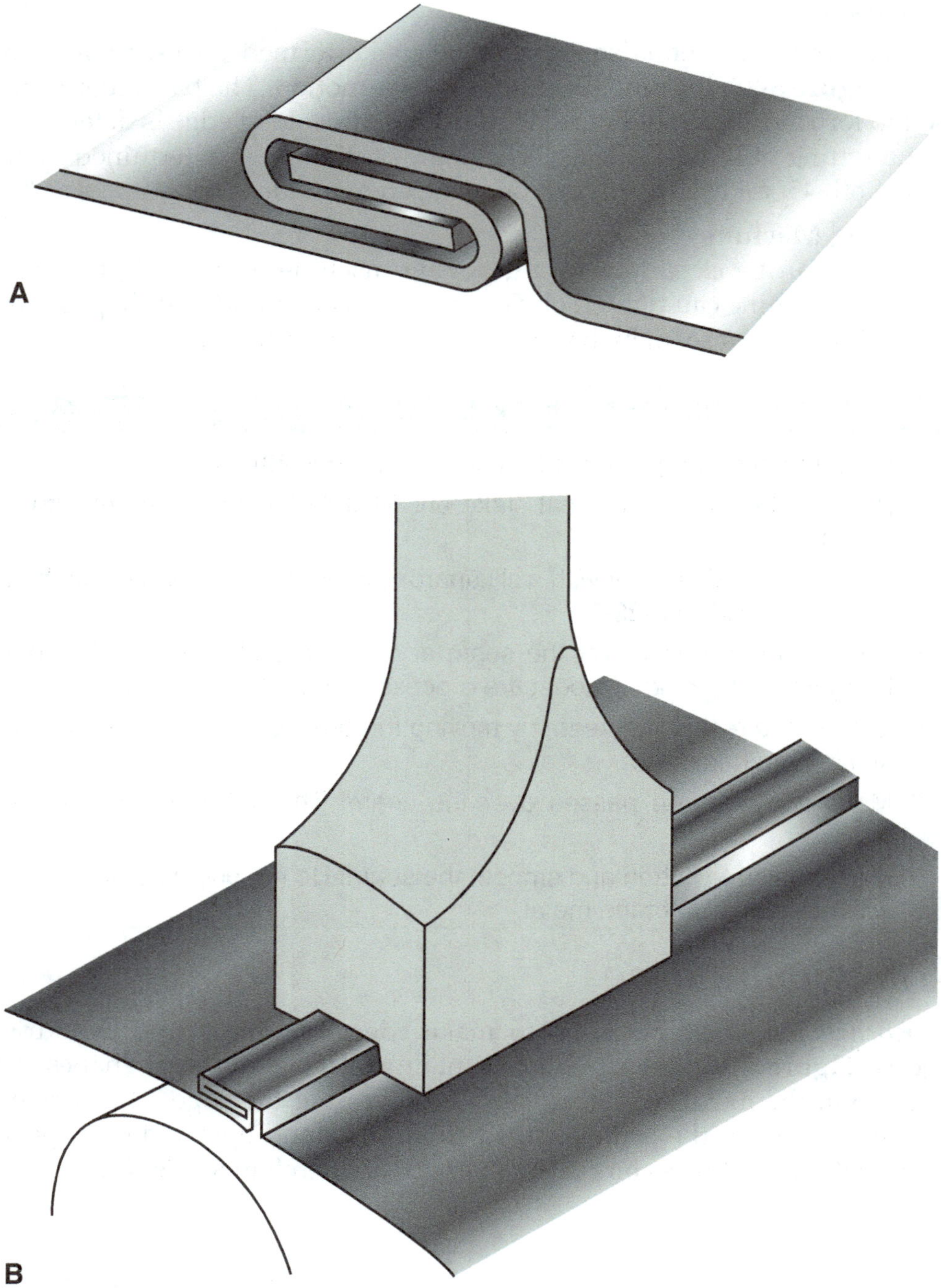

Figure 5-29. Making a hand-grooved seam. A—Hook the pieces together. B—Use the hand groover to secure the seam.

burring machine: A machine used to make small flanges on the edges of sheet metal discs. Flanges are used as part of a seam.

The ***burring machine*** is used to turn the edge on metal cylinders or disks, such as can bottoms. The metal flanges created by the burring process are used for setting down and double seaming. The upper roller on the machine has a sharp edge that creates an angled bend. Sharp, angled bends create stress at the bend, making this machine best for use with light gage metals. Producing uniform flanges requires practice.

Figure 5-30. Rotary machines are used to turn edges on cylindrical work. (Tennsmith, Inc.)

The ***turning machine*** creates a rounded edge that holds a wire. It is used to turn heavier metal than a burring machine can turn. After wire is inserted, the edge is then closed by a ***wiring machine,*** **Figure 5-31.** The seams created to hold the wire can also be made on the bar folder or brake, instead of on the turning machine. They are still closed on the wiring machine.

turning machine: A forming machine that creates a rounded edge for holding wire.

wiring machine: A forming machine that closes the edge surrounding a wire.

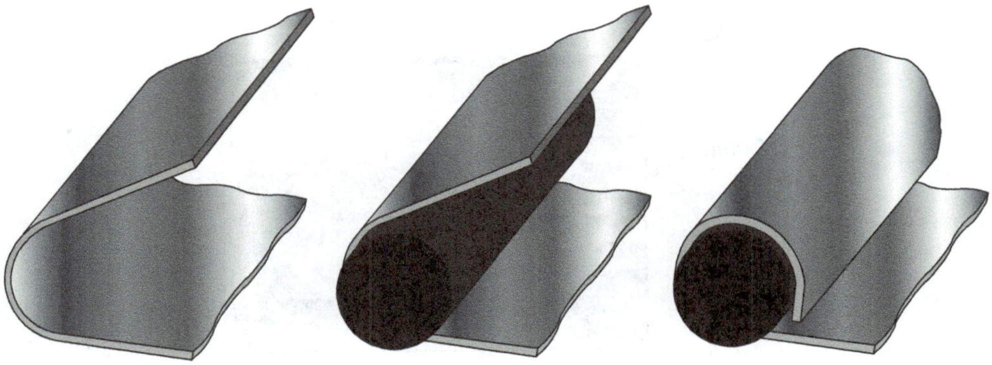

Figure 5-31. Creating a seam and inserting the wire.

setting-down machine: A forming machine that prepares the seams in cans and other vessels for double seams.

double-seaming machine: A forming machine that folds a joint over, securing the seam.

beading machine: A machine used to form beads or impressions on sheet metal.

bead: A decorative impression made on sheet metal that also strengthens and stiffens the metal.

crimper: A machine that reduces the diameter of the end of a sheet metal pipe, allowing it to slide into a pipe of the same diameter.

The ***setting-down machine*** closes the seams left by the burring machine and prepares them for double seaming. This machine has only one adjustment, for the thickness of the material being formed. The machine turns down and compresses the flange of the bottom onto the flange at the end of the body, forming a joint. This joint is then doubled over with a ***double-seaming machine.*** The entire progression is shown in **Figure 5-32.** The double-seaming machine makes an accurate, neat, tight joint.

The ***beading machine*** makes impressions, or ***beads,*** on the work, for reinforcement and decoration. The beads are made using interchangeable rollers. The work is placed between the rollers and then rolled through them. A powered beading machine is designed for deep-throat (up to 13″ or 330 mm) beading, crimping, and other roll operations on material up to 18 gage.

The ***crimper*** reduces, or crimps, the end diameter of one piece, allowing it to slide into another piece of the same diameter. This eliminates the need to build each section with a taper. An ogee bead is often made next to the crimp, preventing the overlapping pipe from sliding beyond the bead, **Figure 5-33.** The crimping and beading machine performs both functions.

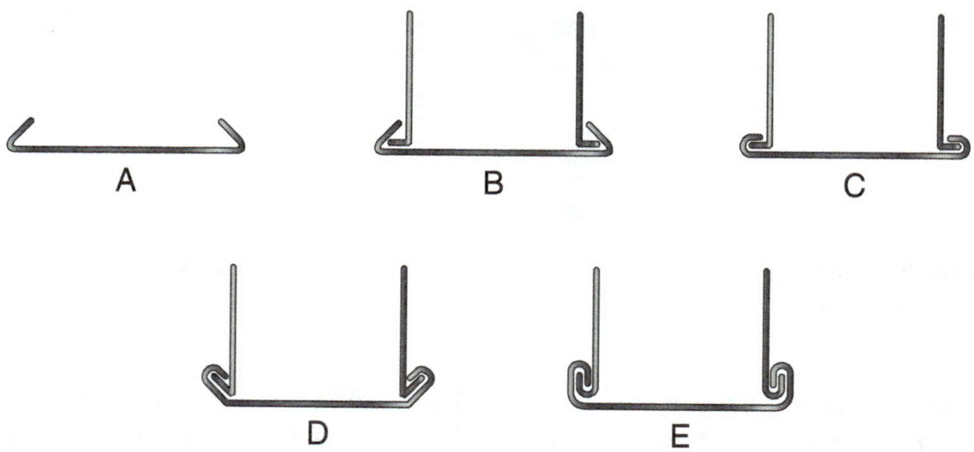

Figure 5-32. Burring, setting down, and double seaming. A—Burred piece that will become bottom of cylinder. B—Burred piece joined with flanged cylinder. C—The compressed seam after setting down. D and E—Seam progressively turned over by double-seaming machine.

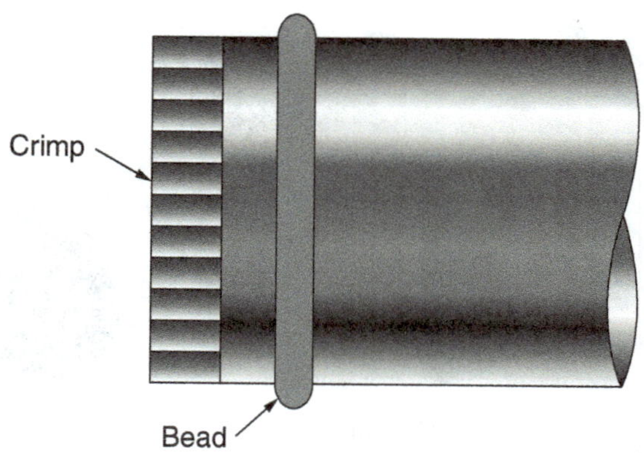

Figure 5-33. Beads keep overlapping pipes in place.

Brakes

Sheet metal is usually bent to desired shapes using a ***brake.*** The universal hand-brake is a staple in the sheet metal shop. It is large and has many parts, **Figure 5-34.** Many of the parts are adjustable, allowing the brake to be used to perform a variety of tasks. It is used to do straight bending on large pieces of sheet metal. It can also be used to close a seam.

brake: A machine used to bend and fold metal. Can be mechanical or hydraulic.

Specialized brakes are used to make other, more difficult bends. These machines include a ***box and pan brake*** and a ***cornice brake.***

The box and pan brake operates in a similar manner as the universal, or standard, brake. However, instead of a solid upper jaw like on a standard brake, the box and pan brake has interchangeable, removable fingers. These fingers can be set up quickly, in a variety of ways. The fingers that are not needed are removed, and only the fingers needed to make the bend are used. Box and pan brakes are used primarily to form four-sided pans and boxes.

box and pan brake: A machine that produces folds and bends on long pieces of sheet metal. Variation of the cornice brake.

cornice brake: A forming machine used to form large pieces of sheet metal. Contains forming blocks that can be arranged in different ways to form different shapes.

The cornice brake is useful for bending large pieces of metal at a set angle. The stop gauge on the brake allows the user to set the angle, limiting the travel of the bending leaf. Cornice brakes can also be used with molds to form metal into a variety of shapes. The cornice brake was originally designed for making decorative cornices on buildings.

Figure 5-34. Hand-operated brake. (Tennsmith, Inc.)

press brake: A powered machine used to bend heavy sheet metal and steel plate. Dies used on the machine produce a large number of different bends and shapes.

Press Brake

A ***press brake*** is a machine found in large shops that performs a wide variety of forming and bending operations. Metal is formed by pressing it into dies. Additional steps are often necessary to complete the work. These machines are not hand operated, but run on hydraulic or electric power. Sizes range from 4′ to 35′ (1 m to 11 m), **Figure 5-35.** Many are computer operated. Force is measured in tonnage.

Powered Punches

Heavy-duty punches are ideal for making many uniform holes in a short amount of time. These machines can punch round and irregular shapes, as well as set rivets. All punching tools have maximum capacities for safe, dependable operation. Do not exceed the rated capacity for the tool.

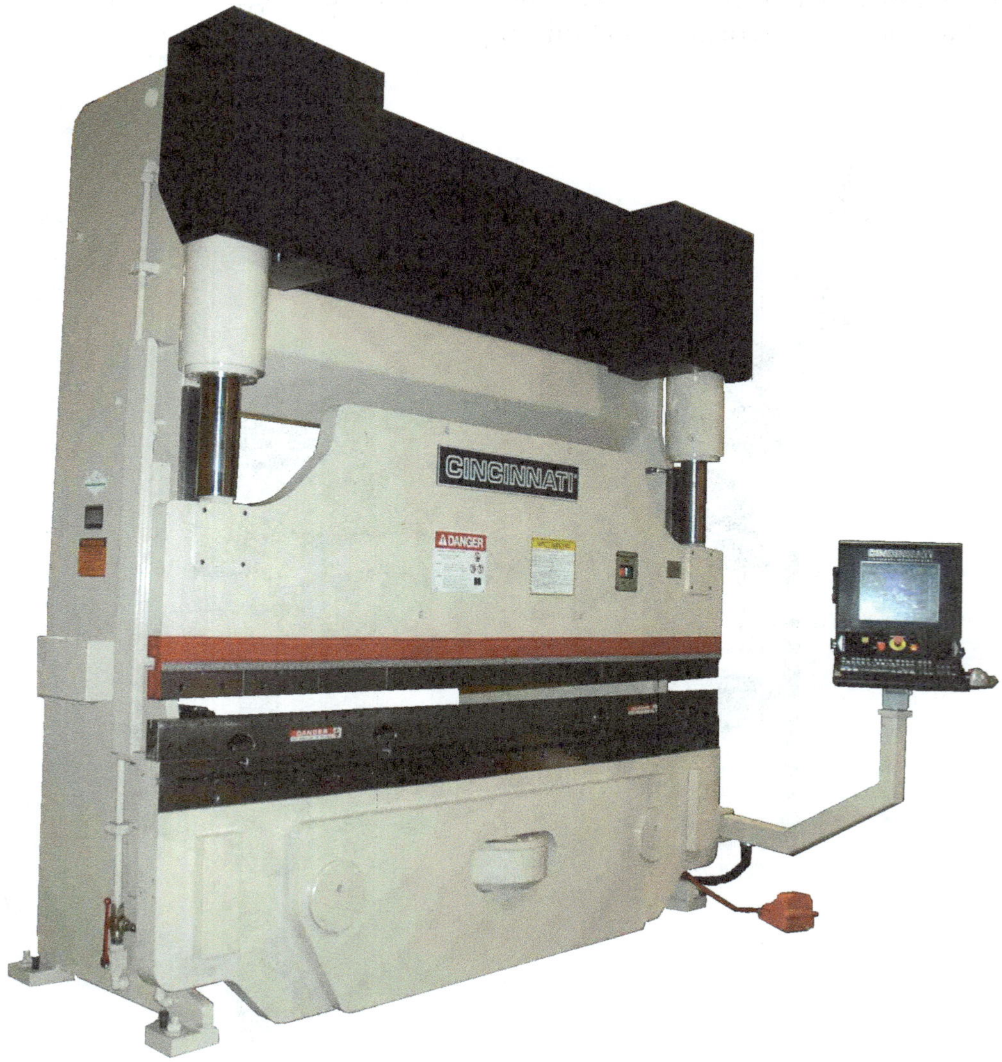

Figure 5-35. Press brake. (Cincinnati Incorporated)

Summary

Sheet metal workers need a wide variety of hand tools and machinery to cut, bend, form, seam, roll, and turn. In general, hand tools and hand-operated machinery is used to complete smaller, individual jobs, such as installing a custom HVAC system in a single house. The large, motorized machines are used in industrial settings, for large production jobs, such as a downspout that is used in thousands of gutter systems.

Hacksaws, files, and chisels are cutting hand tools used by sheet metal workers. Each is used for a particular cutting task and has specific requirements for selection and use. Hammers and mallets are striking tools and often used with other tools to perform a task. For example, the mallet is used to flatten seams without damaging surrounding metal. Screwdrivers, pliers, and wrenches are used to tighten or loosen screws, bolts, and nuts. Pliers can also be used for cutting and forming. There are several types of hand punches and each is used for a certain task. Hand snips are another type of cutting tool. They work much like scissors and come with either a straight or combination blade. Do not use hand snips to cut wire.

While none of the tools mentioned seems particularly dangerous, like all tools they require users to be cautious and aware. Do not use a tool for anything other than its intended purpose.

The sheet metal profession also has a variety of machines available to do some jobs more easily and quickly. Shears, including squaring, ring and circle, and motor driven, are used to cut metal. Slip-rolls and bar folders are used to form sheet metal. Forming tasks includes making arcs, spirals, circles, cylinders, and cones. Rotary machines have interchangeable rollers, allowing the same machine to be used for burring, setting down, beading, crimping, and turning. Finally, bending can be done using hand-operated brakes. Large production runs are done using press brakes.

Internet Resources

Roper Whitney of Rockford, Inc.
www.roperwhitney.com

Tennsmith, Inc.
www.tennsmith.com

Cooper Tools
www.coopertools.com

Klein Tools, Inc.
www.kleintools.com

Outside Activities

1. Using a classroom hacksaw, demonstrate the proper method for inserting and tightening the blade. Locate a piece of metal that can be cut using the blade in the saw. Continue the demonstration by securing the material in a bench vise and showing the proper cutting method.
2. Obtain a single-cut and a double-cut file, with proper handles attached. Use the double-cut file to take a quick cut out of a piece of aluminum mounted in a bench vise. Next, use the single-cut file to smooth the cut surface left by the double-cut file. Note how quickly the single-cut file fills with metal shavings and needs to be cleaned frequently using a card file and brush.
3. Give a presentation to your classmates on chisels. Use the following chisels: flat, cape, round-nose, and diamond-point. Identify each chisel. Use each to cut a groove or remove a chip in a piece of metal that is securely clamped in a vise. Demonstrate proper safety procedures such as wearing protective eyewear and gloves.

Name ______________________ Class ______________ Score ________

Chapter 5
Sheet Metal Hand Tools and Machinery

Review Questions

Write your answers in the spaces provided.

Match each object with the term that identifies it.

A

B

C

D

E

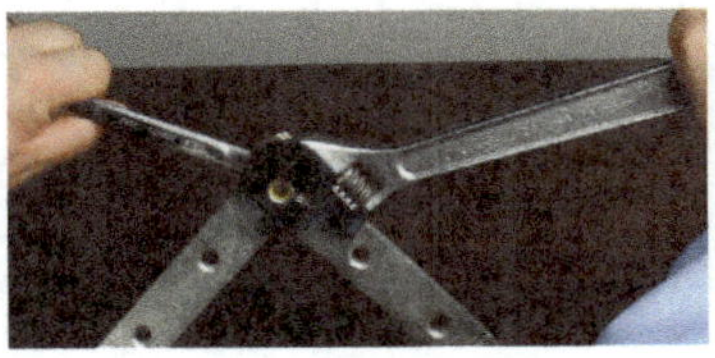

F

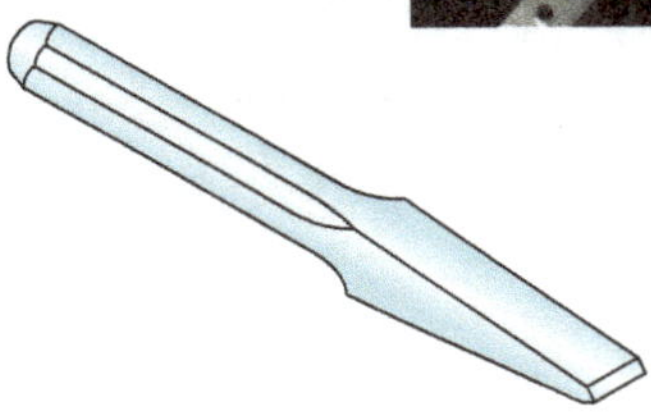

G

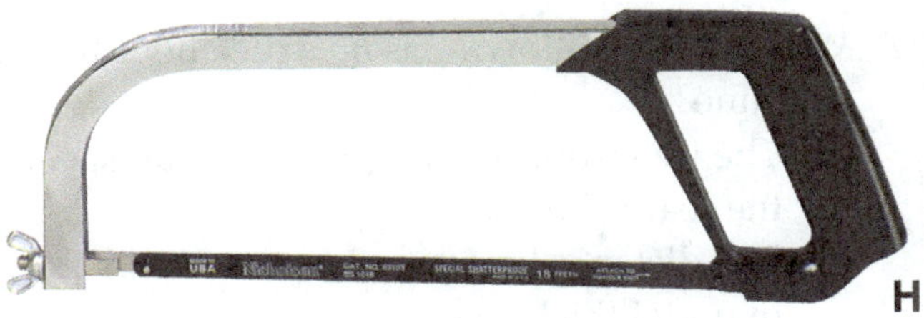

H

1. Flat chisel. 1. ______________
2. Ball peen hammer. 2. ______________
3. File. 3. ______________
4. Allen wrench. 4. ______________
5. Hacksaw. 5. ______________
6. Utility pliers. 6. ______________

7. What is the term for the adjustment made to ensure a blade will cut properly, cleanly, and safely?____

____

8. True or False? The kerf made by an old blade is the same size as the kerf made by a new blade. 8. ____________

9. When choosing a file, consider the _____. 9. ____________
   A. surface being finished
   B. type of metal being filed
   C. amount of material to be removed
   D. All of the above.

10. Name and explain the two most common filing methods. ____

____

____

____

11. Use a(n) _____ to cut, shear, or chip cold metal. 11. ____________

12. Which of the following is not a cutting tool? 12. ____________
   A. File.
   B. Chisel.
   C. Hammer.
   D. Hacksaw.

13. True or False? Pliers should not be used to cut and form metal. 13. ____________

14. A(n) _____ is used to make indentations on metal. 14. ____________

15. Because throatless shears have no handles, there is no limit on the _____ of stock that can be sheared. 15. ____________

16. True or False? A ring and circle shear is used to make cuts that cannot begin on an edge of the material. 16. ____________

17. Which of the following is *not* true about the slip-rolls forming machine? 17. ____________
   A. The two front rollers grip the metal sheet and force it against the rear roller.
   B. The distance between the geared rollers is determined by the material thickness.
   C. Raising the rear roller creates a larger radius.
   D. Lowering the rear roller creates a larger radius.

18. A(n) _____ forms edges, seams, and hems. 18. ____________

19. A turning machine and a burring machine are both types of _____ machines. 19. ____________

20. What is the purpose of beads made on sheet metal? ____

____

Chapter 6
Hems, Edges, and Seams

Objectives

After studying this chapter, you will be able to:

- Discuss the differences between a hem, an edge, and a seam.
- Form several types of hems using the proper equipment.
- Determine the bend allowance needed on a particular fold.
- Fabricate a wired edge on a rotary machine and on a bending brake.
- Explain how to form a Pittsburgh seam on a brake.
- Identify different types of seams, including a grooved seam, a slip joint seam, a standing seam, and a double seam.
- Discuss the differences among the three types of dovetail seams.

Technical Terms

allowance
blind edge
burred edge
capped edge
double seam
dovetail seam
drive clip
grooved seam
Pittsburgh seam
slip joint seam
standing seam
wired edge

Sheet metal pieces, such as ductwork, downspouts, and washing machine housings , begin as flat pieces of metal. Patterns of the stretched out pieces are transferred or made directly on to the flat metal. These patterns include ***allowances***, or additional material, for hems, seams, and edges, **Figure 6-1.** The amount of the allowance is determined by the thickness of the metal, the radius of the bend, and the type of bend being made. Once outlined on the metal piece, the pattern is cut out.

allowance: Extra material added onto the edges of sheet metal patterns to provide for the creation of hems, edges, and seams.

Hems, seams, and edges are important in the fabrication of metal pieces. They join opposite ends of a pattern to make a complete piece. They join entirely separate pieces. They provide a finished appearance to the final product.

Hems, edges, and seams can be made using hand tools or machinery.

Hems

A hem is a folded-over edge on a piece of metal. It can be made in either a single or a double style. It is used to finish off rough, sharp ends of metal sheet and to strengthen and improve the appearance of a project.

Allow about 1/4″ (6 mm) for the hem. When using metal heavier than 22 gage, increase the hem to 5/16″ (8 mm) or 3/8″ (10 mm). Larger edges are easier to bend. Seldom does the edge go beyond 1/2″ (13 mm) wide. Larger edges have a tendency to wrinkle and leave an unsightly finish.

A double hem is a single hem, repeated. Double hems provide much greater strength than single hems. The allowance for a double hem is twice the hem size less 1/16″ (2 mm). Double hems are generally 5/16″ (8 mm) or 3/8″ (10 mm). The outside bend line is made first and is made smaller than the bend allowance. This is done so the outside bend does not cover up the inside bend line, **Figure 6-2.** The inside bend is made to the hem size. The outside bend is made 1/16″ (2 mm) less than the hem size.

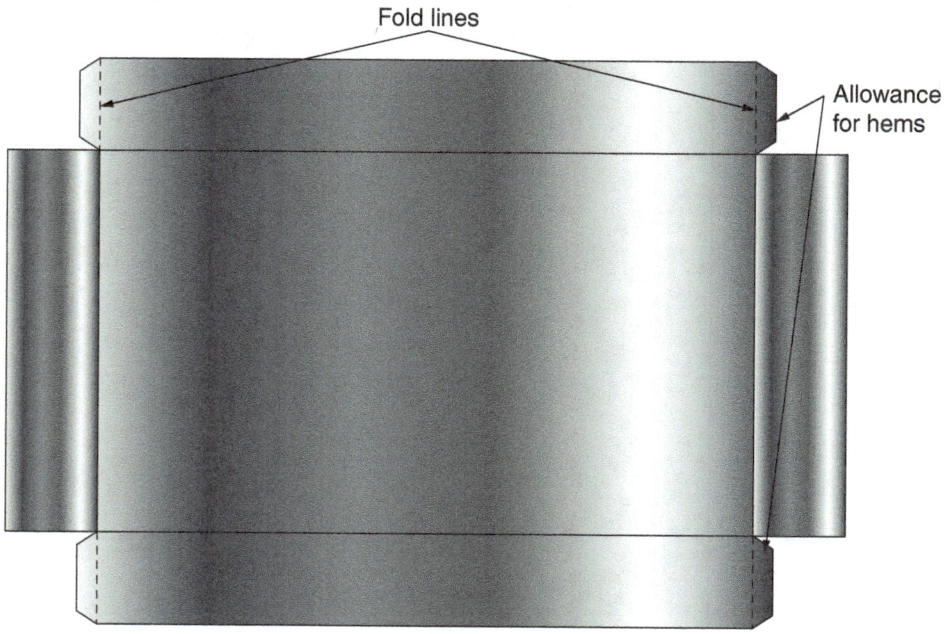

Figure 6-1. A stretched out pattern, showing fold lines and allowances for hems.

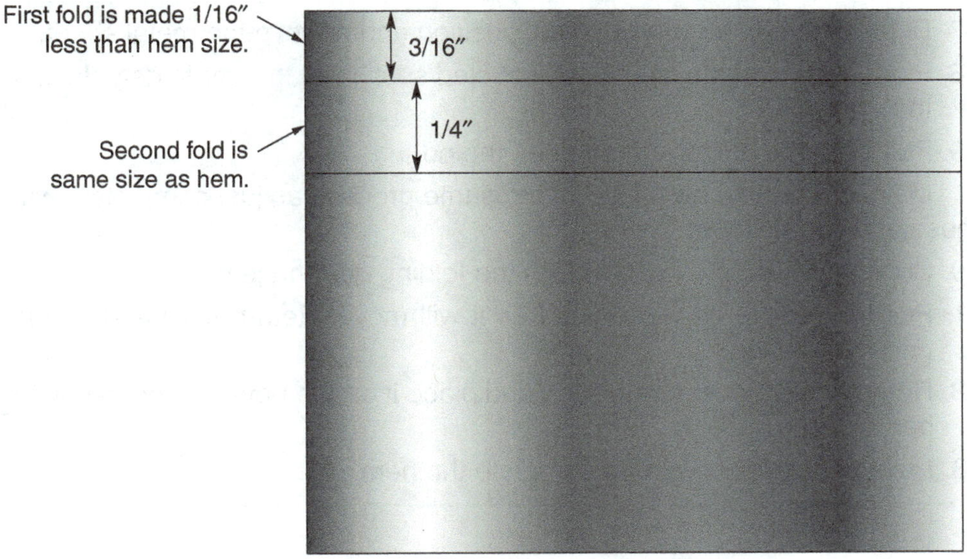

Figure 6-2. Double hem fold allowances for a 1/4" hem.

Hems can be folded using a bar folder or a brake. A bar folder makes the hem in one motion: bending the metal and pressing it flat. Hems made on brakes require two steps: one to bend and one to press flat.

Keep the following points in mind when using a bar folder:

- Make allowances for the thickness of the metal being bent.
- Know the bend sharpness that can be achieved with the wing.
- Know the width of folded edge to be made.
- Know the angle of the fold to be made.
- Know the bending characteristics of the metal being used.

The sharpness of the folded edge is controlled using the wing. Raising the wing even with the bed creates a sharp fold. Lowering the wing below the bed creates a rounded fold. See **Figure 6-3.**

Using the Bar Folder to Make a Hem

1. Turn the adjusting thumbscrew to the desired width. Tighten the lock screw to keep the gage in place.
2. Loosen the wedge lock nut located on the rear of the machine. The wedge lock nut holds the wing in place. Then adjust the wing for the proper bend using the wedge adjustment knob. Tighten the lock nut.
3. Set the adjustable collar stop to the maximum angle.
4. Adjust for the metal thickness using the shoe screws on each side of the machine.
5. Slide the sheet metal under the folding blade and beneath the jaw. Rest the metal against the stops.
6. Hold the metal in place with one hand. With your free hand, pull the handle on the folder as far forward as it will move.

(Continued)

7. Return the handle to its original position and remove the metal sheet.
8. Place the formed edge on the beveled section of the bar folder, close to the wing.
9. Pull the handle forward to flatten the edge.

Making a double hem follows the same process as just completed, with these additional steps:

10. Slide the hem, fold up, between the folding bar and jaw.
11. Pull the handle as far forward as it will move. Return the handle to its original position.
12. Remove the piece, turn it over, and place it on the beveled section of the bar folder.
13. Pull the handle forward to complete the hem.

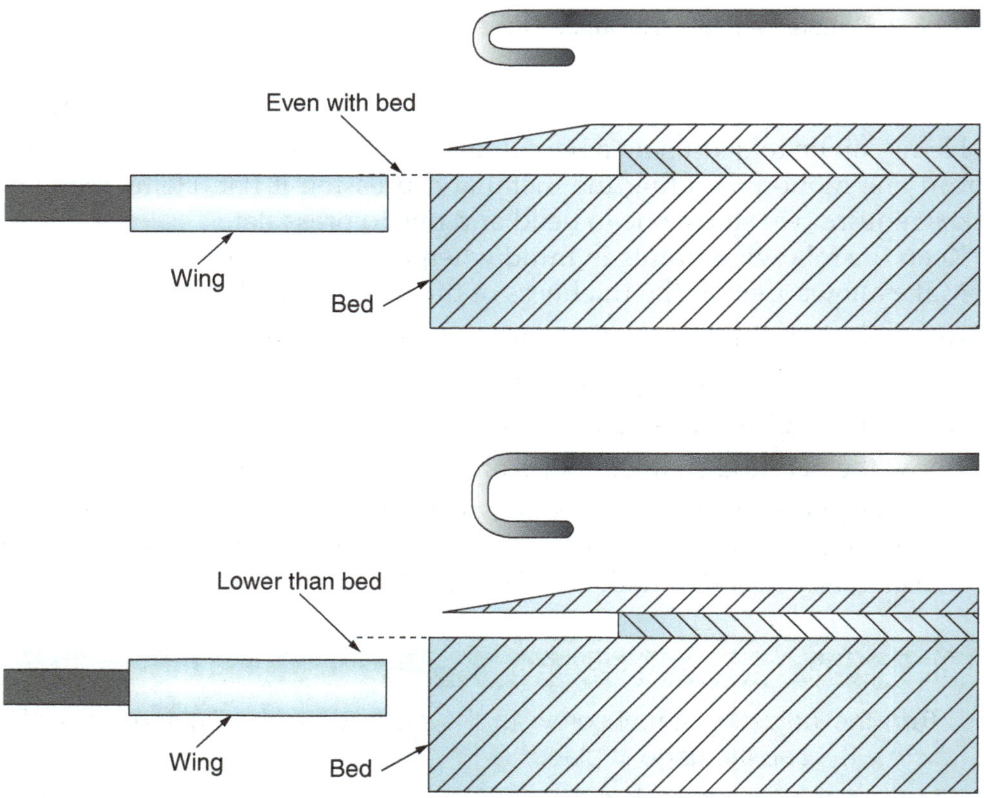

Figure 6-3. The wing position on the bar folder determines the sharpness of the bend.

wired edge: An edge that is additionally strengthened through the use of wire, angle iron, band iron, or other material. The strengthened edge aids the object in keeping its shape.

Wired Edges

Wired edges are used in products that require additional edge strength. Metal pails, for example, have wired edges that provide strength and help the pails to retain their shape.

Like a hem, an allowance is added to the pattern to accommodate the wired edge. The allowance is determined by the diameter of the wire and the gage of the metal. Sheet metal thicker than 26 gage uses an allowance of 2 1/2 times the diameter of the wire. For example, a wire with a 3/8″ (10 mm) diameter is multiplied by 2 1/2 for an allowance of 15/16″ (24 mm). Sheet metal thinner than 24 gage uses an allowance of 2–2 1/2 times the thickness of the metal.

Wire is not the only material used to strengthen edges. Angle iron and band iron are examples of different, stronger edge wraps. Regardless of which material is used as an insert, knowing how to install the material in the edge is an important part of the job.

Recall from Chapter 5 that rotary machines are used to make edges on round or cylindrical objects. The rotary machine most commonly used to make a wire edge is the turning machine. The bend radius made by the turning machine is an excellent shape to hold wire. Turning machines are also capable of turning heavier gage metal, the type of metal used on the sides of objects like a pail.

Making a Wired Edge

1. Adjust the gage. The distance is measured from the face of the gage to the center of the upper roll, **Figure 6-4**. Set the gage to 2 1/2 times the diameter of the wire.
2. Adjust the crank screw to pull the work through the rolls smoothly and evenly. Too much tension will form a groove in the metal.
3. Make one complete revolution of the work.
4. Move the work upward, in a slightly more vertical position. Again, tighten the crank screw just enough to keep the piece moving through the rolls. Make a second pass through the machine.
5. Tilt the work vertically, as far as possible. Tighten the crank screw more, and make one more revolution. The edge is now ready for the wire.
6. Place the wire in the prepared edge.
7. Finish closing the edge using special wiring rolls, **Figure 6-5**.

Making a Wired Edge Using a Bending Brake

Wired edges can also be made using a bending brake, a few hand tools, and a wiring machine.

1. Locate the bend line at 2 1/2 times the diameter of the wire being installed.
2. Bend the edge at a right angle using the brake.
3. Place the proper length of wire against the folded edge. Hold the wire in place with needle-nose pliers.
4. Use a mallet to bend the metal down over the wire.
5. Finish the edge using a wiring machine.

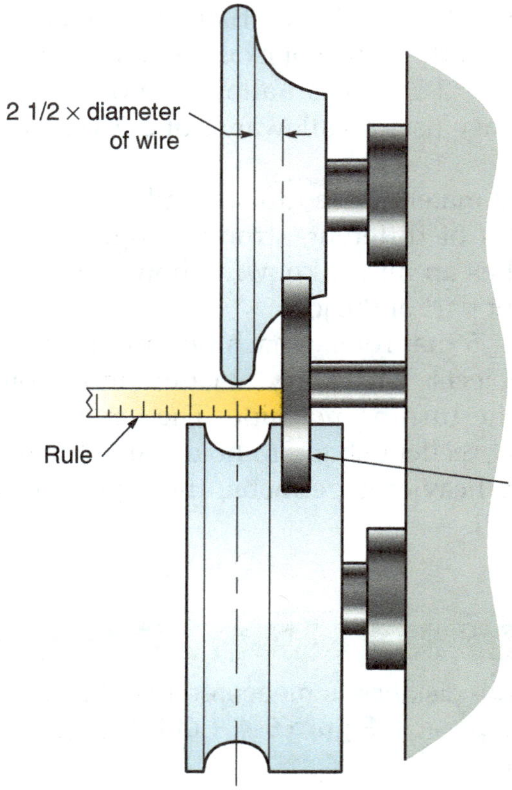

Figure 6-4. Setting the distance on a turning machine when making a wired edge.

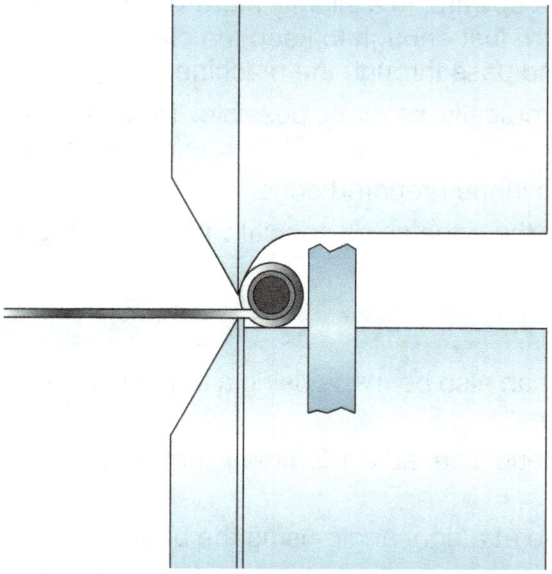

Figure 6-5. The pointed ends on these wiring rolls give the wired edge a crisp, tight finish.

capped edge: An edge made from a strip of metal to cap a raw edge of metal. The metal strip is bent and slipped onto the raw edge. It is held in place by rivets, screws, or soldering.

Capped edges are added to metal edges that are raw or otherwise unfinished, **Figure 6-6.** A strip of metal is bent and slipped over the raw edge of the metal. This cap is held in place by rivets, screws, or tack soldering.

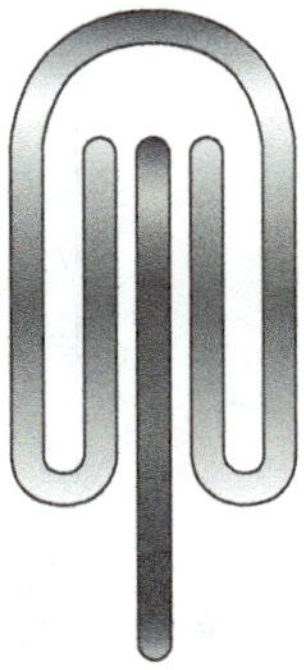

Figure 6-6. Capped edges can be added as an afterthought or as a decorative accent.

Blind edges are used to conceal nail heads and raw edges when metal is used to cover wood. Nails are driven through a strip of metal and the metal sheet and into the wood. The strip of metal will act as the cover for the nail, **Figure 6-7.** This strip forms an upright edge that is carefully pounded down with a mallet. This method is also used on joints where two sheets of metal are joined over wood, providing the same effect.

Elbow edges are made on rotary machines using special rolls, **Figure 6-8.** The angular rolls create the offset angle needed for the elbow.

Burred edges are found on metal disks such as covers and bottoms on cans. Creating uniform burred edge requires a great deal of practice. Learn by turning small disks about 6″ (150 mm) in diameter, until you can turn a good edge.

The process used to make a burred edge is very similar to the process used to make a wired edge. Keep the following in mind:

- Hold your left elbow against your body during the entire sequence.
- Grasp the disk securely between your thumb and index finger.
- Hold some type of material in your hand to act as a guard against cuts.
- Making a gradual bend results in fewer wrinkles and waves on the edge.

blind edge: An edge made of a strip of steel, held in place by a nail to create the appearance of a finished hem. The free edge of the metal strip is hammered into place, covering the nail head.

burred edge: An edge created by rolling two pieces of metal together providing a watertight seal. It is made on a burring machine and is found on cans, buckets, and similar objects.

Seams

Seams are used to fasten sheet metal sections together using locks, rivets, screws, brazing, or soldering. Seams are made by hand, on a brake, or on a bar folder. The type of seam made and the fastening method depend on the type of metal used, its thickness, and the fabricating machinery available to the worker.

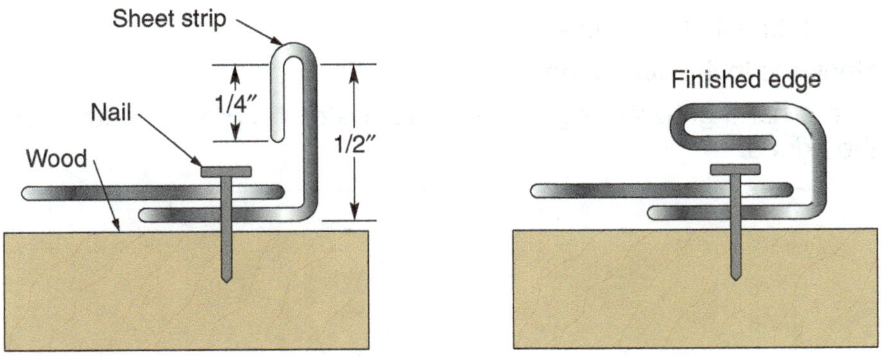

Figure 6-7. Blind edges are also called false edges or Dutchman edges.

Figure 6-8. Fabricating an elbow edge.

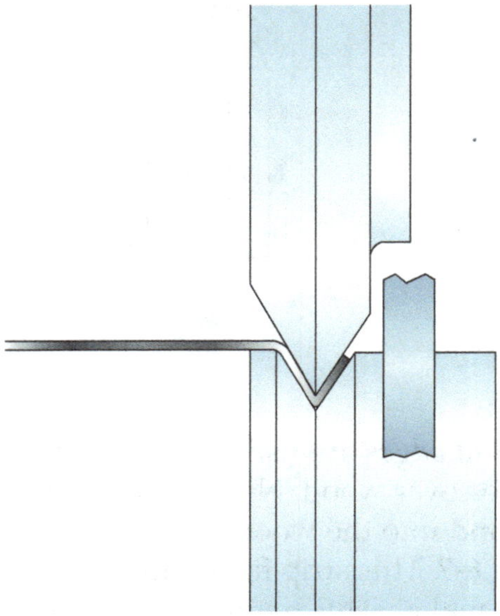

A wide variety of seam types have been designed for use on sheet metal. About 20 types of seams are used in ductwork construction and for other purposes. See **Figure 6-9.** A number of seams commonly used in sheet metal work are discussed in the remainder of the chapter. Of seams fastened together by rivets or solder, the pop (or blind) rivet is the most commonly used. Both rivets and solder are discussed more thoroughly in Chapter 8.

Grooved Seam

grooved seam: A seam with two folded edges locked together with a hand groover. Used on light- and medium-gage sheet metal

The ***grooved seam*** is commonly used when fastening together light or medium gage metals. The material allowance for a grooved seam depends on the width of the locks and the thickness of the metal. In general, however, for metals 24 gage and lighter, the allowance is three times the width of the lock. For metals 22 gage and heavier, the allowance is three times the width of the lock, plus five times the thickness of the metal. Remember to divide the allowance between each side of the pattern.

Making a Grooved Seam

Grooved seams are fairly simple to make. Refer to **Figure 6-10.**

1. Fold edges to form locks.
2. Hook the lock edges together.
3. Use a hand groover or grooving machine of the proper width to secure the edges.

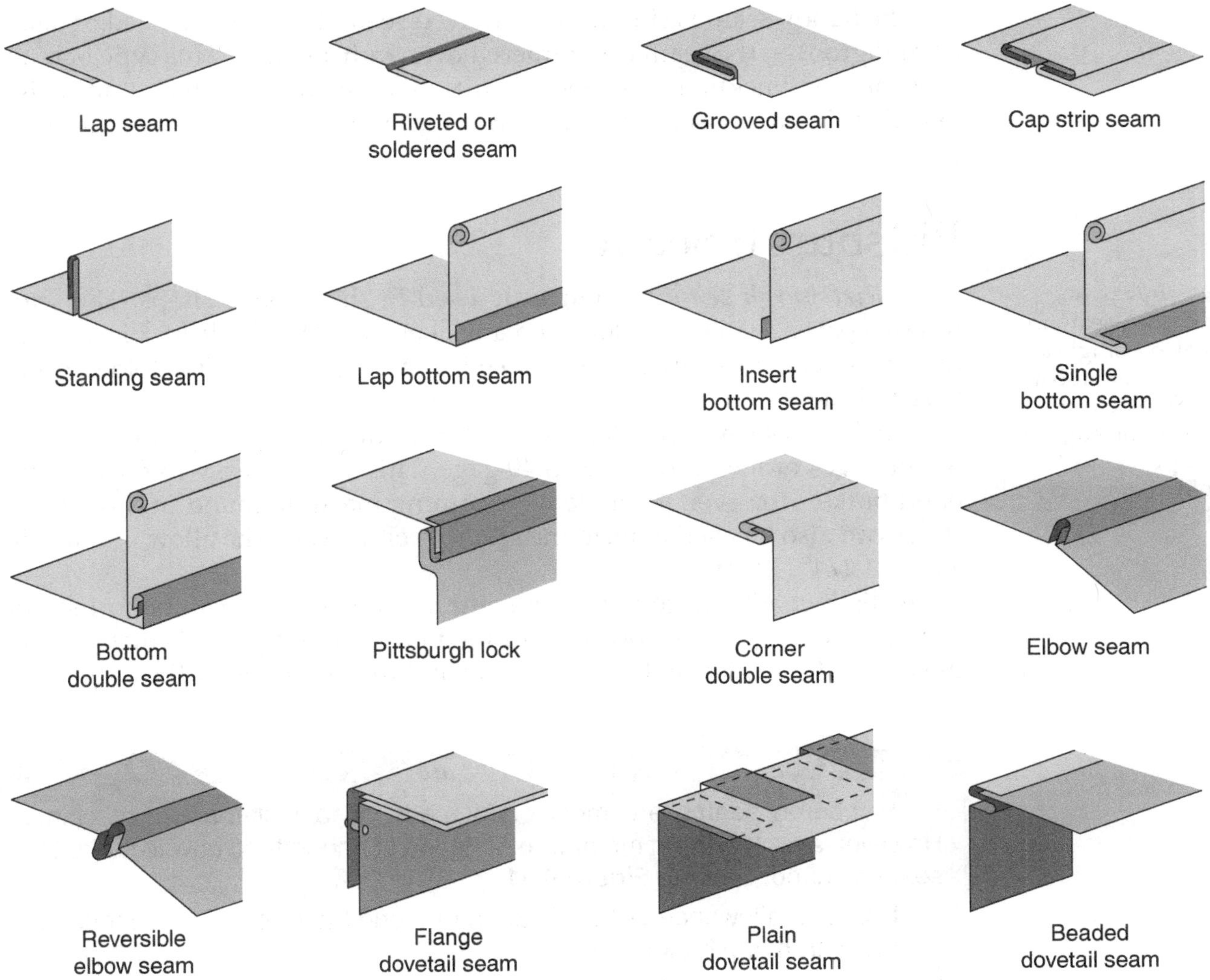

Figure 6-9. There are a wide variety of seams used in sheet metal work.

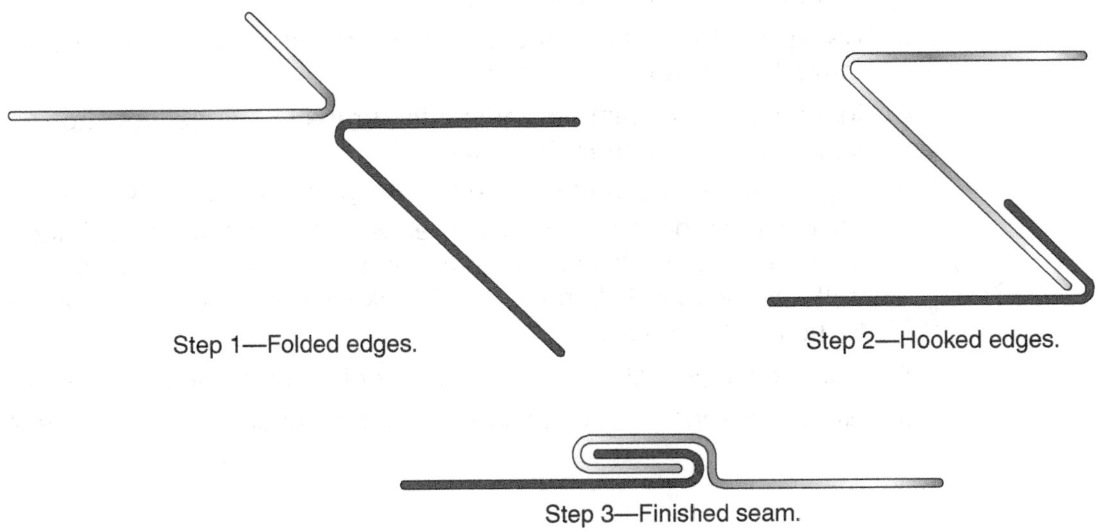

Figure 6-10. Fabricating a grooved seam.

The handy seam is similar to a grooved seam. However, instead of using a hand groover, the seam is hammered over with a mallet. This type of seam prevents the buckling of edges when two large sheets of metal are being joined. It is often used on large ducts that require two or more sheets of metal.

Pittsburgh Seam

Pittsburgh seam: A seam in which a single lock is inserted into a pocket lock, after which the edge of the pocket lock is folded down to seal the seam.

A ***Pittsburgh seam*** is commonly used in sheet metal shops. The seam has two parts: a single, 90° lock and a pocket lock. The single lock is inserted into the pocket lock, after which the edge of the pocket lock is folded down to seal the seam.

Pittsburgh seams are often used in HVAC ductwork. They can be formed on all gages of metal, from 16 to 40 gage. This seam is easy to form on the hand brake. However, a simple lock-forming machine found in most metal shops can also be used to form the seam. In either case, the allowance for the lock is 1 1/4″ (32 mm).

A distinct advantage to the Pittsburgh seam is that it can be formed on curves. The single lock portion is formed on a curved piece of metal, while the pocket lock is formed on a flat sheet and then rolled to fit the curve.

Making a Pittsburgh Seam on a Brake

Pittsburgh seams are most often made on a lock-forming machine. However, a skilled sheet metal worker needs to understand how to make the seam by hand. Refer to **Figure 6-11**.

1. Using an allowance of 1 1/4″ (32 mm), bend the metal 90°. Remove the metal from the brake.
2. Insert the 1 1/4″ (32 mm) bend into the brake, with the long portion of the metal against the bending leaf.
3. Bend the leaf up. Remove the metal from the brake.
4. Reinsert the metal and bend the leaf up. This movement will slightly change the original 90° angle.
5. Using a mallet, pound the 3/4″ (19 mm) piece down as shown. This forms the pocket for the seam.
6. Insert the newly bent pocket into the brake and flatten as shown. Once flattened, remove from the brake.
7. Flip the piece over and reinsert into the brake as shown. Position Point 1 so it sits just beyond the outside edge of the bending leaf. Then clamp the metal behind Point 2. Bend the metal to an approximately 15° angle. With the metal still clamped in the brake, flatten the pocket with a mallet to the position shown.
8. Join the single lock on the other piece of metal to the pocket lock.
9. Gradually knock the flange down along the entire length of the seam.

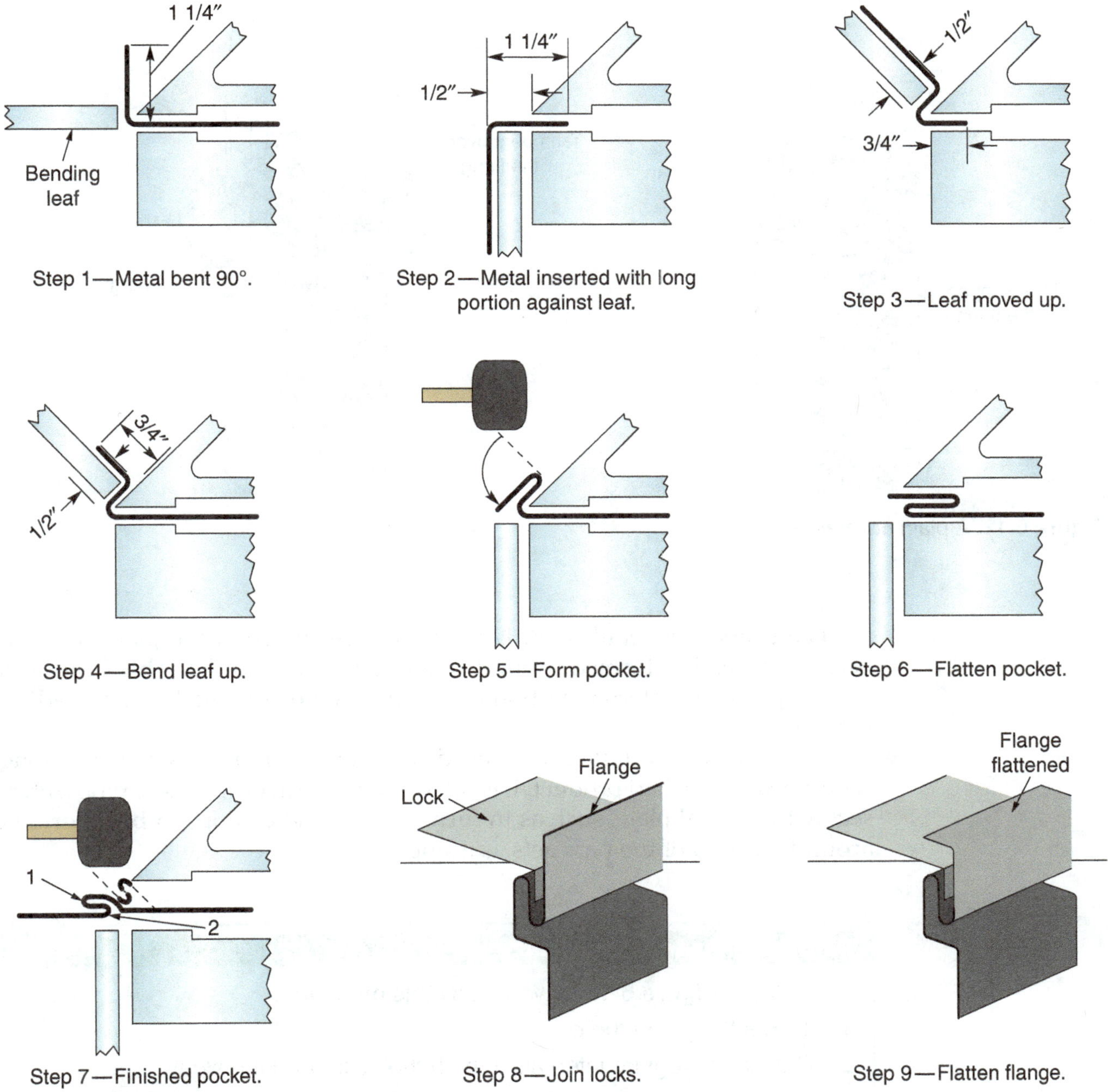

Figure 6-11. Fabricating a Pittsburgh seam.

Dovetail Seams

The ***dovetail seam*** is used primarily on round or elliptical pipes. It is seldom used on rectangular pieces of sheet metal. There are three types of dovetail seams: plain, beaded, and flanged.

dovetail seam: A seam used to join a flange and a collar. It is used almost exclusively on round or elliptical pipes.

The plain dovetail seam is used to join a collar and separate flange. The seam is made without using rivets, screws, or solder. Instead, tabs are cut at the end of the collar and every other tab is then bent down. The separate flange is then added, with the bent tabs holding it in place. The remaining tabs are then bent over the flange to secure it, **Figure 6-12.** This seam is made waterproof by adding solder. If additional strength is needed, the seam may be brazed or the tabs spot-welded and soldered.

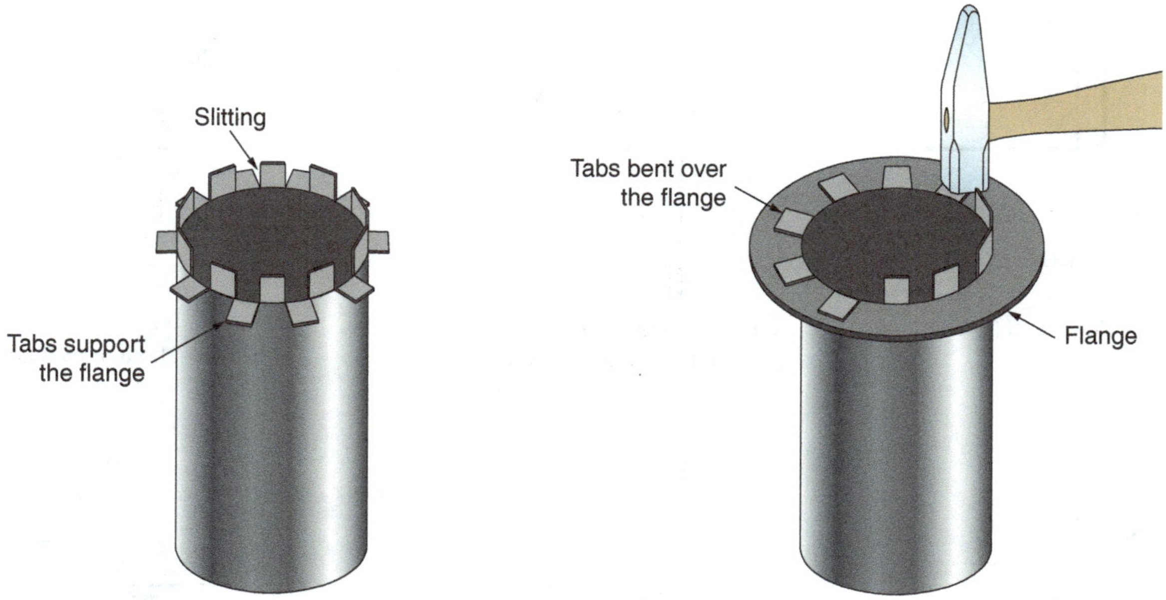

Figure 6-12. A plain dovetail seam.

The beaded dovetail seam is a bit more secure than the plain dovetail because a bead is formed around one end of the cylinder. This bead acts as a stop for the internal section of metal. The bead is made on a beading machine.

The flanged dovetail seam is used when extra strength is needed, along with a neatly finished product. It is often used in situations where pipes intersect with a metal plate, such as in furnace flues and ceilings. A bead formed around one end of the pipe acts as a stop.

Forming a Flanged Dovetail Seam

Refer to **Figure 6-13** as you study this procedure.

1. Turn a flange on the collar.
2. Cut slits at regular intervals in both the collar and the sleeve.
3. Drill rivet holes in both the collar and the sleeve.
4. Rest the metal plate on the collar flange, then insert the sleeve into the collar. Align the rivet holes in the collar and sleeve, then install the rivets.
5. Hammer down the tabs.

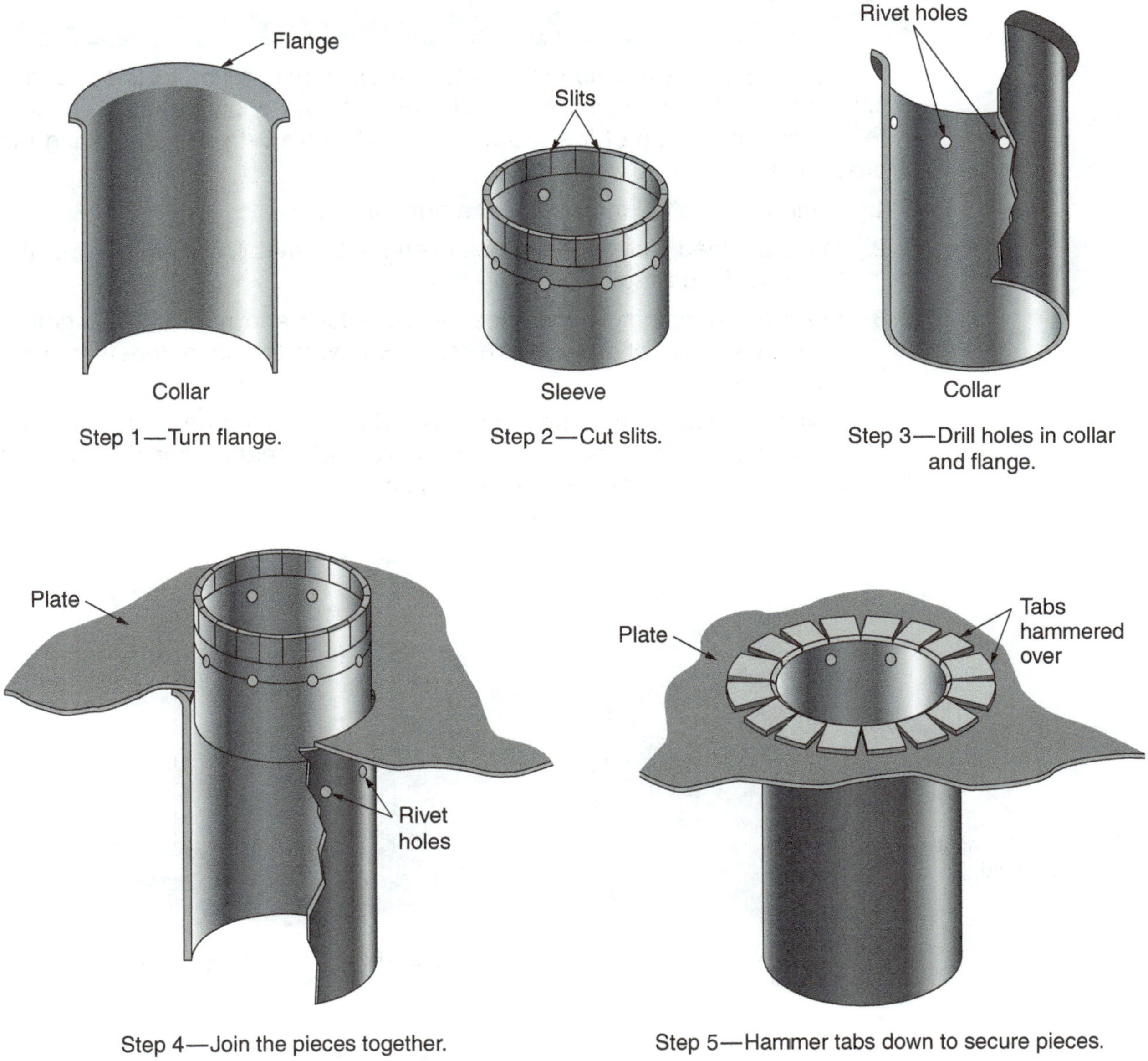

Figure 6-13. Making a flanged dovetail seam.

Double Seam

A ***double seam*** is used on irregular fittings. It can be used on offsets, boxes, and square elbows. A bottom double seam, found on soup cans and water pails, joins cylinders to their bottoms. It is made using a turning machine and a burring machine, as outlined earlier.

double seam: A seam used on irregular fittings and to join cylindrical shaped objects to their bottom pieces.

Making a Double Seam Corner

Before making the double seam corner, punch all bend lines. Leave 1/16″ (2 mm) bending allowance. Allowance for a double seam is equal to three times the width of the seam. Refer to **Figure 6-14** while studying the procedure.

1. Bend the 1/16″ (2 mm) edge at a right angle.
2. Make a folded lock on the opposite end of the metal. This will fit over the bend made in Step 1.
3. Form the remaining bend lines. Join the lock and right angle together and place the duct on a square stake. Squeeze the seam together using a pair of tongs.
4. Using a mallet, flatten the seam over. Square the corner with the seam using the mallet. Keep the square stake underneath the inside edge of the seam to maintain the angle of the corner.

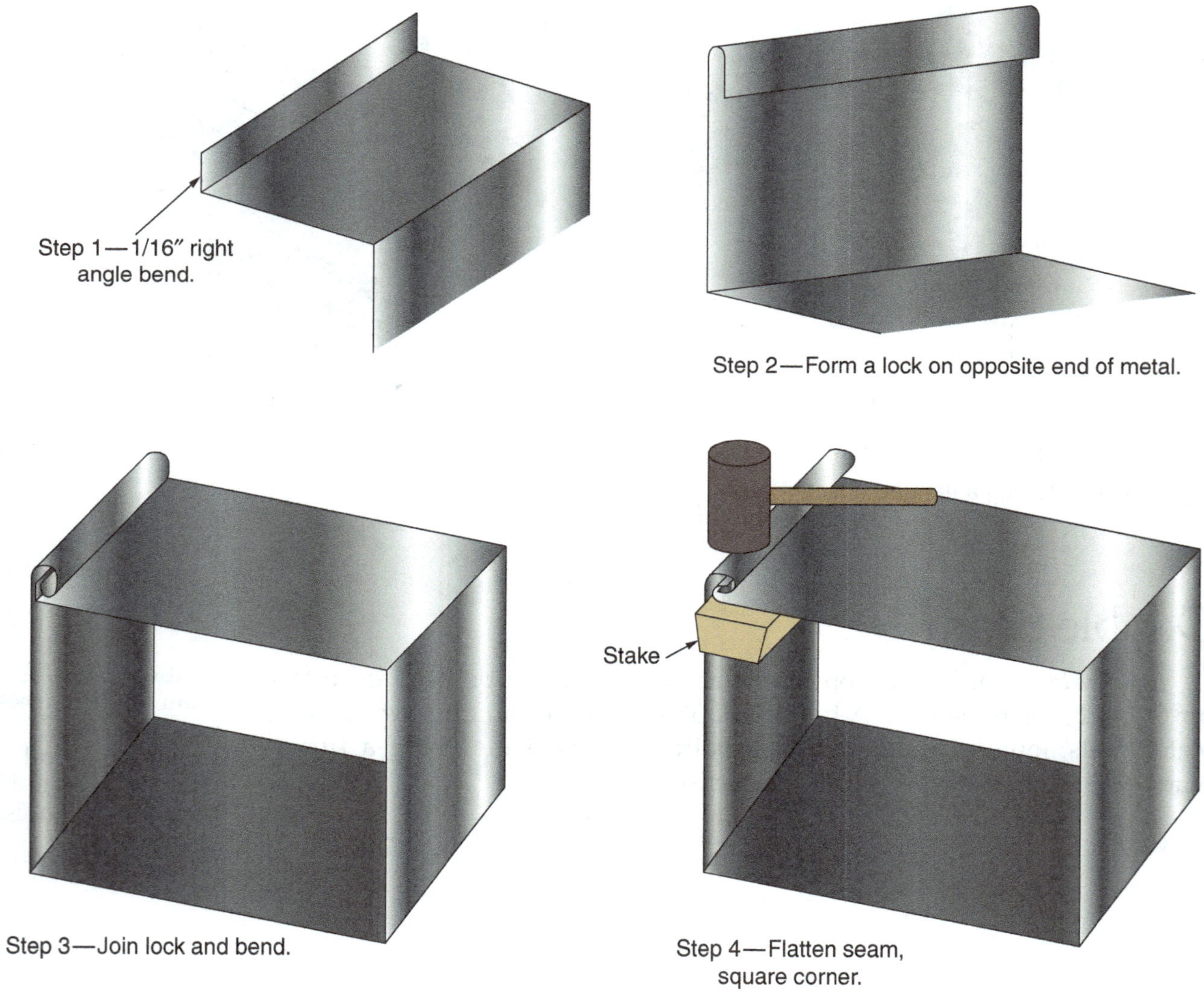

Figure 6-14. Forming a double seam corner.

Standing Seam and Government Clip

A ***standing seam*** is upright and consists of interlocking or overlapping panels, **Figure 6-15.** The upright construction provides its own reinforcement. Standing seams are used on large expanses of metal, such as metal roofs or long pieces of ductwork. On a sheet metal roof, a standing seam provides visual relief and may help manage rainwater and snow.

The government clip is also used on cross-seams of large ducts, **Figure 6-16.** It, however, sometimes requires additional reinforcement from angle bars or bar cleats.

standing seam: An upright seam that provides its own reinforcement. Used on large expanses of metal, where additional strength is needed.

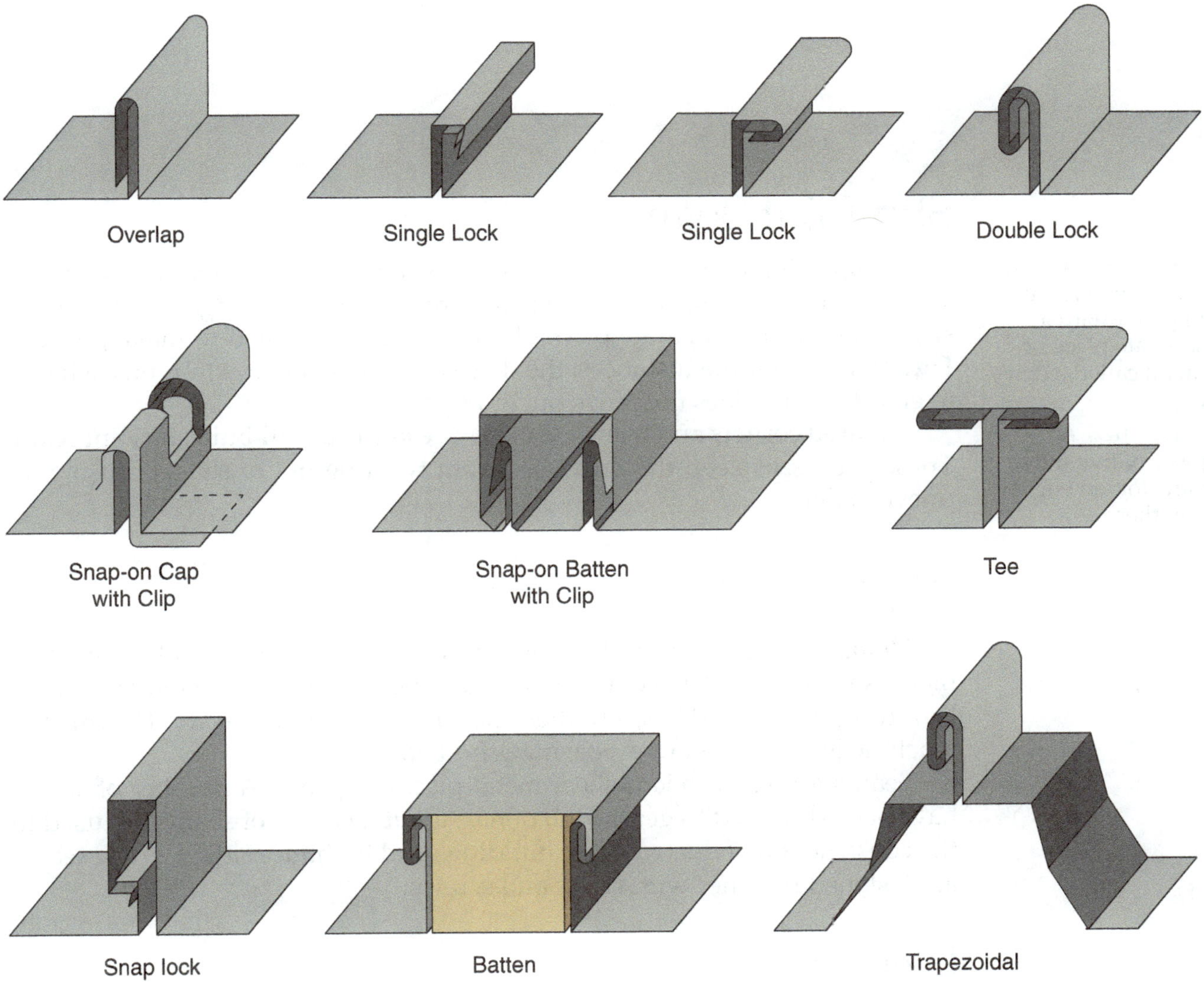

Figure 6-15. Examples of standing seam profiles.

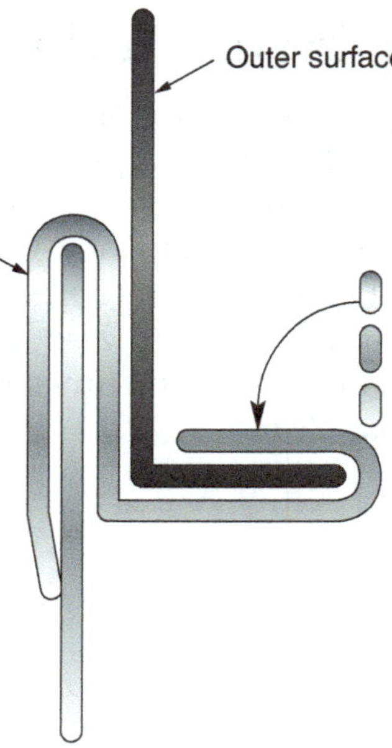

Figure 6-16. The parts of a government clip seam.

Slip Joint Seam

slip joint seam: A seam that is held together with a separate piece of metal called a drive clip.

drive clip: A piece of metal that joins folded edges together.

A ***slip joint seam*** consists of the edges to be joined and a drive clip that joins them. The edges are turned to form pockets. The edges on the ***drive clip***, made from a separate piece of metal, are also turned to make pockets, **Figure 6-17.** In some instances, the drive clip will also have tabs on each end to ensure that it does not move out of place.

Squared corners and trimmed edges are key to a well-built slip joint seam. These processes keep the ductwork from twisting out of shape, causing an uneven edge.

Summary

Hems are used to strengthen objects and to finish off the end of the sheet metal so it does not have a sharp edge. They are made in standard sizes. Wired edges are used to strengthen sheet metal when necessary. The turning machine is often used in preparing wired edges.

Seams are used to join sheet metal pieces together. A number of seams have been designed to get the job done. About 20 types of seams are used in the construction of ductwork for buildings and for house siding. Some seams are fastened together with rivets and screws.

Internet Resources

National Roofing Contractors Association

www.nrca.net

Cincinnati Precision Machinery, Inc.

www.cincinnatiprecisionmachinery.com

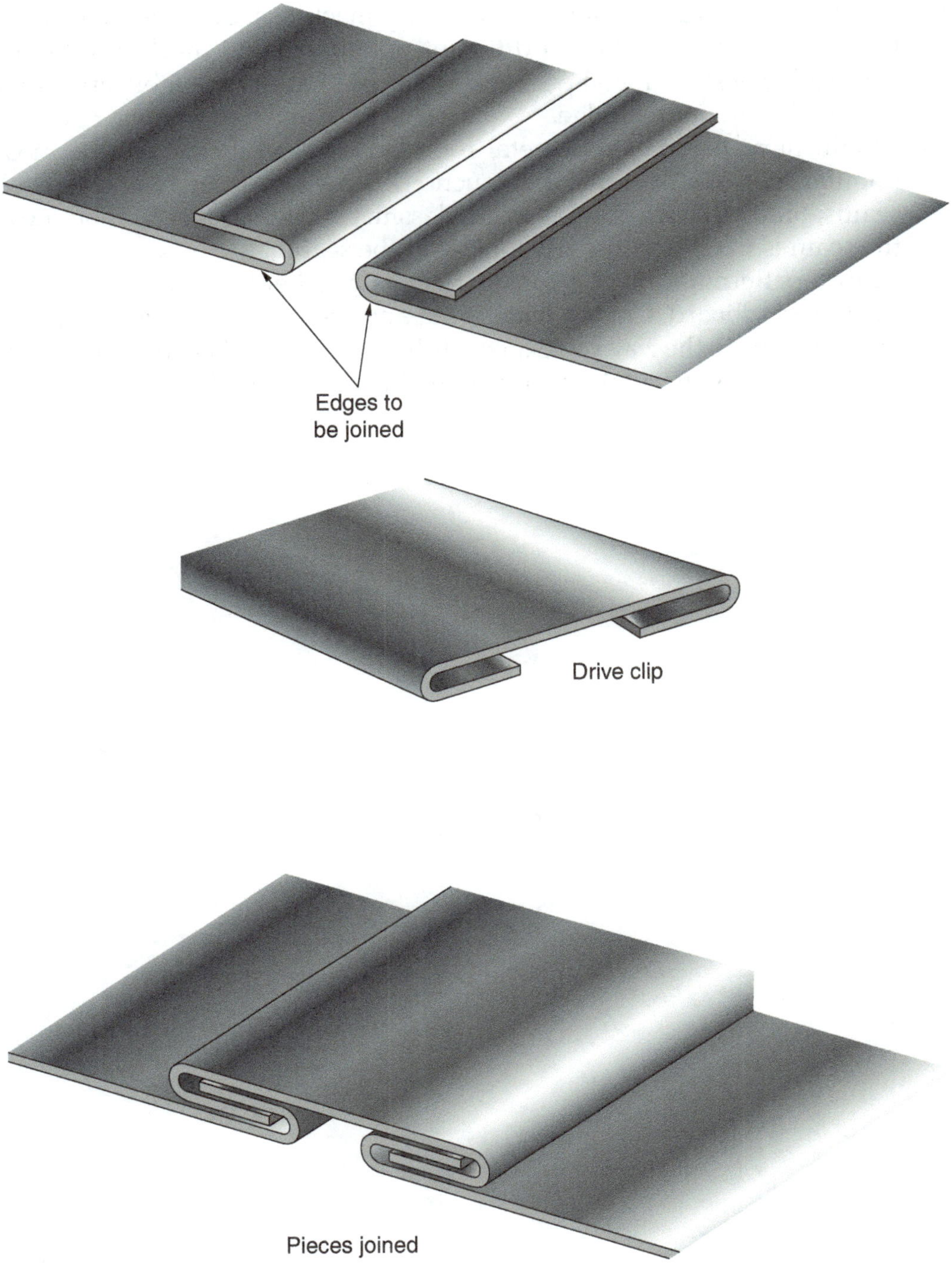

Figure 6-17. The parts of a slip joint seam.

Outside Activities

1. Research bar folders on the Internet. Determine what companies manufacture them and then study the descriptions and abilities of their machines. Make a list of projects that a sheet metal worker can do on these machines.

2. Locate manufacturers' literature on rotary machines. Compile a list of the types of rollers available on different manufacturers' machines. Brainstorm a list of sheet metal projects that would require the fabrication of a rolled edge or a wired edge.
3. Obtain literature on sheet metal roof construction. What are the most popular seams in steel roof construction? Draw a diagram of one of these seams. Show this diagram to your classmates and explain to them how the seam is made and the advantages of the seam.
4. Draw a pattern for a box, a cookie sheet, or another basic shape on index stock or cardboard. Cut the pattern out and then assemble them. Did the pattern turn out the way you expected? If not, try again. When your pattern is correctly drawn, cut it out, and assemble it for your classmates.

Name ______________________________ Class ______________ Score ________

Chapter 6
Hems, Edges, and Seams

Review Questions

Write your answers in the spaces provided.

1. The amount of _____ for a hem is determined by metal thickness, bend radius, the type of bend being made. 1. ______________
2. A hem _____. 2. ______________
   A. adds strength to a sheet metal piece
   B. improves the appearance of a finished product
   C. can be folded over once or twice
   D. All of the above.
3. True or False? A fold made on a bar folder with the wing positioned below the bed will be sharp. 3. ______________
4. The allowance on a pattern for a wire edge is determined by what two considerations? ________

   ____

   ____
5. What rotary machine is most commonly used to make wired edges? ______________

   ____

   ____

Match each of the following descriptions with the correct term.

6. Used to conceal nail heads and raw edges on metal used to cover wood. 6. ______________
7. Found on metal disks such as can bottoms. 7. ______________
8. Added to raw or unfinished edges. 8. ______________
9. Made using special rolls to create an offset angle. 9. ______________

A. Burred edge.
B. Blind edge.
C. Capped edge.
D. Elbow edge.

10. True or False? A seam is the same thing as a hem. 10. ______________
11. Explain the difference between a grooved seam and a handy seam. ______________

    ____

    ____

12. Pittsburgh seams are used on metals _____. 12. ____________
    A. 24-gage and lighter
    B. 16–40 gage
    C. 22-gage and heavier
    D. of any gage

13. Name three types of dovetail seams.____________________________

    ____

14. An upright seam consisting of interlocking or overlapping panels is called a(n) _____. 14. ____________

15. The edges of a slip joint seam are held together by a(n) _____. 15. ____________

Chapter 7
Drilling and Punching

Objectives

After studying this chapter, you will be able to:

- Identify the parts of a drill bit.
- Explain the method for sizing drill bits.
- Mount a drill bit in a portable drill and in a drill press.
- Explain the difference between cutting speed and drill press speed.
- Use a sheet metal punch.

Technical Terms

body clearance
chisel edge
chuck
cutting speed
drill bit
drill press speed
drill walking
feed
flute
lands
lip
margin
peripheral speed
point
shank
sleeve
socket
straight-fluted drill bit
taper mount
twist drill bit
web

Fabricated sheet metal products consist of many separate sheet metal pieces fastened together. The pieces can be fastened using mechanical methods (seams, rivets, screws and bolts) or by welding.

Welding is the most common fastening method used in sheet metal work. Welding creates a strong bond between parts. But there are instances in which a weld is not the best way to join separate pieces. Lighter metals can be joined using seams. You learned about several of these seams and how to make them in Chapter 6. Heavier materials may not be held securely with either of these methods, however, and that is when a riveted or screwed bond is best.

Holes must be cut in material to accommodate rivets or screws. Holes are made using either a drill or a punch. Generally, punches are used to make holes in lightweight sheet metal, while drills are used on heavier sheet metal. A drilled hole will usually have cleaner edges than a punched hole.

Determining what fastening method is best for a particular task is one of a sheet metal worker's responsibilities. Study this chapter to improve your ability to make this decision.

Drills and Drilling

Drilling holes in sheet metal is a common task for sheet metal workers. Holes are drilled using either a handheld drill or a drill press. The handheld drill is a portable tool that holds and twists the drill bit. A drill press is a fixed machine. It can be floor mounted or bench mounted, **Figure 7-1**. It has several advantages over a handheld drill:

- Less effort is needed for drilling.
- Work can be securely locked into place, using the vise or clamp attached to the table.
- A fixed angle on the drill results in uniform holes.

Parts of the Drill Bit

Regardless of which machine is used, the ***drill bit*** is the portion of the drill that does the actual cutting. It is a pointed cutting tool that makes cylindrical holes. ***Twist drill bits*** are the most commonly used drill bits in sheet metal work. The parts of a twist drill bit are shown in **Figure 7-2**.

The ***shank*** of the drill bit is either straight or tapered. Straight shanks are held in place by a ***chuck***. Tapered shanks are used exclusively on drill presses. They fit into a taper holder on the drill press spindle. The body of a drill bit extends from the shank to the point. Body lengths vary. ***Flutes*** are helical grooves running along opposite sides of the drill.

The drill bit ***margin*** extends along the entire length of the flute. This determines the diameter of the hole. The ***body clearance*** is immediately in back of the margin. This reduced diameter decreases the friction between the drill and the wall of the hole so that the drill does not bind.

The cutting end of the twist drill bit is the ***point***. The point includes ends of the ***lands***, the ***web***, and the ***lip***. The extreme tip of the drill bit, which forms one sharp edge, is the ***chisel edge***. It acts as a flat drill, cutting its own hole in the work.

drill bit: A metal cutting tool that makes cylindrical holes.

twist drill bit: A drill bit with helically grooved flutes.

shank: Portion of the drill bit that fits into the spindle or the chuck.

chuck: Portion of a drill that both holds and drives the drill bit. Drill bit is held using a movable jaw.

flute: Groove that runs the length of the drill body.

margin: Part of a drill bit that determines the diameter of the hole. It runs the entire length of the flute.

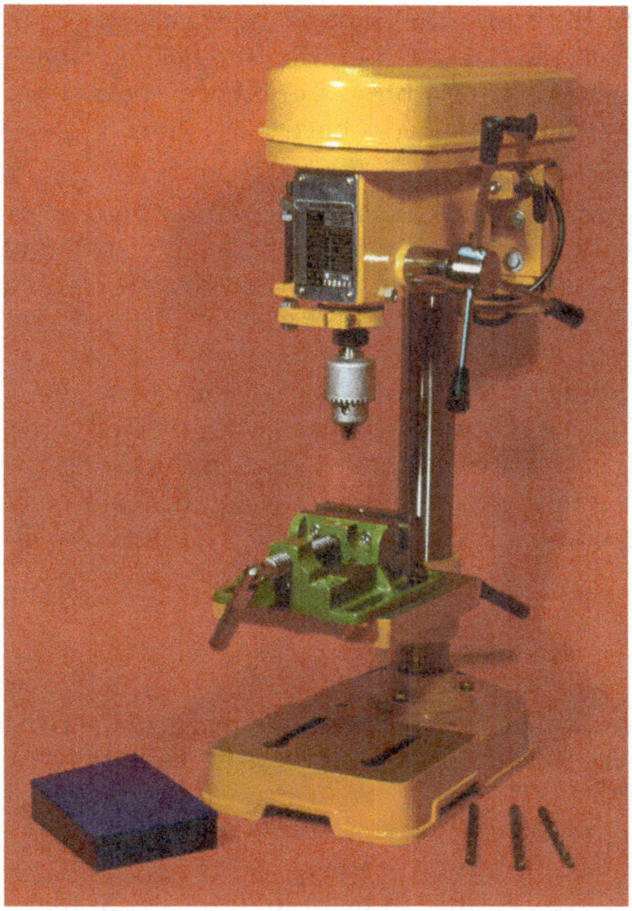

Figure 7-1. Drill press.

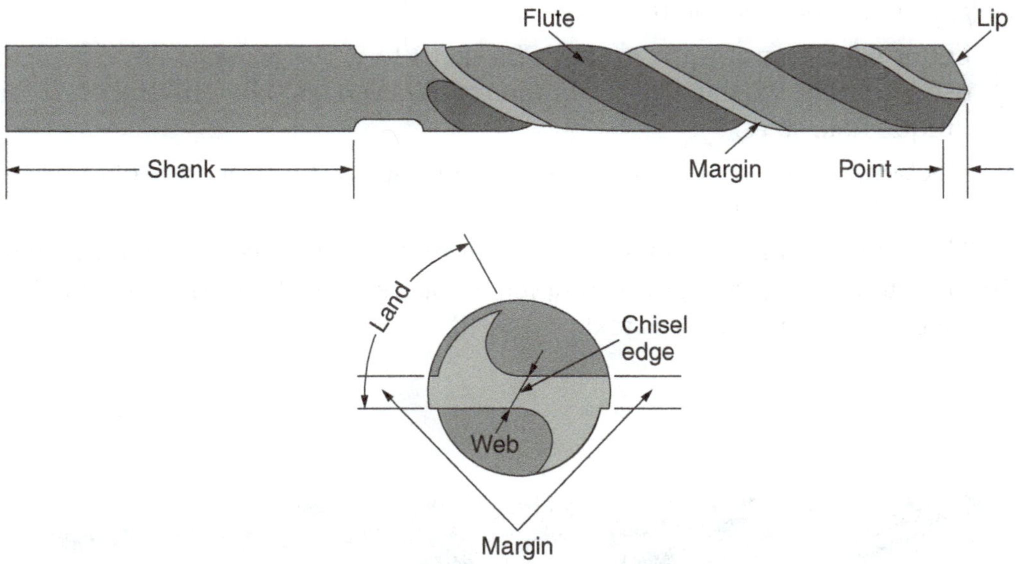

Figure 7-2. Familiarize yourself with the parts of a twist drill bit.

body clearance: Part of a drill bit directly behind the margin. It has a smaller diameter than the margin, which decreases the friction between the drill and the wall of the hole so the drill does not bind.

point: Cone-shaped end of the drill bit that does the actual cutting.

lands: The outermost portion of the drill bit body between adjacent flutes.

web: The center part of the drill bit body where the lands meet.

lip: The cutting edges of the drill bit.

chisel edge: The portion of a drill bit that connects the lips.

straight-fluted drill bit: A drill bit with unbent flutes.

As the drill bit is fed into the work, the borings, or waste materials, are passed up through the flutes and deposited outside the hole. ***Straight-fluted drill bits*** are sometimes used on brass, bronze, or other soft materials. However, most drill bits used in sheet metal work have spiral flutes.

Drill Bit Point Construction

Drill bit points are constructed based on how they will be used, **Figure 7-3**. For example, a general-use drill bit is made with a point angle of 118° and moderately spiraled flutes. This construction is useful for general drilling purposes, such as shallow depth in soft metals and wood. A flatter point angle, such as 135°, is found on drill bits used to drill tough metals.

The force needed to drill on sheet metal can cause denting on the metal. For this reason, the point of the bit is sometimes ground in the web area. Grinding the webs reduces the width of the chisel edge, minimizing the amount of force needed to push the drill bit into the metal. This split-point design reduces the chance of denting the metal surfaces and minimizes ***drill walking***.

drill walking: Tendency for an uncentered drill bit to move away from a marked hole location.

Drill Bit Sizes

Drill bit sizes have been standardized over the years. The size of the twist drill bit is usually stamped on the shank. Very small bits are not identified by size, but must be measured with a drill gauge. The drill gauge is also useful when measuring bits whose stamps have been worn away through use.

Standard drill bit sizes are classified in three ways:

- Number sizes, ranging from No. 80 (0.0135 inch) to No. 1 (0.228 inch). Number sizes correspond approximately to the Stubbs Steel Wire Gage size, an English and American standard. Number series drill bits fill the gaps between fractional sizes.
- Letter sizes, ranging from *A* (0.234 inch) to *Z* (0.413 inch) and are also used to fill in the gaps between the fractional sizes. They fill the need for unique hole sizes.
- Fractional sizes are drills with diameters defined in fractions of an inch. They are the most commonly specified.

The table in **Figure 7-4** shows some of these bit sizes and how they compare to one another. For a complete table, refer to the Drill Size Chart in the Reference Section of this text.

Figure 7-3. Drill bits are constructed with various point angles. This drill bit has a 118° point angle and is used for general sheet metal work. (Klein Tools)

Inch	mm	Drill	Decimals of an inch	Inch	mm	Letter sizes	Decimals of an inch
13/64			.2031		5.9		.2323
		6	.2040			A	.2340
	5.2		.2047	15/64			.2344
		5	.2055		6		.2362
	5.25		.2067			B	.2380
	5.3		.2087		6.1		.2402
		4	.2090			C	.2420
	5.4		.2126		6.2		.2441
		3	.2130			D	.2460
	5.5		.2165		6.25		.2461
7/32			.2188		6.3		.2480
	5.6		.2205	1/4		E	.2500
		2	.2210		6.4		.2520
	5.7		.2244		6.5		.2559
	5.75		.2264			F	.2570
		1	.2280		6.6		.2598
	5.8		.2883			G	.2610

Figure 7-4. Drill bit sizes are designated by numbers, fractions, and letters.

Metric size drill bits are also available but not as widely used. The diameters correspond to commonly used metric bolt sizes. They are sized in millimeters and tenths of a millimeter.

Types of Drill Bits

Drill bits are manufactured in a variety of sizes and shapes for a variety of uses. General-use drill bits are commonly used by homeowners. Variations on a general-use bit, however, make the bit more useful to the professional sheet metal worker. See **Figure 7-5.**

Using Twist Drill Bits

Choose the twist drill bit to be used based on the metal being drilled. Check the bit to make sure it is sharp and straight. Mount the drill bit into the machine being used, either the portable drill or the drill press. Mounting in a portable drill is straightforward. Open the chuck with the chuck key. Insert the drill bit, then tighten the chuck around the bit, using the chuck key. Remove the chuck key before drilling the material. Briefly run the drill to make sure the drill bit is operating properly.

Types of Twist Drill Bits	
Name	**Purpose**
Screw machine length (stubby)	Name applied to the shortest drills commonly available. Designed primarily for use in automatic drilling machines Short length allows for maximum rigidity.
Jobbers length	Applies to the length of any standard drill bit up to approximately 11/16″ diameter. Drill bit length most often used by both do-it-yourselfers and professionals.
Extended length (aircraft extension)	Longer than jobbers length bits. Used to drill in hard-to-reach places (between studs) or through extra-thick material.
Silver and Deming	Applies to 1/2″ shank drill bits only. Features a general-purpose flute design and is recommended for use in portable electrical drills and drill presses. Name refers to a time when drill bits were made specifically for unique machining applications.
Double-ended	Both ends are fluted for a short distance and have a solid center shank for secure chucking. Drills are designed to be reversed quickly in machine tools to reduce changeover time. Often feature a split-point design for fast starting in sheet metal.
Low-helix	Have wider flutes than those on regular drill bits. This increases the space available for chips to be pushed out of the hole, preventing catching or clogging. Require less pressure, penetrate more easily, and stay cooler than regular drill bits.
High-helix	Have a higher cutting rake and improved chip-conveying properties for use on aluminum and some plastics.
Three- or four-fluted	Specially adapted for enlarging punched, cored, or drilled holes.
Step drill bits	Have two or more diameters on a single bit. The bit looks like a set of stairs and can cut a number of hole diameters. Not used often in sheet metal work because they dull quickly when cutting it.
Left-hand	Used where the spindle of the machine rotates in a left-hand direction.
Aircraft	Designed for drilling light sections of high-strength, high temperature alloys and similar materials.
Die	Used for drilling hard steel.
Metalworking	Has a 1/4″ (7 mm) diameter shank adapted to fit the chucks of electric and hand drills.
Clearance	Used to drill holes large enough to allow a bolt or screw to pass through. The *clearance* is the difference between the drill size and the thread.

Figure 7-5. The drill bits listed here are designed for specific applications.

taper mount: Portion of a drill press that both holds and drives the drill bit. Tapered opening holds the drill bit in place through friction.

sleeve: A drill press holding tool used to hold drill bits that are too small for the spindle.

Mounting a drill bit in a drill press may require using either a permanently mounted chuck or a ***taper mount***. A permanently mounted chuck has a tapered shank that fits the taper on the drill press spindle. This chuck may also be used to hold straight-shank twist drill bits up to 1/2″ (13 mm).

Tapered-shank twist drill bits can be inserted either directly into drill press spindles or into various holding devices having a taper hole. The tang of the twist drill bit fits into a small, slotted recess inside the drill press spindle, above the tapered hole. This keeps the drill bit from slipping.

A steel ***sleeve*** is used to hold twist drill bits that are too small for the spindle, **Figure 7-6**. The sleeve has a Morse taper on the outside so that it will

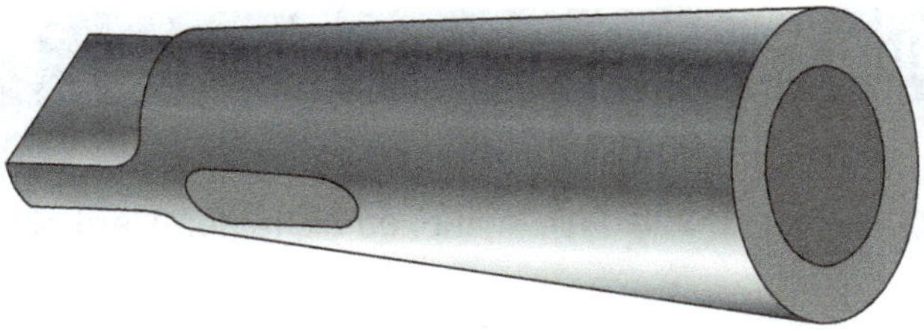

Figure 7-6. A drill bit sleeve is used when the shank of a drill bit is too small to fit securely in the spindle.

fit into the tapered hole in the spindle. The tang on the end of the sleeve fits the slot in the spindle. The taper on the drill bit shank fits the tapered hole on the inside of the sleeve.

A steel ***socket***, **Figure 7-7**, is used on tapered-drill bit shanks that are too large to fit in the drill press spindle. The socket is then inserted in the spindle or sleeve.

socket: A drill press holding tool used to hold drill bits that are too large to fit in the spindle.

On the drill press, drill bits are either pushed or twisted into position and held there by friction. Use a drift to safely remove the drill bits.

As with the portable drill, once the drill bit is mounted in the drill press spindle, run it briefly. Make sure it is running straight and is properly secured.

Secure the work piece to avoid bodily injury or property damage. Hold the drill bit perpendicular to the work. Wear safety goggles to minimize the chance of eye injury.

Place the drill bit on the mark for the hole. Apply pressure to the drill and drill briefly. Remove the drill bit to ensure the hole is being drilled in the correct spot. Correct, if necessary. Apply pressure and continue drilling until the hole is complete. With experience, you will learn the amount of pressure needed to create a properly drilled hole.

Use cutting compounds when drilling metals to increase the life of the drill bit. Cutting compounds keep drill bits cool and reduce friction, resulting in lower surface temperatures. Compounds that aid in drilling are listed in **Figure 7-8**.

Speeds and Feeds

The cutting speed of a drill bit and the rate of feed determine the time needed to drill a hole. ***Cutting speed***, or ***peripheral speed***, expressed in surface feet per minute (sfpm), is the speed difference between the drill bit and the surface of the work piece it is drilling into. Cutting speeds have been calculated by industry, based on the material being drilled, **Figure 7-9**. Use this chart, or a similar chart, to choose the drill bit that will cut a hole in the material.

cutting speed (peripheral speed): The speed difference between the drill bit and the surface of the work piece it is drilling into. Measured in surface feet per minute.

Drill press speed is expressed in revolutions per minute (rpm) of the drill press spindle. The amount of material surface area coming in contact with the drill bit changes depending on the diameter of the bit. Therefore, the spindle revolutions will have to change to maintain the suggested cutting speed. For

drill press speed: Measurement of the revolutions per minute made by the drill press spindle.

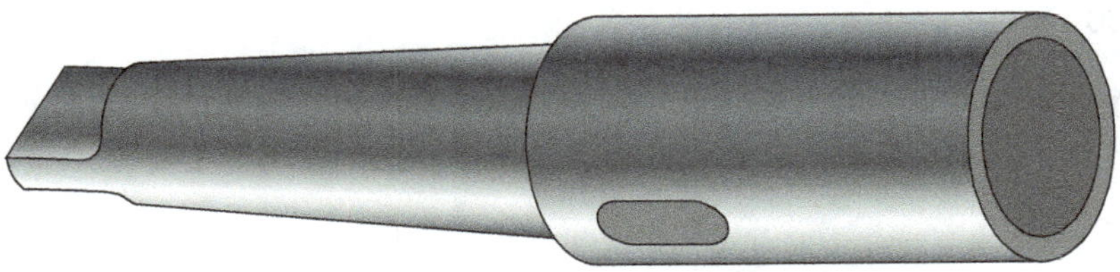

Figure 7-7. A drill bit socket is used for oversized drill bit shanks.

Cutting Compounds	
Material	**Compound**
Tool steel Forging steel Manganese steel Soft steel Wrought iron	Soluble oil, sulfurized oil, mineral lard oil
Aluminum Aluminum alloys	Kerosene
Copper	Soluble oil, dry, mineral lard oil, kerosene
Brass, soft	Dry, soluble oil, kerosene, lard oil
Bronze	Soluble oil, lard oil, mineral oil, dry
Cast iron	Dry, air jet, soluble oil

Figure 7-8. Using cutting compounds increases drill bit life.

Recommended Cutting Speeds			
Material	**Recommended speed in feet per minute (fpm)**	**Material**	**Recommended speed in feet per minute (fpm)**
Aluminum and alloys	200–300	Steel, annealed	60–70
Bakelite	100–150	Steel, forgings	50–60
Plastics	100–150	Steel, machine	80–110
Brass and bronze, soft	200–300	Steel, manganese	15–25
Bronze, high tensile	70–100	Steel, soft	80–100
Cast iron, chilled	30–40	Steel, stainless (free machining)	60–70
Cast iron, hard	70–100	Steel, stainless (hard)	30–40
Cast iron, soft	100–150	Steel, tool	50–60
Magnesium and alloys	250–400	Slate, marble, and stone	15–25
Malleable iron	80–90	Wrought iron	50–60
Monel, metal	40–50	Wood	300–400
Nickel	40–60		

Figure 7-9. Cutting speeds vary widely, depending on the type of material being drilled.

example, a 4″ (100 mm) diameter drill bit removes twice as much material as a 2″ (51 mm) diameter drill bit, thereby doubling the amount of surface feet of material that the drill bit contacts. Use the following formula to calculate the spindle speed of the drill press to compensate for the size of the drill bit:

$$\text{rpm} = \frac{4 \times \text{cutting speed}}{\text{drill bit diameter}}$$

As a general rule, the larger the drill bit, the slower the drill press spindle speed.

Feed is the downward motion into the work during each revolution of the drill press spindle. Feed rates can also be found in charts and are also based on the material being worked.

feed: Downward motion of the drill bit into the material made in each revolution of the drill press spindle.

Speed and feed rates are commonly needed in assembly-line situations. In that environment, speed and feed rates result in clean, uniform holes made quickly and efficiently. Speeds and feeds are not specific when drilling with a portable drill. They cannot be controlled accurately in this situation. Different users apply different pressures to their drills. These differing pressures will cause the drill bit to work differently in each case.

Drill Bit Troubleshooting

Twist drills withstand more strain in proportion to their size than any other small cutting tool. When a drill bit chips, breaks, burns, or does not cut properly, it is often due to misuse. Refer to **Figure 7-10** for suggestions on ways to eliminate many drilling problems and ensure maximum drill life.

Drill Bit Troubleshooting

Problem	Cause
Drill will not cut	• Dull drill bit. • Work piece too hard. • Drill speed too fast.
Hole rough	• Too much pressure on the drill. • Work not secure. • Dull or improperly sharpened drill. • Cutting oil needed.
Hole oversize	• Drill bit bending. • Improper size drill. • Work not secure. • Drill not sharpened properly.
Drill bit walks	• No center punch. • No pilot hole to lead drill into place. • Improper drill point.
Drill bit grabs work	• Incorrect drill point. • Work not secure. • Too much pressure or speed.
Drill bit burning	• Too much speed or pressure. • Flutes clogged. • Work too hard. • Cutting oil needed. • Dull drill bit.
Drill breaks	• Flutes clogged with chips. • Work not secure. • Incorrect drill type. • Too much pressure or pressure not applied on centerline of bit.

Figure 7-10. Many problems encountered while drilling sheet metal can be solved with these simple solutions.

Punches and Punching

A punch is a tool used on sheet metal to create or start holes and to mark the location of holes to be drilled. Punches fall into two categories: hand punches and machine punches. Hand punches commonly used in sheet metal work include the solid, pin, prick, and center punches. The prick punch and the center punch are commonly used to mark layout work, such as fold lines and points. The center punch is also used to make indentations on sheet metal heavier than 18 gage.

Making a Hole with a Hand Punch

1. Locate the position of the hole. To locate the center point on circular work pieces, use a center head or hermaphrodite calipers, **Figure 7-11**.
2. Select the correctly sized punch.
3. Attach the work piece to be punched to the end grain of a block of hardwood, **Figure 7-12**.
4. Hold the punch in one hand against the metal and strike the top of the punch with a hammer.
5. Turn the metal over, with the punched hole projecting outward. Smooth this section with a mallet.

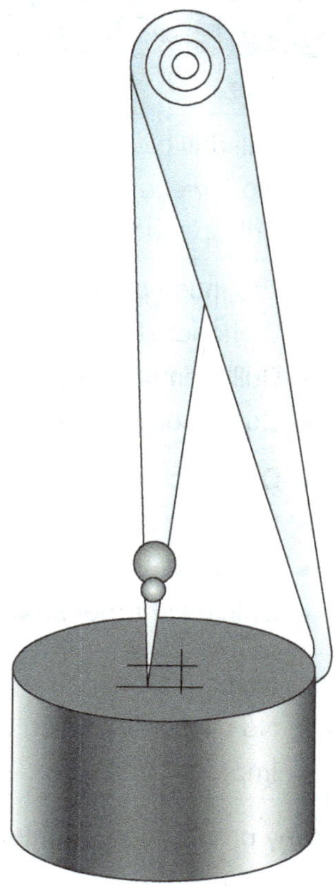

Figure 7-11. Finding the center of a work piece, using hermaphrodite calipers.

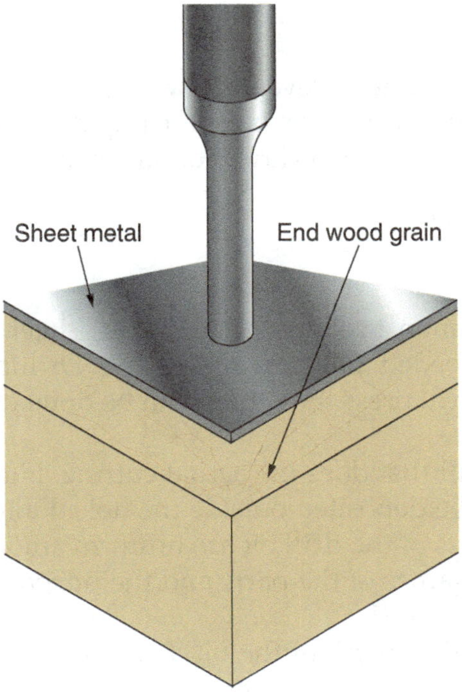

Figure 7-12. Preparing the work piece to be punched.

Hand-Lever Punches

A hand-lever punch uses handles to apply pressure to a work piece, creating a hole, **Figure 7-13**. Squeezing the handles forces the male punch through the sheet metal into the female die. Hand-lever punch sets come with various sizes of punches and dies. Each punch has a matching die. They can be used to produce different sizes and shapes of holes.

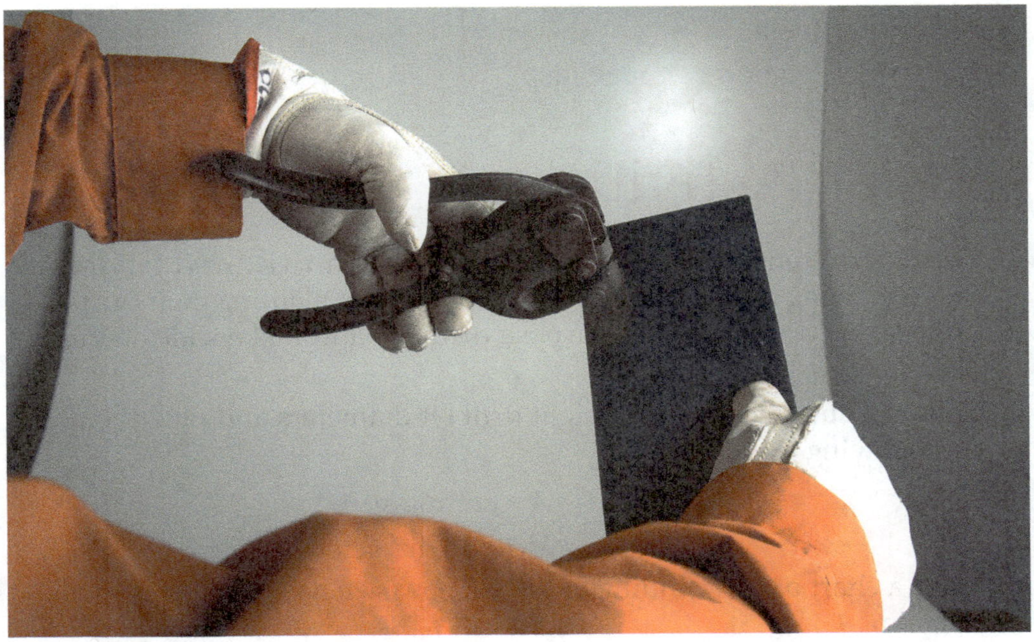

Figure 7-13. A hand-lever punch is useful when making a large number of punches.

Machine Punches

Machine punches are powered by either a lever, screw, or gear. They are used on very heavy material or when many uniform holes are required. Like the hand-lever punch, machine punches come with a variety of punch and die sets.

Summary

Drilling and punching are common tasks associated with sheet metal work. Both are used to create holes needed for rivets and screws. Drilling can be done with a handheld drill or a drill press. Punching can be done using a variety of hand and machine punches.

The drill bit is the part of the drill that does the actual cutting. Flutes on the bit allow the waste materials to be deposited outside the hole. Flutes can be either twisted or straight. The parts of the drill bit are uniform among the many sizes available. Knowing the names of the parts and the purpose they serve can help when choosing a bit for a job.

Drill bit sizes are stamped onto the shank of the bit. For very small bits or bits that have a great deal of wear, size can be determined by using a drill gauge. Size is expressed using numbers, letters, and fractions.

Before using the drill bit, it is important that it be mounted properly in the machine you will be using. Mounting in a portable drill is straightforward. Mounting in a drill press, however, requires a few steps to make sure the fit is correct. Once mounted, determine what speed and feed rates are necessary for drilling the hole. Then drilling can begin.

Punches are used to create or start holes, or to mark the locations of holes to be drilled. Most punching is done on lightweight metals and is done using hand punches. Machine punches are only used in industrial settings.

Internet Resources

Morse Cutting Tools

www.morsecuttingtools.com/data.asp

Outside Activities

1. Create a presentation illustrating the parts of a twist drill bit. Include slides showing different types of drill bits and explaining their purposes. Choose one or two special purpose drill bits and make slides outlining their construction and uses.
2. Choose a cutting speed and eight drill bit diameters and plug them into the following formula.

   $$\text{rpm} = \frac{4 \times \text{cutting speed}}{\text{drill bit diameter}}$$

   Make a chart of your findings. Explain how size has an effect on drill bit rotation.

Name ______________________ Class ______________ Score ________

Chapter 7
Drilling and Punching

Review Questions

Write your answers in the spaces provided.

1. Name two machines that are used to drill holes. ____________________

____

____

2. The _____ is a pointed cutting tool that makes cylindrical holes. 2. ____________

3. True or False? There are two types of shanks: straight and tapered. 3. ____________

Match each part of the twist drill bit with the term that identifies it.

4. Point. 4. ____________

5. Shank. 5. ____________

6. Lip. 6. ____________

7. Flute. 7. ____________

8. Margin. 8. ____________

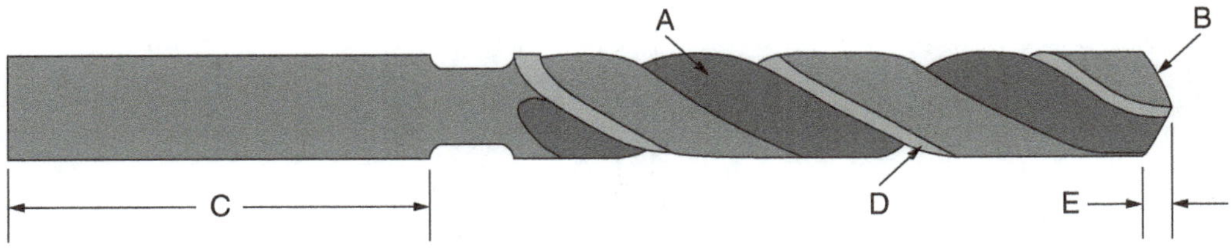

9. What is meant by drill walking? ____________________

____

____

____

10. Name three methods for sizing standard drill bits. ____________________

____

____

11. A steel _____ is used to hold twist drill bits that are too small for the spindle. 11. ____________________

12. Cutting speed _____. 12. ____________________
    A. is measured in revolutions per minute
    B. measures downward motion
    C. must always be calculated using a formula
    D. is also known as peripheral speed

13. Explain why speed and feed rates are not specific when drilling with a portable drill. ________

____

____

____

14. What procedure is being done in the following drawing?____________________

____

____

____

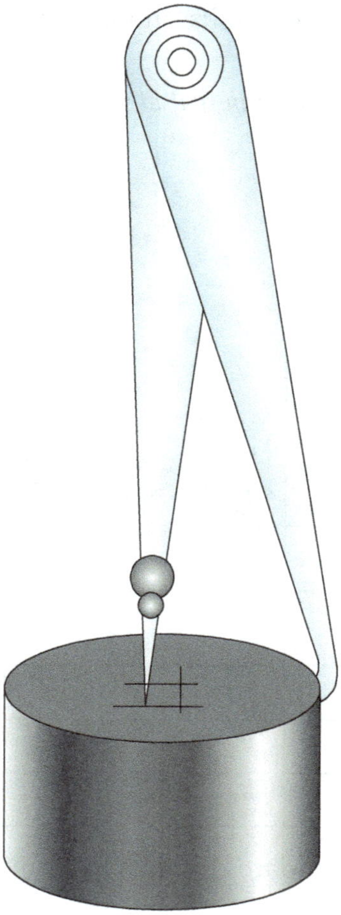

15. True or False? Both hand-lever and machine punches come with a variety of punch and die sets. 15. ____________________

Chapter 8
Fastening Sheet Metal Components

Objectives

After studying this chapter, you will be able to:

- Identify different types of screws and explain in which assemblies they should be used.
- Demonstrate how bolts, nuts, and washers fit together.
- List the characteristics of fasteners that can be used for identification.
- Use riveting tools properly to install rivets.
- Assemble the tools and materials needed and successfully solder a joint or seam.
- Explain the oxyacetylene welding, arc welding, inert gas welding, and spot welding processes.
- Discuss soldering and welding safety.

Technical Terms

blind rivet
bolt
brazing
capillary action
corrosive flux
flux
flux core solder
hard solder
lock washer
machine screw
noncorrosive flux
nut
resistance spot welding
rivet set
rivet
self-drilling sheet metal screw
self-locking nut
sheet metal screw
soft solder
solder
soldering
spot welding
thread pitch
tinning
washer
welding

Assembly is the final step in sheet metal production. The pieces that have been planned, measured, and cut now need to be joined together into a finished product. There are several methods for securing pieces to each other. Fasteners are used to join these pieces to each other or to another structure. Commonly used sheet metal fasteners include screws, bolts, and rivets. Sheet metal may also be joined by soldering or welding. Deciding which fastener or fastening method to use is based on job specifications, costs, and availability of materials.

Fasteners

Screws, bolts, and rivets are used on metals that cannot be exposed to the high temperatures used in welding. In addition, screws and bolts are used in assemblies that may need to be separated from time to time, for repair or service. The fastening action of screws and bolts can be easily reversed using the proper tools. Soldering and welding are, essentially, permanent ways of assembling pieces.

Screws

A screw has an externally threaded shaft that is able to fit through holes in assembled parts, to mate with preformed threads, or to make its own threads. Screws are tightened or released by turning the screw heads. Screws are used in a variety of applications to join a variety of different materials. Screws are made from a variety of metals depending on the required strength, corrosion resistance, and cost.

Machine screws have threads made to fit into a nut or into a tapped hole. ***Sheet metal screws*** have fully threaded shafts ending in a point. These screws are used on sheet metal that has existing holes. A ***self-drilling sheet metal screw*** has a point capable of drilling a hole into the material before the threads engage. The friction created by both types of sheet metal screws eliminates the need for a fastening device, such as a nut, to hold it in place. See **Figure 8-1.**

machine screw: Fastening device that has threads made to fit into a nut or into a tapped hole.

sheet metal screw: A type of machine screw that has a fully threaded shaft, ending in a point. This screw is used on sheet metal that has existing holes.

self-drilling sheet metal screw: A machine screw that has a point capable of drilling a hole into the material before the threads engage.

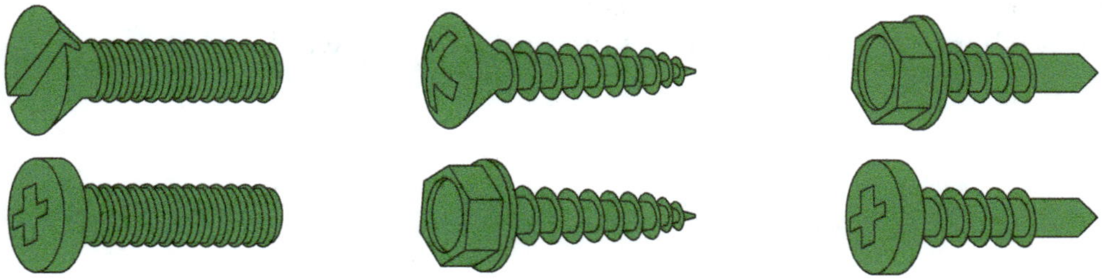

Figure 8-1. Three types of screws used on sheet metal work.

Bolts

Bolts have an externally threaded shaft designed to fit through a preformed hole and into an internally threaded nut, **Figure 8-2.** During assembly, the bolt is prevented from turning as the nut is tightened. A bolt must be used with a nut to do its intended job.

Nuts

Nuts are used with bolts and some screws to fasten and lock parts together. The size of nut used is determined by the diameter of the screw or bolt. ***Self-locking nuts*** are used in assemblies that experience a great deal of vibration or stress, **Figure 8-3.** These nuts have a mechanical feature that prevents the nut from loosening.

Washers

Screws are often inserted into flat discs called ***washers*** before being driven into an assembled part. Washers serve several purposes. They are used to distribute the load created by the screw, to protect the finish of the part, or to create a finished look to the part. ***Lock washers*** are used to secure nuts and bolts to the assembled part.

bolt: A fastener with an externally threaded shaft designed to fit through a preformed hole and into an internally threaded nut. A bolt must be used with a nut to do its intended job.

nut: A fastener used with bolts and some screws to fasten and lock parts together.

self-locking nut: Type of nut with a mechanical feature that prevents the nut from loosening. It is used in assemblies that experience a great deal of vibration or stress.

washer: A flat disc used to distribute the load created by a screw, to protect the finish of a part, or to create a finished look to a part.

lock washer: A type of washer used to secure nuts and bolts to assembled part.

Fastener Identification

Fasteners have several characteristics that are useful for identification. They can be identified by their category, head shape, drive type, length, diameter, and thread pitch.

Category designations are based on a fastener's function or design. For instance, wood screws are used on wood and hex bolts have a hexagonal head. Drive-type designations refer to the type of tool used to drive the fastener into the material. Head shape is used to further identify fasteners,

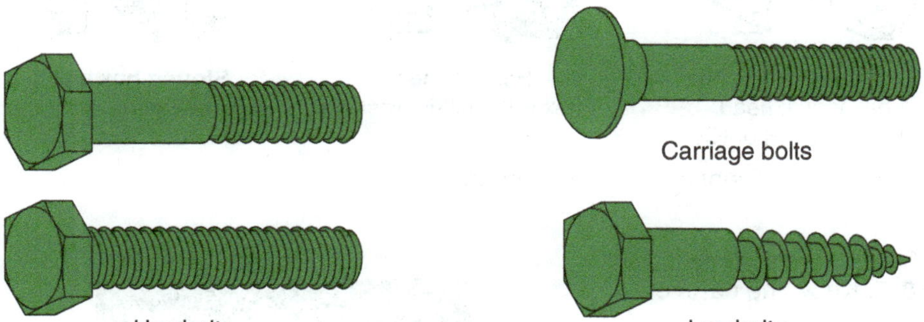

Figure 8-2. Bolts come in a variety of shapes and sizes.

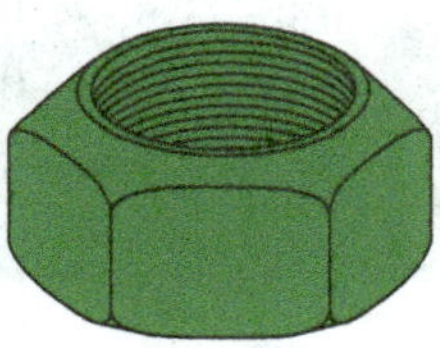

Figure 8-3. This is a nonreversible self-locking nut.

beyond their category or drive type. Head types are shown in **Figure 8-4.** Length is generally considered to stretch from the point at which the fastener head touches the material to the end of the thread. For countersunk heads, length measurement includes the head, as it is flush with the material once it is in place. Diameter is generally the widest point across the thread, given in fractions of an inch or decimal. ***Thread pitch*** is a measurement of the number of threads per inch and is used only on machine screws. The measurement is divided into either Unified National Coarse (UNC) or Unified National Fine (UNF) designations. Coarse fasteners have a lower number of threads per inch than fine fasteners.

thread pitch: Measurement on a threaded fastener that identifies the number of threads per inch.

rivet: A metal fastener made up of a head and a shaft. It is attached to a work piece using a riveting hammer, other riveting hand tools, or a riveting machine.

Rivets

Rivets are metal fasteners made up of a head and a shaft, **Figure 8-5.** They are attached to work pieces using a riveting hammer, other riveting hand tools, or a riveting machine. Rivets commonly used in sheet metal work include Tinner's, flat head, round head, and countersunk, **Figure 8-6.**

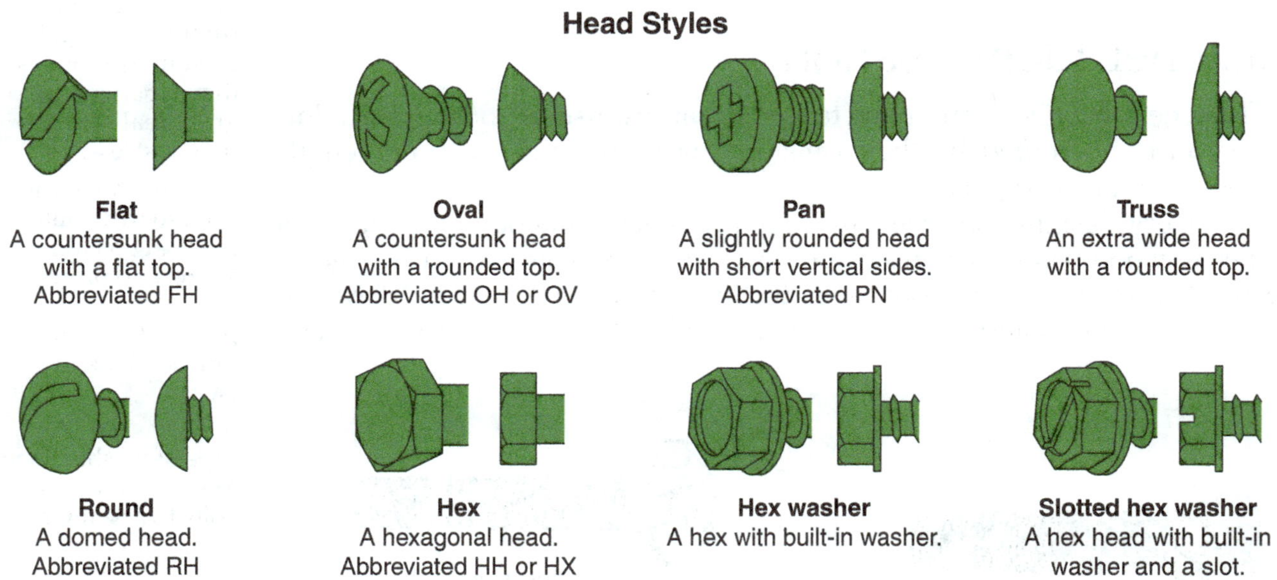

Figure 8-4. Note the abbreviations used to identify a particular head.

Figure 8-5. Know the parts of a rivet.

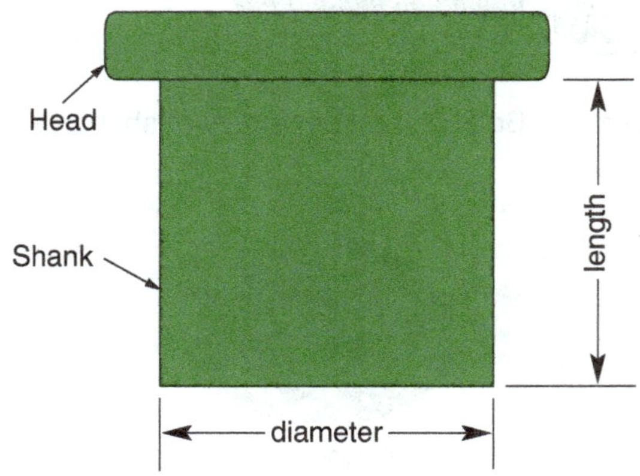

Rivets		
Type	**Appearance**	**Use**
Tinner's		Fabrication
Flat head		Fabrication
Round head		Applications needing extra strength.
Countersunk		Applications requiring a smooth surface finish.

Figure 8-6. Rivets come in a variety of styles, for use in varying circumstances.

Rivet sizing or designations vary. Tinner's rivets are referred to as 1 pound, 2 pound, and so on. This refers to the weight of 1000 rivets. Therefore, the weight of 1000 5-pound rivets is 5 pounds. Flat, round, and countersunk rivets are sized based on shaft diameter and length.

Blind rivets are used in situations where only one side of the work piece can be seen or reached. The rivet is inserted from one side. Using a specially designed riveting tool, the rivet is driven into place. Blind rivets are sometimes called pop rivets. However, POP® is a trademarked term and refers specifically to those rivets made by the Emhart Teknologies Company.

blind rivet: A type of rivet used in situations where only one side of the work is reachable or visible.

Rivets must be placed properly in order to securely hold work pieces, **Figure 8-7.** Drill holes for rivets at least two times the rivet diameter away from the edge of the work. The distance between rivets should be at least three times the rivet diameter but no more than 2″ (51 mm).

When laying out a riveted seam, create a template for any jobs that will be laid out repeatedly. Install rivets in the end holes first, then install the center rivet. Finish the seam by installing rivets in the remaining holes.

Installing Rivets

The tools needed to install rivets include a rivet set and a riveting hammer. The ***rivet set***, **Figure 8-8,** uses friction to draw the work pieces together and to set the rivet.

1. Insert the rivet through the predrilled holes in the work pieces.
2. Place the deep hole of the rivet set over the head of the rivet. Strike the top of the rivet set with the riveting hammer.
3. Remove the rivet set.
4. Strike the rivet directly with the hammer to set the rivet.
5. Place the shallow hole of the rivet set over the head of the rivet. Strike the top of the rivet set with the hammer to give the head a smooth, finished quality.

rivet set: A tool used to seat and lock a rivet into place.

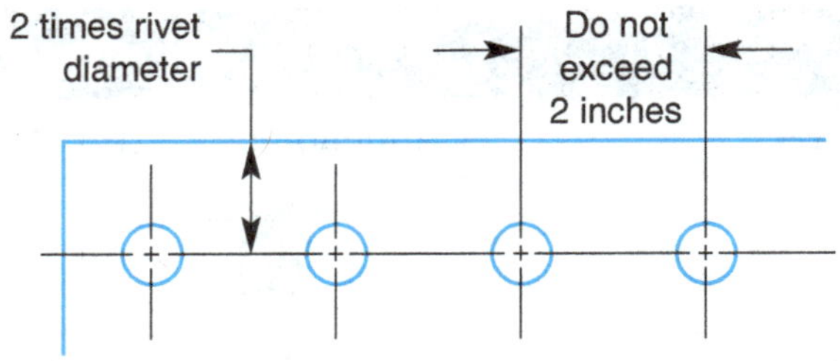

Figure 8-7. Rivets need to be properly spaced to ensure the integrity of the assembly.

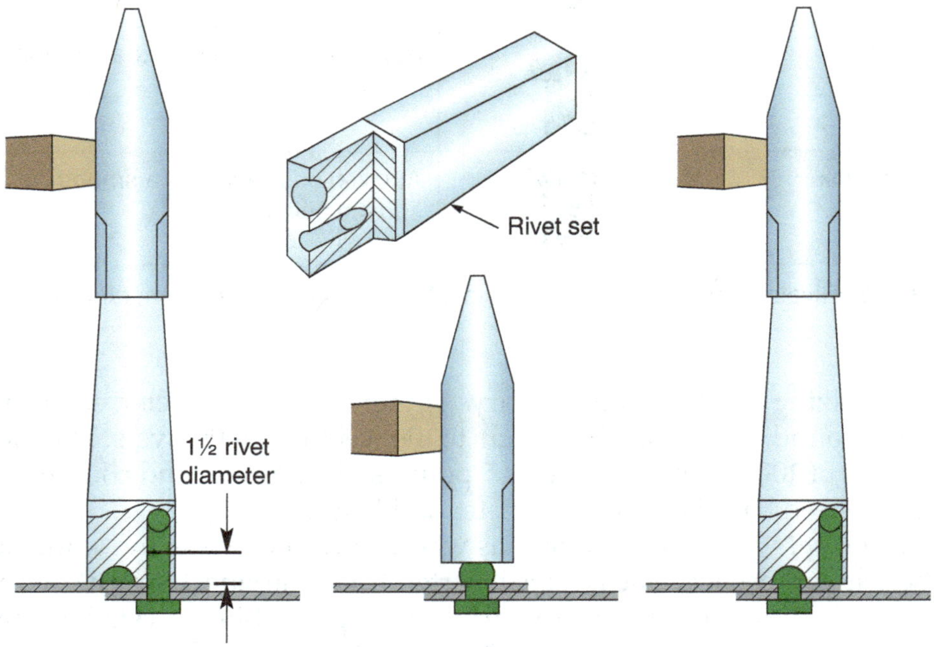

Figure 8-8. Installing a rivet with a rivet set.

soldering: The process of joining metals by heating a filler material (solder). The filler material melts, joining the base metals without melting them.

capillary action: A physical reaction between a solid and liquid that causes liquid solder to flow into air spaces found between pieces of tightly fitted metal.

solder: Filler material used in the soldering process. It is an alloy of several materials.

soft solder: Solder that melts at temperatures below 750°F (400°C).

hard solder: Solder that melts at temperatures above 750°F (400°C). It is usually a copper/zinc or copper/silver alloy.

brazing: A process of joining metals with a nonferrous filler at temperatures exceeding 840°F (450°C).

Soldering

Soldering is a nonmechanical fastening method used with metals. Using heat, several metal pieces can be joined together by melting a filler metal into the joint. The metal pieces do not melt together. Instead, the solder is pulled into the joint through ***capillary action*** and bonds it.

Solder

The filler material, called ***solder,*** is an alloy of several materials. Each solder alloy has a specific melting point. ***Soft solder*** has a relatively low melting point, below 750°F (400°C). The melting point of ***hard solder*** is above 750°F (400°C). Hard solder is usually a copper/zinc or copper/silver alloy. Using hard solder tends to melt the pieces being joined, so it is technically a ***brazing*** process.

A lead/tin alloy is commonly used in sheet metal work. The higher the concentration of tin, the stronger the tensile and shear strength of the solder. A 50/50 mixture of lead and tin is most workable for sheet metal applications. In lead/tin mixtures, the first number is the percent of lead and the second is the percent of tin. For example, 80/20 solder is 80% lead and 20% tin. Solder is available in wire or bar form. Choose a solder that will bond readily with the metal pieces being joined.

Solder is used in a variety of ways. It is used in copper tubing to join seams. It is used in the automobile industry to fill holes and deep scratches in car bodies. Bottoms of buckets, tea kettles and tin cans are also soldered.

Flux

All metals will develop a thin film of rust (metallic oxide) or tarnish from exposure to the air. It is necessary to remove this film so the solder can cling to the metal. ***Flux*** has the following characteristics:

- It removes tarnish and rust from the metal.
- It prevents further oxide formation while the metal is being heated to soldering temperature.
- It lowers the surface tension of molten solder, enabling the solder to spread and penetrate properly.

Fluxes are either corrosive or noncorrosive. Use a ***noncorrosive flux*** for permanent results. Rosin is a noncorrosive flux. It works best on clean tin plate, solder-coated surfaces, and clean copper. Other noncorrosive fluxes are available for most metals. These fluxes usually have a rosin base with small quantities of other materials.

Corrosive flux contains acid. Clean corrosive fluxes from metal surfaces immediately after soldering to stop further corrosion of the metal. Scrub with a wire brush, then clean with water or a damp rag. Commonly used corrosive fluxes include muriatic acid and a zinc/ammonium chloride mixture.

Flux core solder is a solder that has a flux core. As the solder is melted, the flux core cleans the surface of the joint.

Remove flux from the finished joints to prevent corrosion and for a better appearance of the joint. Water-soluble fluxes can be removed with hot water and a brush. Alcohol will remove most rosin fluxes. A flux containing some form of grease, as in most paste fluxes, requires a grease solvent, followed by a hot water rinse.

flux: Material (acid or rosin) used in soldering that prevents oxidization of the soldering iron tip and the heated metal. It can also remove the oxide film that forms on metals.

noncorrosive flux: A type of flux that will not continue to corrode metal after soldering.

corrosive flux: A type of flux that contains acid. The work piece must always be cleaned after soldering with corrosive flux to prevent residue from corroding the work.

flux core solder: A solder that has a flux core. As the solder melts, the flux core cleans the surface of the joint.

Soldering Tools

A number of devices are made for soldering, including soldering irons and soldering guns. Soldering irons can be nonelectric or electric. Nonelectric soldering irons have no heating element but instead are heated to the proper temperature in a gas furnace. Butane furnaces are portable and easily moved from job to job. Electric soldering irons generate uniform heat, making them useful for continuous soldering. Soldering guns are used on jobs requiring high degrees of heat over a large area.

Soldering tips are made of copper or iron and come in a variety of shapes. These tips need to be properly filed and tinned to prevent corrosion and solder build-up. Excess corrosion or build-up will harm the heating ability of the iron. ***Tinning*** involves rubbing the tip of a heated soldering iron on a sal ammoniac block and then covering it with a thin layer of solder, **Figure 8-9.** After tinning, remove excess solder from the tip of the iron.

tinning: A process done before soldering. It involves rubbing the tip of a heated soldering iron on a sal ammoniac block and then covering it with a thin layer of solder.

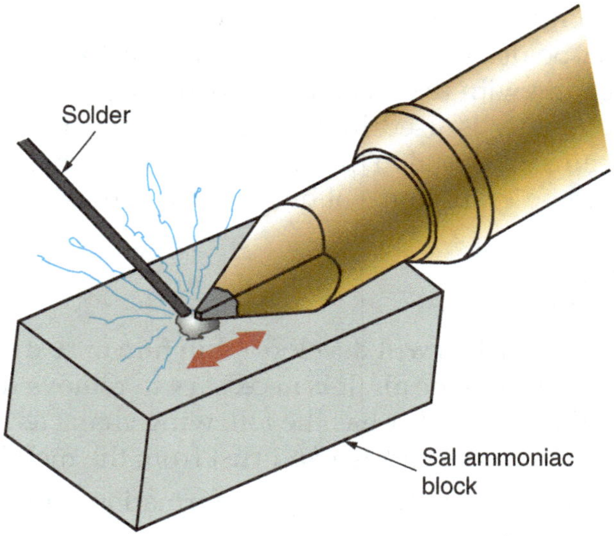

Figure 8-9. A properly tinned iron makes soldering jobs go smoothly.

Soldering Process

1. Clean the metal pieces being joined.
2. Fit the pieces together to form a joint or seam.
3. Heat the soldering iron and the parts. Hold the soldering iron with the tip in full contact with the metal, **Figure 8-10**.
4. Apply flux.
5. Apply solder directly in front of the tip of the soldering iron, not on the tip. Keep the iron moving along the joint or seam as the solder melts and flows into place.
6. Remove the soldering iron from the work.
7. Hold the assembled pieces still until solder has solidified, to prevent cracks.
8. Clean the corrosive flux off the assembly if necessary.

Soldering Safety

- Burns and fires are always a danger. Learn how to properly operate soldering equipment before working on any job.
- Handle soldering iron carefully. Avoid the tip. The tip of the soldering iron can reach 500°F (260°C).
- Use a piece of metal or an insulated board as a rest for the soldering iron.
- Work in a well-ventilated area. Soldering fumes are dangerous.
- Do not handle soldered materials until cool.

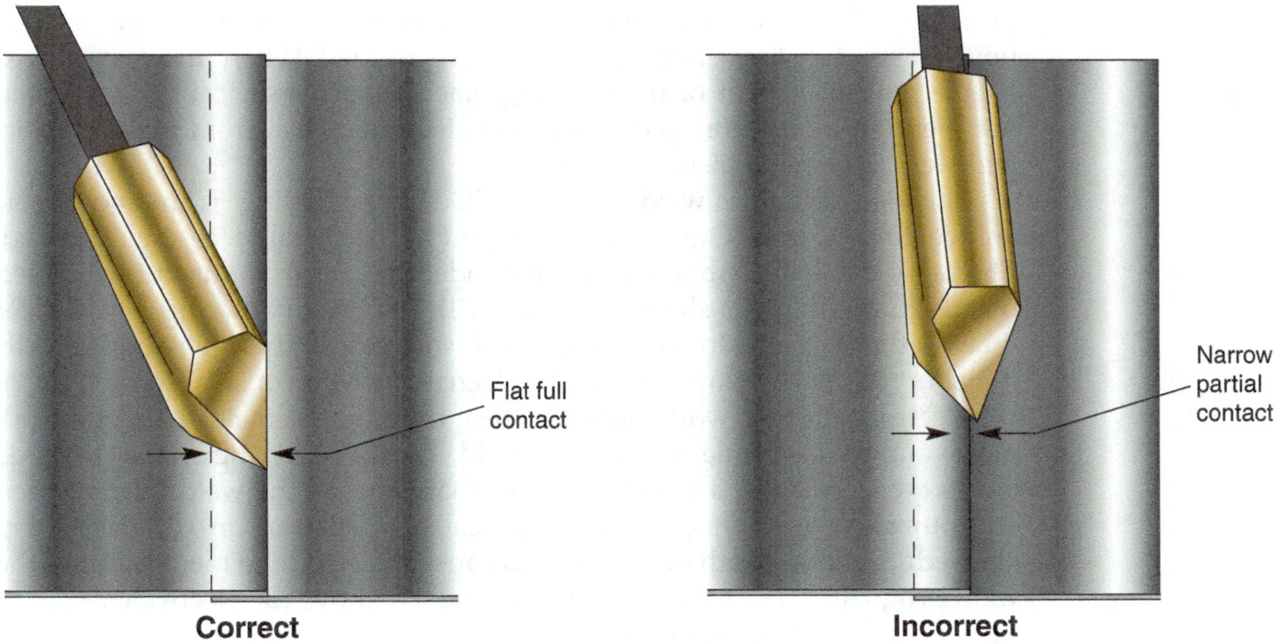

Figure 8-10. In order to make a securely soldered seam, hold the soldering iron in flat, full contact with the metal.

Welding

Welding is used in many sheet metal applications, although the work is most often done by welders and not sheet metal workers. It is important, however, for sheet metal workers to have some understanding of welding.

Welding refers to the permanent joining of two metal parts by melting and fusing them together. It is used with pieces too heavy to be joined using screws, bolts, or soldering. Common welding methods used with sheet metal include oxyacetylene welding, shielded metal arc welding, gas metal arc welding, gas tungsten arc welding, and resistance spot welding.

welding: A nonmechanical fastening method that permanently fuses metal pieces together by melting the metal.

Oxyacetylene Welding

Oxyacetylene welding uses two gases, oxygen and acetylene, to create a high-temperature flame. Each gas is stored in its own tank. To create the flame, the gases are mixed together in a torch and released through the torch tip. The gases exiting the torch tip are ignited, creating a very hot flame. The flame is used to heat the metal work pieces up to their melting temperature, and then the welder feeds a metal welding rod into the weld pool. The welding rod melts, adding extra metal to the weld. Oxyacetylene torches are also often used to heat heavy steel so it can be bent more easily.

Arc Welding

Arc welding is a group of welding processes that use the heat of an electric arc to melt and weld metal. All arc welding machines have one positive cable and one negative cable. One of these cables is attached to the metal work pieces and the other is in contact with an electrode. The electrode is positioned a short distance from the metal being welded. When the welder turns on the output current of the welding machine, an electrical arc is created between the metal work pieces and the electrode. Heat from that arc melts and welds the metal pieces together.

Shielded metal arc welding, also called stick welding, is the most basic arc welding process. One cable is clamped to the metal pieces being joined. The other cable is clamped to an electrode. The electrode used in shielded metal arc welding is a short metal rod with a solid flux coating. The welder touches the electrode to one of the metal work pieces and then quickly pulls it a short distance away, causing an electric arc to form between the tip of the electrode and the work piece. The heat from the arc melts the metal work pieces and the electrode, creating the weld. The flux covering on the electrode burns away, creating a cloud of smoke that protects the weld from oxygen and other contaminants, which would cause brittleness or oxidation (rust) and reduce the strength of the weld. Because the electrode melts away as the weld progresses, the welder must continuously move the electrode closer to the work pieces. See **Figure 8-11.**

Shielded metal arc welding is a quick and dirty welding process. It is typically used to weld mild steel plate. This process generates a great deal of smoke, spatter, and slag. Therefore, any welds created with this process must be thoroughly cleaned.

Gas metal arc welding, also called metal inert gas (MIG) welding, is a semiautomatic arc welding process. It is called a semiautomatic welding

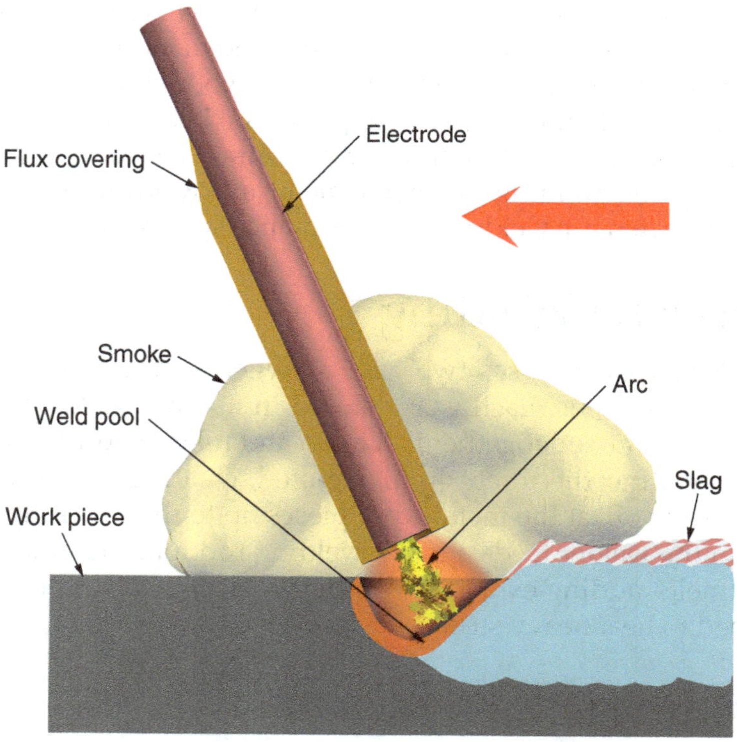

Figure 8-11. In the shielded metal arc process, the arc melts the work piece and the electrode. The flux coating burns off the electrode and forms smoke that protects the weld area from oxidation.

process because the welding equipment automatically feeds the electrode into the weld. One of the cables from the welding machine is attached to the metal work pieces and the other is connected to a welding gun. When the welder points the welding gun at the work pieces and pulls the trigger, a continuous wire is fed off a spool mounted on or inside the welding machine. The wire is fed through the welding gun, where it passes through a contactor tip that electrically charges it. As the charged wire electrode emerges from the end of the gun, an electrical arc is created between the wire electrode and the metal work pieces. The heat from the arc melts the work pieces and the electrode wire, creating the weld.

A supply of inert gas is also connected to the gun. When the trigger on the welding gun is pressed, a gas valve inside the gun is opened, allowing the gas to flow out the end of the gun. This gas completely envelopes the welding area and protects the weld from oxygen. See **Figure 8-12.**

The gas metal arc welding process creates clean welds. Different types and sizes of electrode wires are available, so the process can be used to weld mild steel and aluminum in a wide range of thicknesses.

In gas tungsten arc welding, also called tungsten inert gas (TIG) welding, a non-consumable tungsten electrode is mounted in a torch body. A variable switch on the torch body or a foot pedal allows the welder to control the intensity of electrical current through the electrode. A supply of inert gas is also connected to the torch body.

To start welding, the welder positions the torch close to the metal work pieces. Then, the welder activates the switch on the torch body or depresses the foot pedal. An arc is formed between the electrode and the metal work pieces.

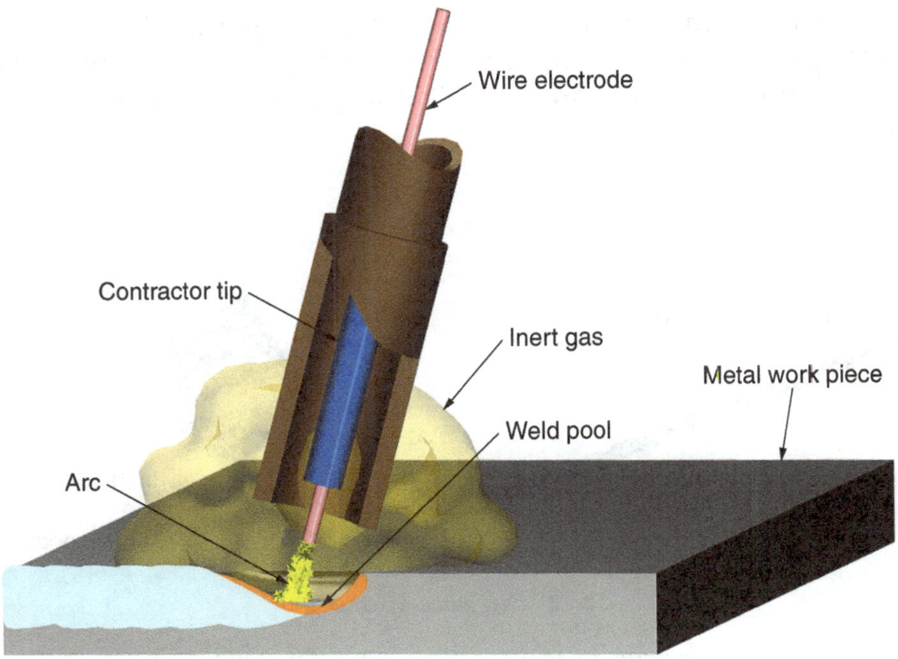

Figure 8-12. In the gas metal arc welding process, the welding machine automatically feeds wire electrode into the weld at the same rate that it melts. Inert shielding gas flows through the end of the welding gun and protects the weld area from oxidation.

Activating the switch also causes shielding gas to flow out of the torch, enveloping the weld area. The heat from the arc melts the metal work pieces, but not the electrode. The welder must operate the torch with one hand and feed a metal rod into the weld pool with the other hand. See **Figure 8-13.** The inert gas flowing out of the torch prevents oxygen from reaching the weld area.

Gas tungsten arc welding is a clean welding process that can create very high quality welds. This process is commonly used to weld aluminum, steel, and stainless steel. However, the process can be used on most types of metal as long as the right combinations of electrode, shielding gas, and filler metal are chosen. It is a slow but precise welding process, and therefore well-suited for sheet metal work.

Resistance Spot Welding

spot welding (resistance spot welding): A welding technique in which the pieces of sheet metal to be welded are held in contact between two copper electrodes. Current from the electrodes passes through the metal to fuse the pieces together.

Another welding process called ***resistance spot welding,*** or simply ***spot welding,*** is also widely used in sheet metal work. In spot welding, the pieces of sheet metal to be fastened are held tightly between two copper electrodes. Heavy electric current then passes through the metal. The resistance between the electrodes is very low, so a large amount of current flows through a very limited spot. The heavy current flow causes the point of contact to heat to a high temperature very quickly, and melts the metal so it runs together, **Figure 8-14.** When spot welding is done correctly, it creates a stronger bond than riveting.

Nearly all commercial metals can be spot-welded. Most of them can be spot-welded to each other. In only a few cases do the welds tend to be brittle. The majority of spot welding is done with wrought sheet. Sheets can be spot welded to rolled shapes and steel castings, and some types of die castings can be welded without difficulty.

The practical limit of thicknesses that can be spot-welded is about 1/8″ (3 mm). That is, of course, in cases where each piece is the same thickness. However, thinner pieces can be welded to pieces more than 1/8″ (3 mm) thick.

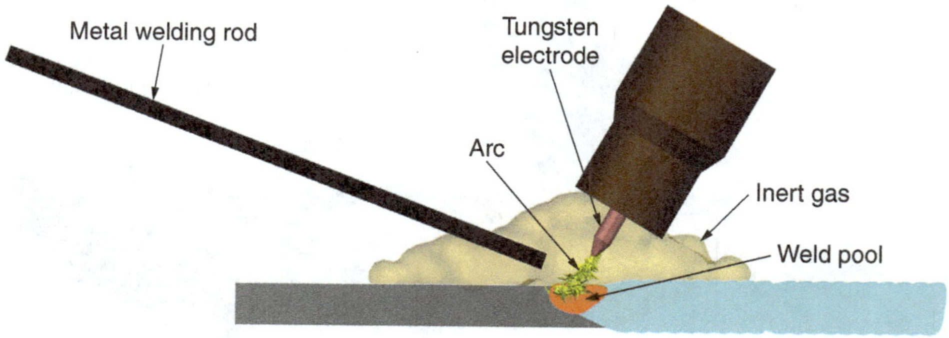

Figure 8-13. In gas tungsten arc welding, an arc is created between the work piece and a tungsten electrode. The work piece melts, but the electrode does not. The welder must manually feed metal welding rod into the weld pool to create the weld. Inert gas from the welding torch protects the weld.

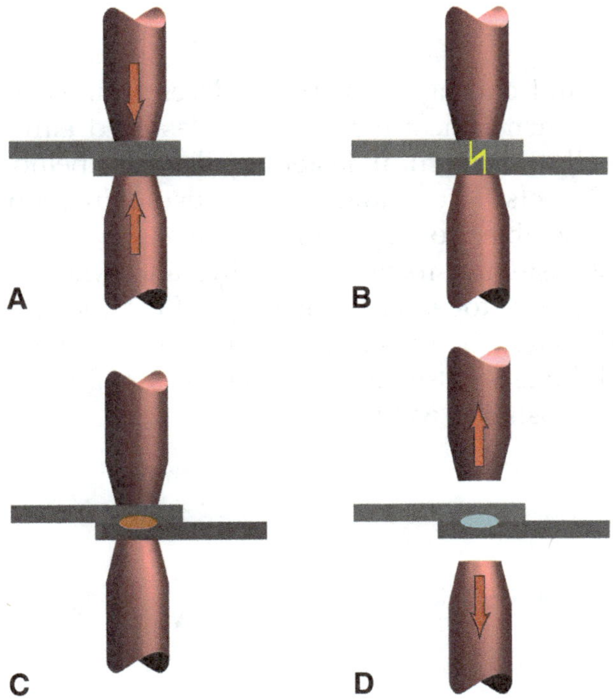

Figure 8-14. The resistance spot welding process is shown here. A—Two copper electrodes are clamped down on the metal to be welded. B—Current passes from one copper electrode, through both pieces of metal, to the other electrode. C—Increased resistance where the two pieces of metal touch causes that area to heat up and melt. D—The current is shut off and when the weld solidifies, the electrodes are removed.

Welding Safety

- Cover all skin, including your neck and forearms. Wear a leather jacket, pants, and gloves when welding. Wear a leather welding cap if welding overhead.
- Do not carry a lighter or matches.
- Never look at an arc with the naked eye. This can cause permanent damage to your eye.
- Wear a helmet with a proper shade to protect your eyes from light and your face from sparks.
- Do not weld metals containing cadmium, zinc, beryllium, or lead without a fume extractor or respirator.
- Weld in a well-ventilated area only. If the area lacks proper ventilation, wear a respirator to avoid toxic fumes.
- If shielding gas is being used in an enclosed area, good ventilation is necessary to replenish the air supply.
- Keep the work area clear of combustible materials.
- Properly maintain welding equipment.
- Keep a fire extinguisher handy.
- Post a fire watch if you are working on one side of a large assembly and the other side is hidden from view.
- Post a fire watch when work *must* be done around combustible materials.

Summary

After planning, measuring, and cutting the various pieces for a sheet metal job, it is time to assemble them. Sheet metal can be fastened either mechanically or nonmechanically. Mechanical fastening devices include screws, bolts, nuts, washers, and rivets. Screws and bolts are useful in applications that may need to be disassembled for repair or service.

Nonmechanical methods of fastening include soldering and welding. Soldering bonds separate metal pieces together by melting a filler material between them. Welding actually melts the metal pieces together, creating a bond. While several welding methods are used for sheet metal assembly, the most commonly used is spot, or resistance, welding.

Internet Resources

Curious Inventor

www.curiousinventor.com/guides/Metal_Working/Screws

Bolt Depot

www.boltdepot.com/fastener-information/

Outside Activities

1. Using several pieces of scrap metal, practice inserting various fasteners into the metal. Use screws, bolts, and rivets. Which fastener was the most difficult to insert? Which was the easiest? Try to remove the screws and bolts. Was one type easier to remove than another? Discuss your findings with your classmates.
2. File all four sides of a nonelectric soldering iron. Tin only one side with solder. Gather several pieces of scrap metal and the soldering iron. Heat the iron and then try soldering with the soldered edge. Note the way the solder melts and moves. Observe the process of capillary action. Now try the same thing with the untinned tip. Does it take more time to heat the joint? How does the solder react with the untinned tip?
3. Try to pull a soldered joint apart using pliers. Were you able to break the bond? How difficult was it to pull the pieces apart? Do you think the solder did its job?

Name ______________________ Class ______________ Score ________

Chapter 8
Fastening Sheet Metal Components

Review Questions

Write your answers in the spaces provided.

1. True or False? Bolts and screws create a permanent bond between parts and should not be used on assemblies that may need to be disassembled. 1. ______________

Match each of the following descriptions with the correct term.

2. Threads fit into a nut or into a tapped hole. 2. ______________
3. Fully threaded shaft ends in a point. 3. ______________
4. Externally threaded shaft is designed to fit into an internally threaded nut. 4. ______________
5. Mechanical feature prevents fastener from loosening once it is in place. 5. ______________
6. Distributes the load created by a screw. 6. ______________

A. Self-locking nut.
B. Washer.
C. Machine screw.
D. Bolt.
E. Sheet metal screw.

7. _____ designations are either Unified National Coarse or Unified National Fine. 7. ______________
   A. Drive type
   B. Head shape
   C. Thread pitch
   D. All of the above.
8. Use _____ rivets in situations where only one side of the work piece can be seen or reached. 8. ______________
9. True or False? Predrilled holes are not required when using rivets. 9. ______________
10. Which of the following statements in *not* true about soldering? 10. ______________
    A. Soft solder has a melting point above 750°F (400°C).
    B. Solder is pulled into a joint through capillary action.
    C. Solder is available in wire or bar form.
    D. The higher the concentration of tin in solder, the stronger its tensile and shear strength.

11. Name three characteristics of flux.______________________________

____

____

____

12. What are the two types of flux used in metal?______________________________

____

13. Rubbing the tip of a heated soldering iron on a sal ammoniac block and then covering it with a thin layer of solder is called _____. 13. ______________

14. Which common welding method uses two gases to create a high-temperature flame? 14. ______________
    A. Resistance spot welding.
    B. Gas tungsten arc welding.
    C. Gas metal arc welding.
    D. Oxyacetylene welding.

15. The process in which sheet metal pieces are fastened together using two copper electrodes is called _____ welding. 15. ______________
    A. arc
    B. spot
    C. oxyacetylene
    D. inert gas

Chapter 9
Drafting and Pattern Development

Objectives

After studying this chapter, you will be able to:

- Discuss the three types of pattern development: parallel line, radial line, and triangulation.
- Use visualization techniques to determine how a sheet metal object is constructed.
- List the drafting equipment needed to draw patterns for sheet metal work.
- Identify the shapes, angles, and lines encountered when developing sheet metal patterns.
- Demonstrate parallel line developments of a rectangular duct, a round duct, and a mitered round duct.
- Use radial line development to create a cone-shaped duct.
- Develop a duct that changes from a square shape to a round shape using the triangulation pattern development method.
- List the advantages of using fabrication software in sheet metal work.

Technical Terms

acute angle
centerline
dimension line
extension line
fabrication software
flexible curve
French curve
hidden line
irregular curves
mitered duct
multiview drawing
object line
obtuse angle
parallel line development
pattern
radial line development
right angle
stretchout
template
triangulation
true length
vertex
visualization

pattern: A flat outline of an object that has only width and depth, and in which all sides of the object can be seen.

parallel line development: Method of creating patterns for objects in which the sides run parallel to each other.

radial line development: Method of creating patterns for objects that radiate from a common center. Arcs are used to create the pattern.

triangulation: Method of creating patterns for objects that gradually transition from one shape to the other. Patterns for these pieces are created using triangles.

visualization: The ability to see an object as a collection of shapes and how the shapes fit together.

Finished sheet metal products are three dimensional. They have width, depth, and height. These three-dimensional objects are made using a pattern. The ***pattern*** is a flat outline of an object and has only width and depth, **Figure 9-1**. All sides of the object can be seen on a pattern. Hems, edges, and seams are included on the pattern.

Other pieces are more detailed, however, and require more work to develop. Three pattern development methods used in sheet metal work include ***parallel line development***, ***radial line development***, and ***triangulation***. Parallel line developments are used for pieces in which the sides run parallel to each other, including straight ducts, elbows, and tee joints. Radial line developments are used for pieces that radiate from a common center, using arcs. Triangulation is used for transition pieces. These pieces have differently shaped openings on either end, gradually transitioning from one shape to the other. Patterns for these pieces are created using triangles.

Visualization

The ability to visualize is necessary in order to develop patterns. The ability to see the piece as a collection of shapes and the way the shapes fit together is called ***visualization***. Observe the finished piece from all angles: top, front, and sides. Then imagine disassembling the part. How does the part look when pulled apart at one seam? Now, look at a flat layout. Can you visualize the completed piece? How will you assemble it? Being able to answer these questions is key to developing patterns.

Drawing Patterns

Simple patterns, such as a sheet metal drain pan, can be developed directly on the metal sheet. They can also be developed on paper and then transferred

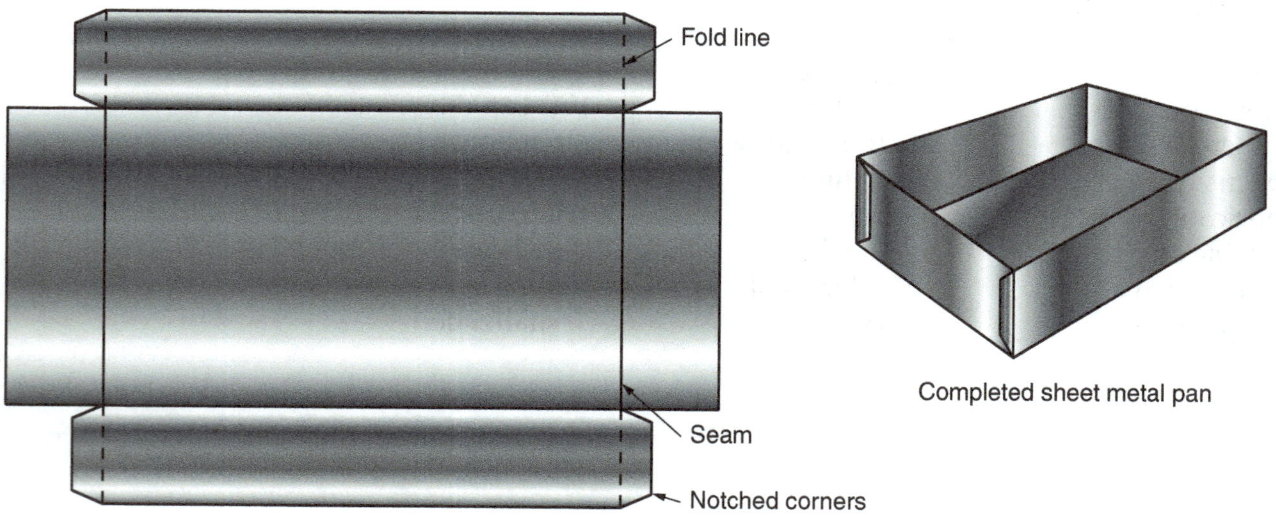

Figure 9-1. This simple pattern is for a rectangular metal pan.

to the sheet metal, **Figure 9-2**. This paper pattern is called a ***template***. It is usually made of metal if it will be used repeatedly. Laying out a pattern on thick construction paper and folding it to the desired shape is an inexpensive way of testing a new pattern. Brazing or welding a piece of sheet metal that has been cut too short is time consuming, expensive, and usually warps or ruins the appearance of the piece.

template: A pattern developed on paper or metal, that is then transferred onto a piece of metal.

Drafting Equipment

The following drafting equipment is needed to draw patterns for sheet metal work, **Figure 9-3**. Variations of this equipment or additional equipment are useful but not necessary for pattern development.

Drawing Board

A drawing board is a flat board made of wood with metal edges on the right and left sides. The metal edges improve the board's strength and enhance the T-square's ability to slide along the edge. A drawing board must be larger than the size of the drawing to be produced and is used to hold the paper in place while drawing.

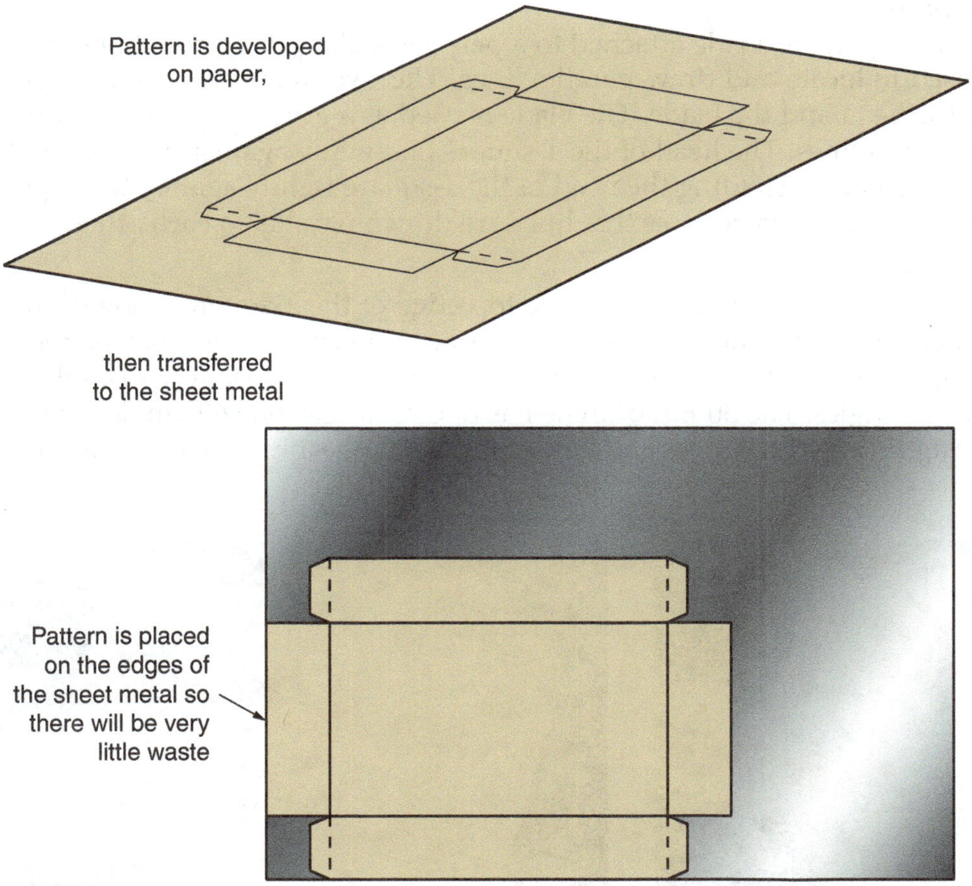

Figure 9-2. When transferring a paper pattern to metal, place the pattern in the lower left-hand corner of the metal. This will minimize waste.

Figure 9-3. The drafting equipment needed for drawing patterns.

T-square

A T-square is a rule attached to a perpendicular crosspiece at one end. It is used to locate and draw parallel lines. The two main parts of a T-square are the head and the blade. The blade is used as a guide for making straight horizontal lines. The head of the T-square fits snugly against the edge of the drawing board. Holding the head of the T-square tightly against the edge of the drawing board ensures that lines are drawn parallel to each other.

Triangles

Triangles are placed against the top edge of the T-square blade to draw vertical or angled lines, **Figure 9-4**. Commonly used triangles include the 45° triangle and the 30-60° triangle. The 45° triangle has one right (90°) angle and two 45° angles. The 30-60° triangle has one 90° angle, one 60° angle, and one 30° angle. These triangles can be used interchangeably to make a variety of

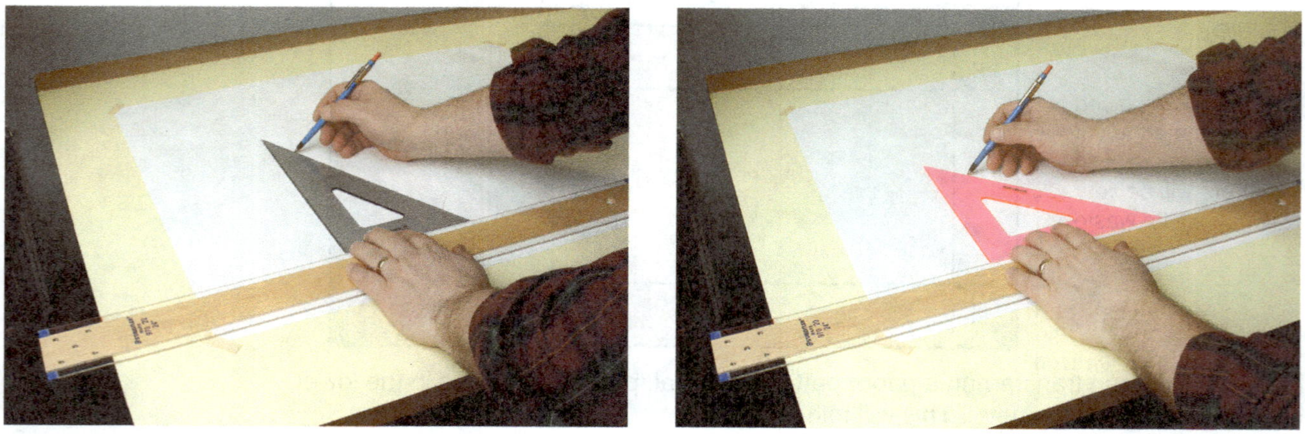

Figure 9-4. When using triangles, place them firmly against the T-square to ensure your lines are parallel.

other angles. For instance, a 45° triangle placed on a 30-60° triangle will make a 75° angle. Most angles needed to create a pattern can be drawn using these two triangles. Use a protractor or adjustable triangle for any other angles that may be needed.

Compass

A compass is used to draw circles and arcs. Most circles and arcs can be drawn with a bow-type compass while larger details usually require the use of a beam compass. The pencil point of most compasses can be interchanged with a steel point so the compass can be used as a divider for transferring measurements.

Protractor

Protractors are used to measure or lay out angles. Most protractors are shaped in a half-circle, with degree graduations on the outside edge. The graduations start at 0° at either end and meet at 90° on the top.

Irregular Curves

Sheet metal patterns often have irregular curves, or curves that are not a uniform shape. These curves are created by plotting a series of dots where the curve is to be located. These dots are then connected, using tools called ***irregular curves***. The ***French curve***, **Figure 9-5**, is typically a 1/8″ (3 mm) thick piece of clear plastic that has a variety of curves on the outside and inside edges. The ***flexible curve*** is a thin piece of bendable plastic that contains a thin metal wire. This wire allows the curve to retain its shape once it is bent. The curve is bent to go through all the plotted points and then the points are connected.

irregular curve: A plastic drafting tool used to connect plotted dots that designate the outline of a curve.

French curve: A type of irregular curve that has a variety of curves on the outside and inside edges.

flexible curve: A thin piece of bendable plastic that contains a thin metal wire. The curve is bent so it goes through all plotted points and then the points are connected.

Rules

Rules are used to measure length. Traditional, flat rules are useful for basic measuring tasks. A flat rule with 1/16″ (2 mm) graduations is used for most small sheet metal projects. A triangular rule has a variety of scales on its three sides.

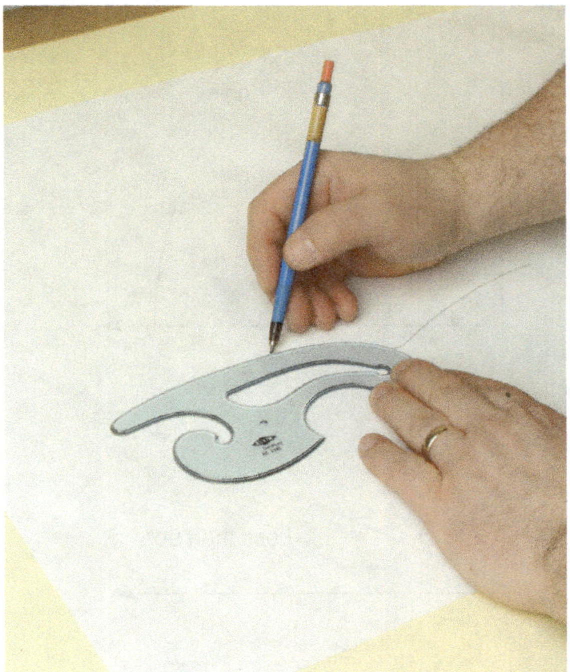

Figure 9-5. The French curve is useful for connecting the points on a curve whose shape is not standard.

Angles and Shapes

While visualizing and drawing, you will encounter many angles and shapes. Some of the common angles and shapes are listed in the following paragraphs.

Angles are formed when two straight lines meet. The point at which these lines meet is called the ***vertex***, **Figure 9-6**. There are three groups of angles:

- ***Right angles*** are formed when the two intersecting lines are 90° from each other, **Figure 9-7**.
- ***Acute angles*** are formed when the two intersecting lines are less than 90° from each other, **Figure 9-8**.
- ***Obtuse angles*** are formed when the two intersecting lines are more than 90° from each other, **Figure 9-9**.

vertex: The point at which two straight lines forming an angle meet.

right angle: A shape formed when the two intersecting lines are 90° from each other.

acute angle: A shape formed when the two intersecting lines are less than 90° from each other.

obtuse angle: A shape formed when two intersecting lines are more than 90° from each other.

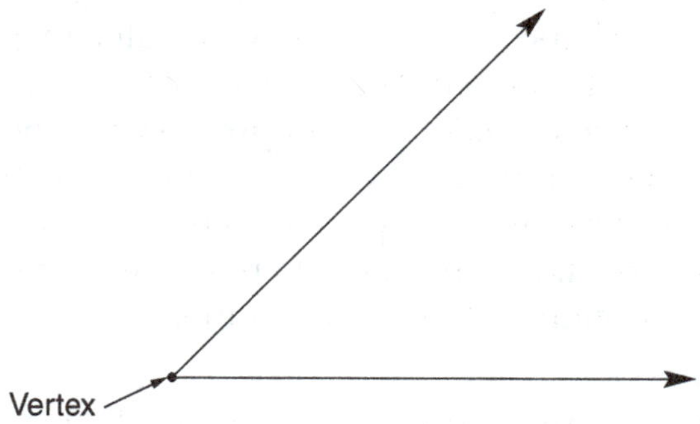

Figure 9-6. The point at which two intersecting lines meet is the vertex of an angle.

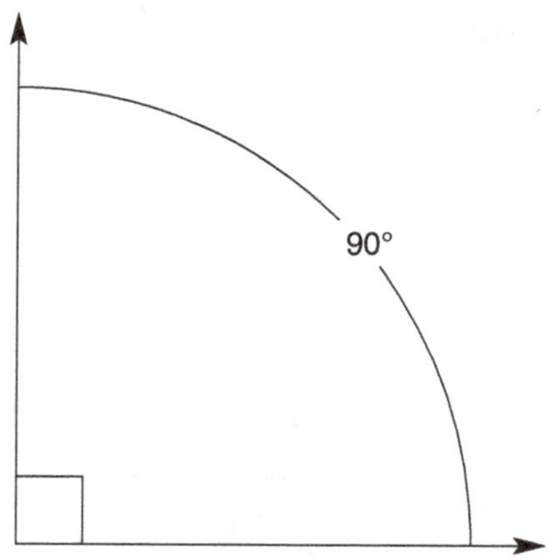

Figure 9-7. The intersecting lines of a right angle are 90° apart.

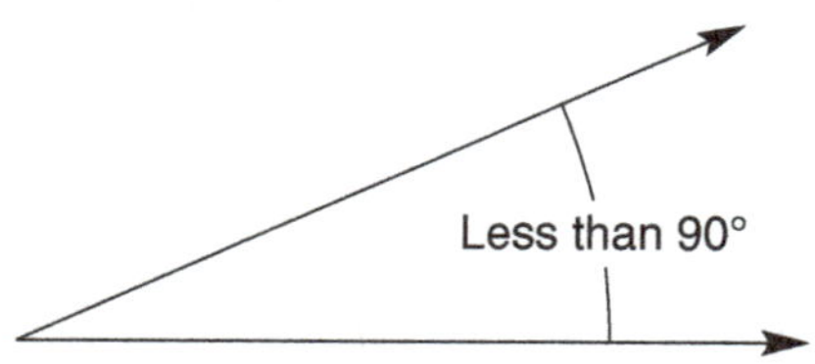

Figure 9-8. One type of acute angle.

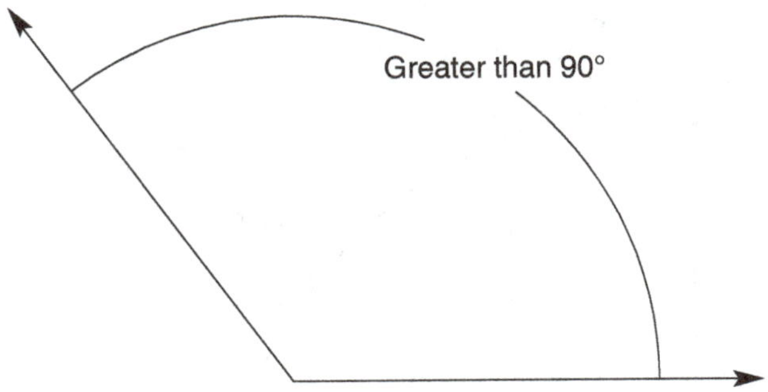

Figure 9-9. An obtuse angle.

In addition to angles, many geometric shapes are encountered when developing patterns. Most are two-dimensional polygons. Polygons commonly used in sheet metal patterns include:

- Squares. They have four sides of equal length that are at right angles to each other.
- Rectangles. All sides are at right angles, but only opposing sides are of equal length.
- Triangles. A three-sided figure whose interior angles add up to 180°. A right triangle has one 90° angle. An acute triangle has all acute (less than 90°) angles. An obtuse triangle contains one obtuse (more than 90°) angle.
- Pentagons are five-sided figures. All sides are of equal length.
- Hexagons are six-sided figures. All sides are of equal length.
- Octagons are eight-sided figures. All sides are of equal length.
- Circles. This is a closed plane curve in which every point is equidistant from the center.

Additionally, there is terminology specific to circles, **Figure 9-10**. It includes:

- Circumference, or perimeter, is the length of the outside boundary of a circle.
- Diameter is the length of a straight line through the center of a circle, ending at each side.
- Radius is half the distance of the circle diameter or the length of a straight line from the center of a circle to an outside edge.
- An arc is any section of the curved outside boundary of a circle.
- A chord is a straight line located on the interior of a circle that ends at two points on the outside edge of a circle.
- A tangent is a straight line that touches the outside edge of a circle at one point.

Lines

The following are lines commonly used when developing patterns for sheet metal. Refer to **Figure 9-11**.

- An ***object line*** represents the visible edges and details of a part. The lines are solid and thick so they stand out from dimension, hidden, center, and other lines that can clutter the drawing of a complicated part.

object line: Solid, thick line that represents the visible edges and details of a part.

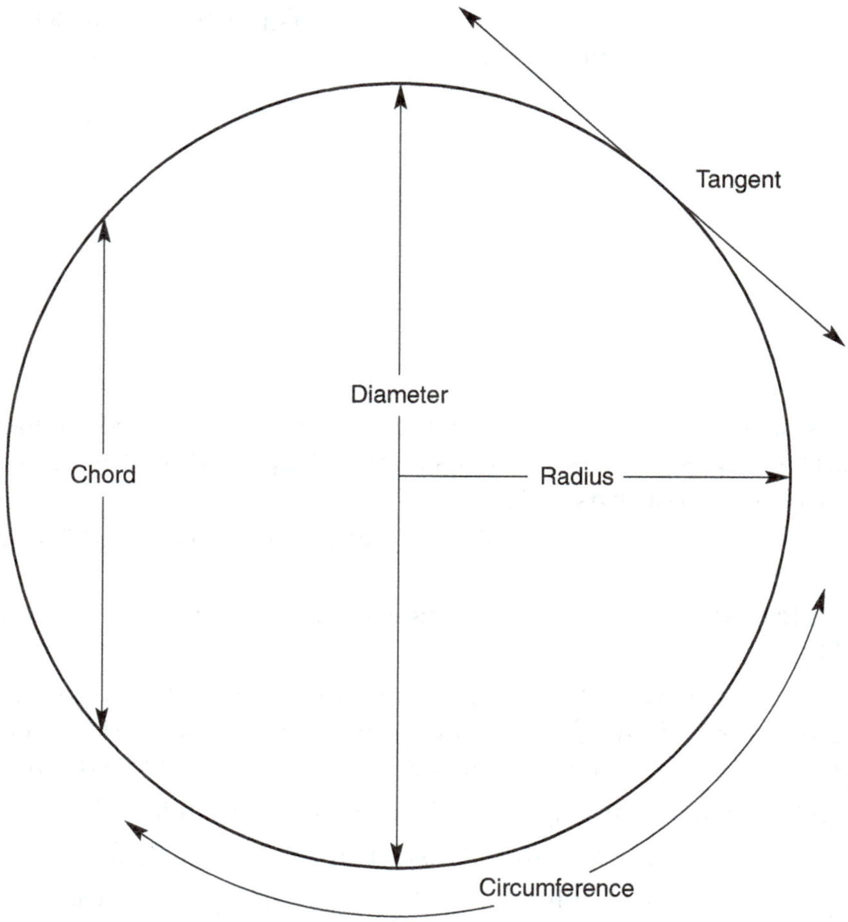

Figure 9-10. The terminology used in relation to a circle.

Figure 9-11. An alphabet of lines commonly found in sheet metal drawings.

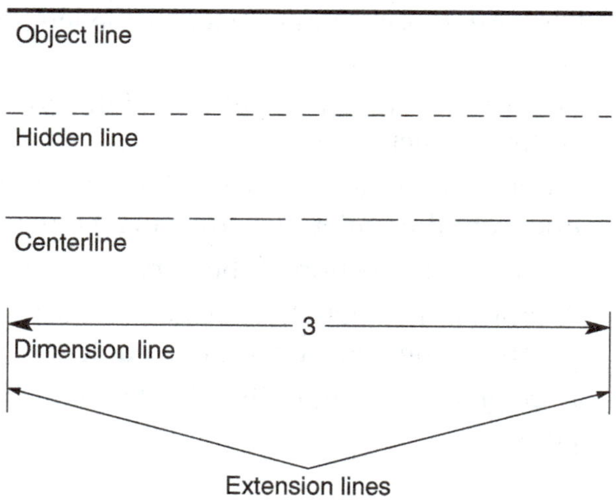

hidden line: A thin, broken line that represents internal features of a part that are not visible from its exterior.

centerline: A thin, broken line that indicates the location of the middle of an object.

- A ***hidden line*** represents internal features of a part that are not visible from its exterior. It is a thin line made with dashes.
- The ***centerline*** represents the location of the middle of an object. These lines are used for locating the center of holes that need to be punched or

drilled. A centerline is a thin line, drawn in this order: long dash, space, short dash, space, long dash, and so on. If the distance is not great, the short dash is drawn in the middle of the feature and long dashes are drawn on either side an equal distance from the short dash.

- The ***dimension line*** displays the measurement of a particular feature in it. It is usually terminated at each end with thin arrowheads, 1/8″ (3 mm) in length.
- ***Extension lines*** are perpendicular to dimension lines. Extension lines project out from an object's feature and are connected by the dimension line. Extension lines are used so the drawing is easier to read. An excess of dimension lines inside the pattern edges makes the pattern difficult to read and can result in incorrectly manufactured parts.

dimension line: A line that displays the measurement of a particular feature in it and terminates at each end with thin arrowheads.

extension line: A line that projects out from an object's feature to a dimension line.

Developing Patterns

In order to begin a development, an exact representation of the object is needed. Most pieces require at least two views in order to include all sides. This is called ***multiview drawing***. The views used most often in multiview drawings are the top view and the front view, **Figure 9-12**. The top view is found by looking down at the top of an object. The front view is found by facing the front of the object.

multiview drawing: An exact representation of an object that usually consists of at least two views in order to fully describe the object.

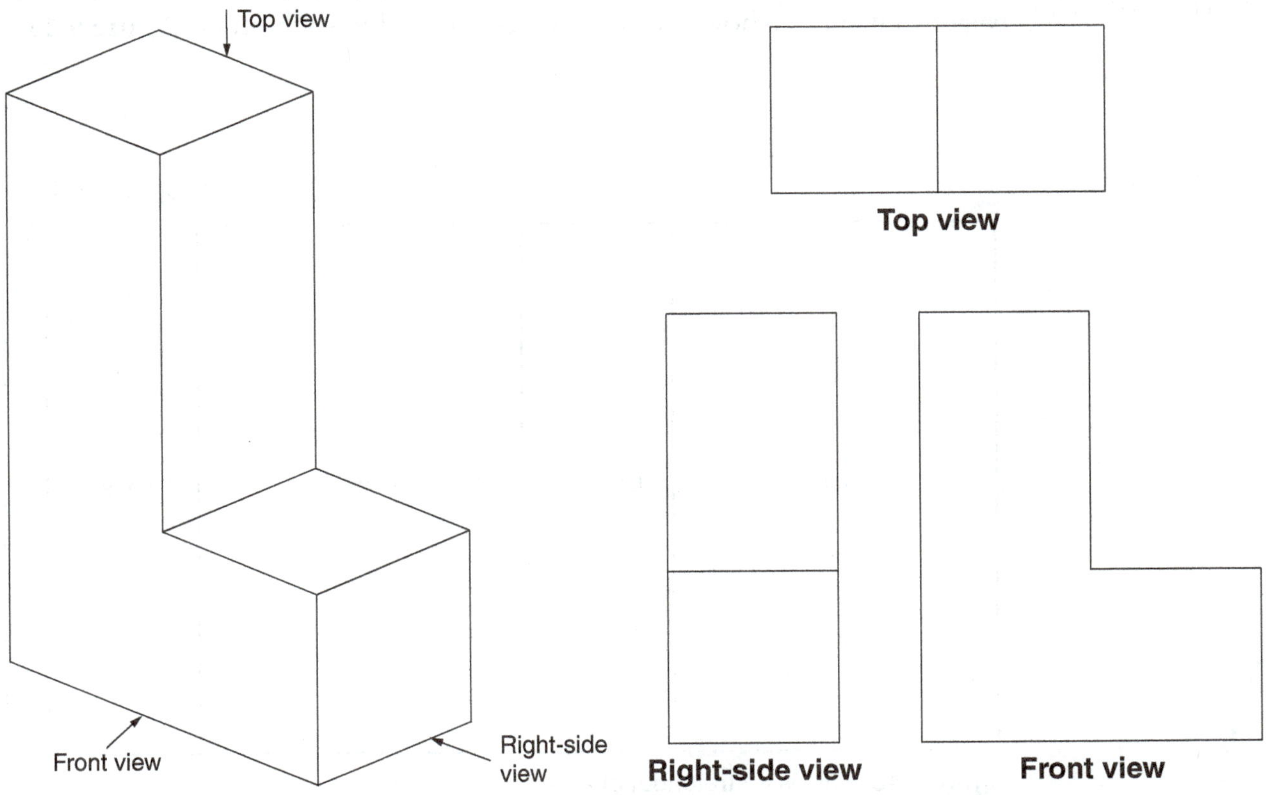

Figure 9-12. This multiview drawing includes a front view, top view, and right-hand view. Additional views, if needed, can include a bottom view and a left-hand view.

stretchout: The distance from each end of a flat pattern development. It also refers to the line projecting from the right of the front view when drawing a pattern.

A ***stretchout*** is the distance from each end of a flat pattern development. It also refers to the line projecting from the right of the front view when drawing a pattern.

Parallel Line Development

Parallel line development is used for objects in which the sides run parallel to each other, such as in ducts, elbows, and tee joints. This type of development is used for most sheet metal work. Start by drawing the front and top views of the object.

Rectangular Duct

Ductwork has an opening at each end for the flow of forced air. Rectangular ducts have four sides, requiring a minimum of three 90° bends. More bends may be required based on the type of seam used to fasten the duct together. Remember, the piece of sheet metal to be folded must be long enough for the width of the four sides, plus the metal used in the seam, **Figure 9-13**.

Label each corner of the top view, **Figure 9-14**. Then project down to the front view. Project the stretchout line to the right of the front view of the object. This is where the parallel line development of the pattern is laid out. Notice that all four sides of the object are drawn connected by the vertical lines that represent where the sheet metal will be folded. An additional tab may be added to form a seam.

mitered duct: A sheet metal object with a beveled end.

Laying out rectangular duct is fairly simple and straightforward. A more difficult layout is on ***mitered ducts***. Mitered ducts are cut to go around corners. Begin by sketching a pictorial drawing of the duct. This makes it easier to visualize how the sides are connected to each other, **Figure 9-15**.

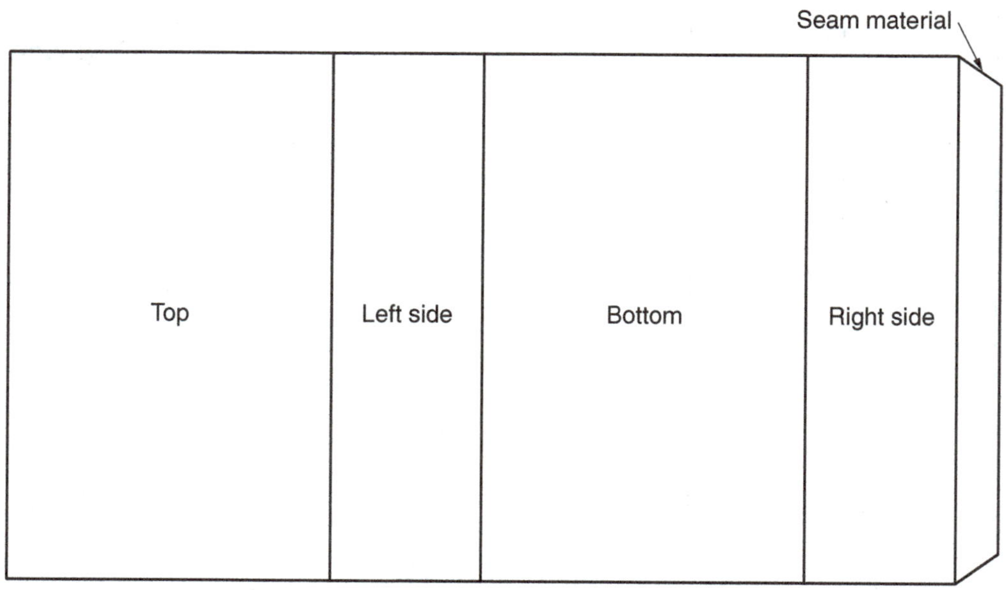

Figure 9-13. This is a stretchout of a rectangular duct.

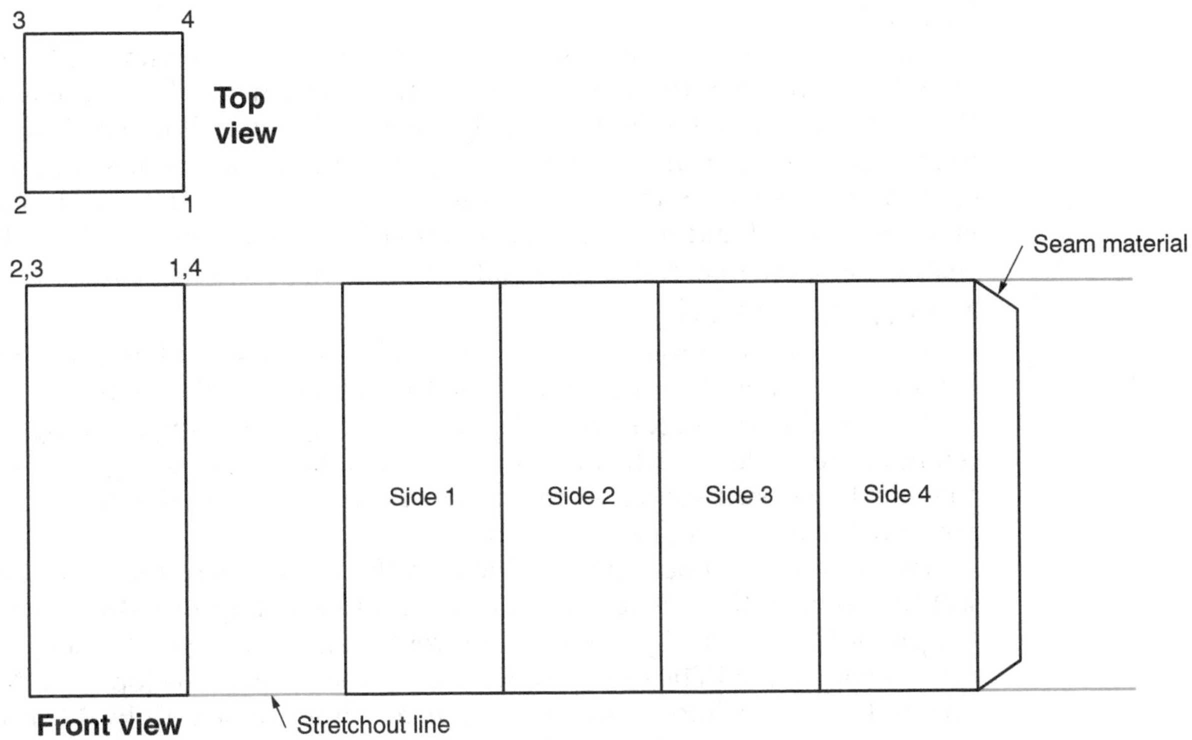

Figure 9-14. Projecting the top view to the front view to the stretchout.

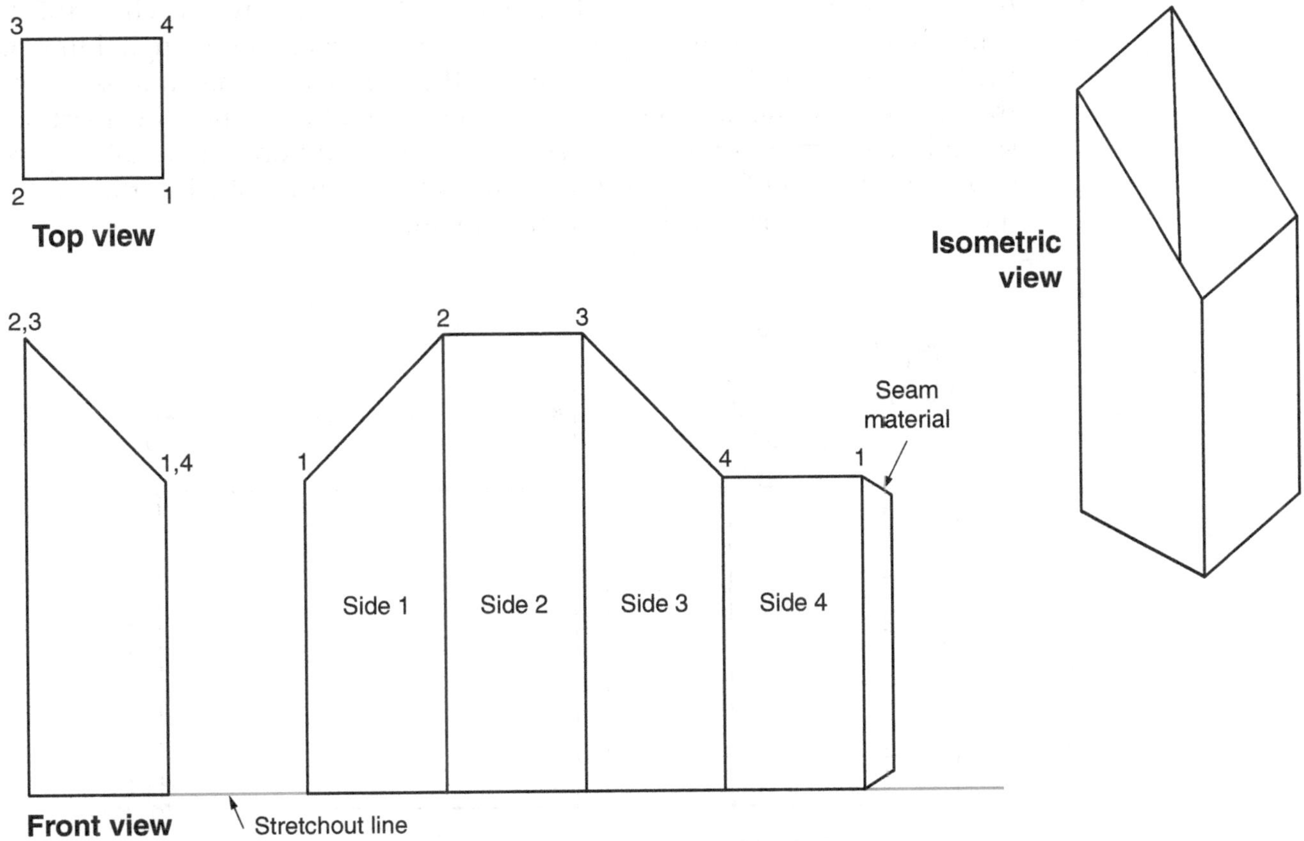

Figure 9-15. Pattern development for a mitered rectangular duct.

Round Duct

Parallel line development of a round duct requires a top view and a front view. In the top view, the diameter of the duct is drawn and then the length of the duct is drawn in the front view, **Figure 9-16**. The stretchout line is drawn to the right side of the front view. The length of the development is simply the circumference of the circle (3.14 × diameter). In other words, if the round duct is unraveled and laid out flat, this would be the circumference of the duct. Again, add extra material at each end of the development for seams.

Mitered Round Duct

It can be difficult to determine how a flat pattern should be cut to form a perfect 45° angle. The trick to making that angle is to divide the top view circle into 12 parts, **Figure 9-17**. Begin by making a 30° angle through the center of the circle to each side. Next make two 60° angles using the 30–60° triangle. Label each section from 1 to 12, starting with 1 in the three o'clock position. Label in a clockwise rotation.

Project each number straight down to the place where they will intersect the angle of the mitered duct on the front view, **Figure 9-18**. Draw the stretchout line to the right side of the front view. Remember, the length of the development will be the circumference of the circle in the top view. Now divide this length into 12 equal parts. If dividing the length by 12 results in an uneven number, draw a 45° line beneath Point B. From Point A, select a scale slightly smaller than Line AB which can be divided into 12 equal units. Swing the scale at Point A until the twelfth unit intersects the angled line drawn to Point B. The 45° lines should be drawn from each of the 12 units to divide the development evenly. The numbers on the angled line on the top view should then be projected to the right where they intersect with their corresponding number labeled on the development. These intersections should be connected with an irregular curve. Additional material may be added to each end for seams. Once this flat pattern is rolled to the correct diameter, it will form a cylinder with a 45° miter.

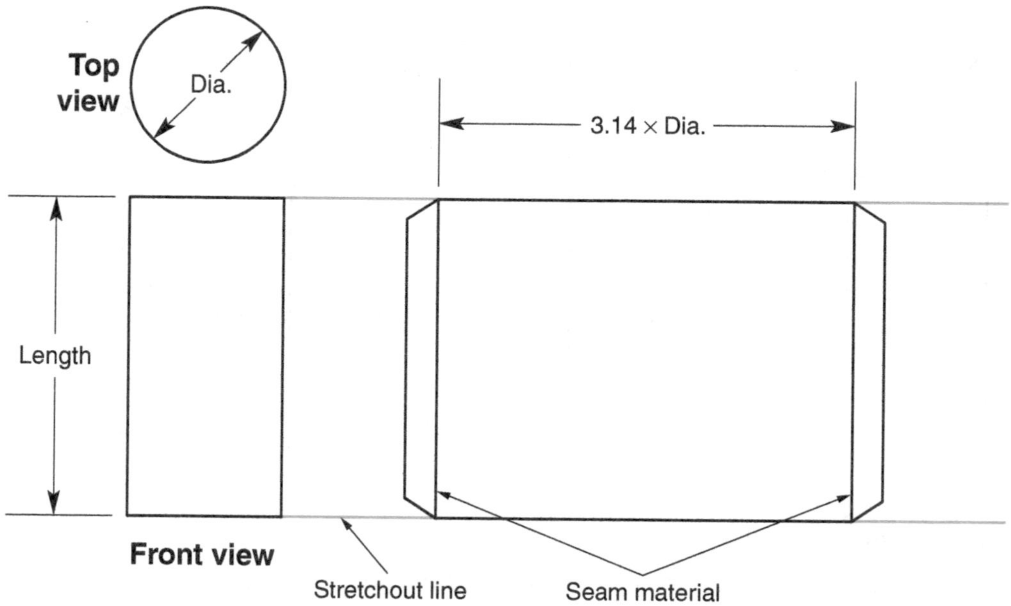

Figure 9-16. Pattern development for a round duct.

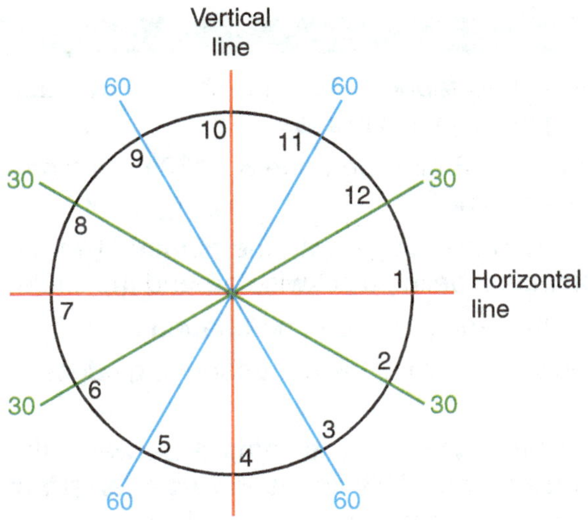

Figure 9-17. Dividing a top view circle to ease projection of the view.

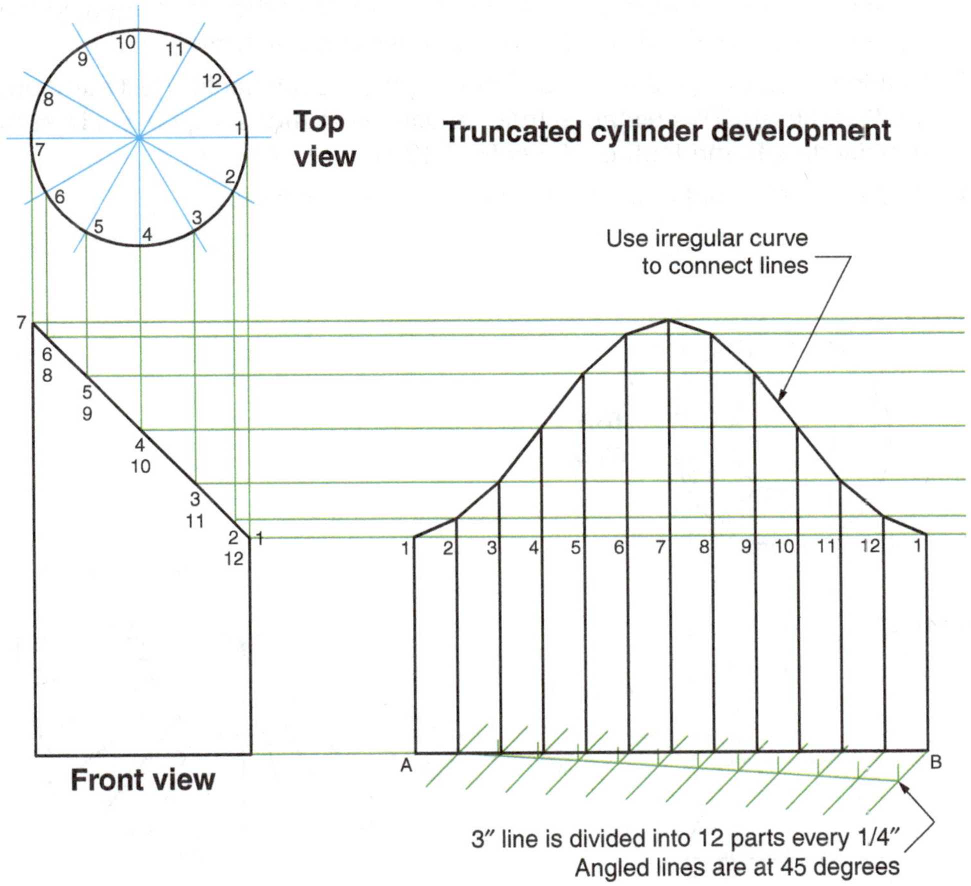

Figure 9-18. Mitered round duct pattern development.

Radial Line Development

Radial line developments are used for cone-shaped and pyramid-shaped ductwork. These types of ducts have a gradual change in diameter. Radial line developments use a curved stretchout line that is typically made with a compass.

Making a Radial Line Development

The following steps illustrate how to fabricate a cone-shaped cylinder. Refer to **Figure 9-19** while working through this process.

1. Draw the top view, making one 3″ (76 mm) and one 6″ (152 mm) diameter circle using the same center point.
2. Divide each of the circles into 12 parts. Begin with the number 1 at the three o'clock position and continue labeling clockwise around the circle.
3. Draw the front view illustrating the length of the cone-shaped duct.
4. Project the numbers on the top view down to the corresponding edges of the front view.
5. Measure the length of the inclined line that is formed between the 3″ (76 mm) and 6″ (152 mm) diameters. This line is the ***true length*** of the sheet metal that will be formed into the cone-shaped duct.
6. Draw an arc to the right of the front view with a radius equal to the length of the inclined line mentioned in Step 5. The length of the arc is equal to the circumference of the 6″ (152 mm) diameter duct, or 18.84″ (478 mm).
7. Divide the 18.84″ (478 mm) duct into 12 parts. Each section will measure 1.57″ (40 mm). This distance may be difficult to measure, so it is best to use dividers to mark off each section, 12 times.
8. Add seam material to the ends of the arc if necessary.

true length: A line on a plane that is perpendicular to a viewer's line of vision.

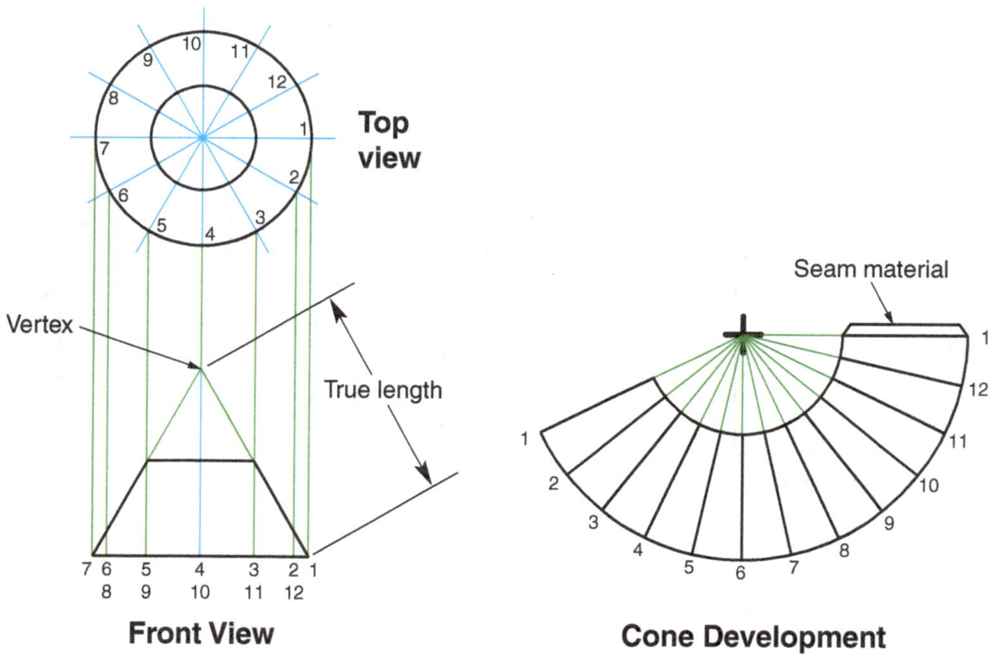

Figure 9-19. Pattern development for a cone.

Triangulation

Parallel and radial line development techniques do not work on a duct that changes shape. Triangulation is the best method for developing these surfaces. Using this method, the surface is divided into a series of triangles. The true length of the sides of each triangle must be found. Then the triangles are drawn next to each other on a flat surface. An example of this method is the development of an oblique cone.

Oblique Cone

Oblique cones can have a circular or elliptical base, **Figure 9-20.** Either type can be developed approximately using triangulation.

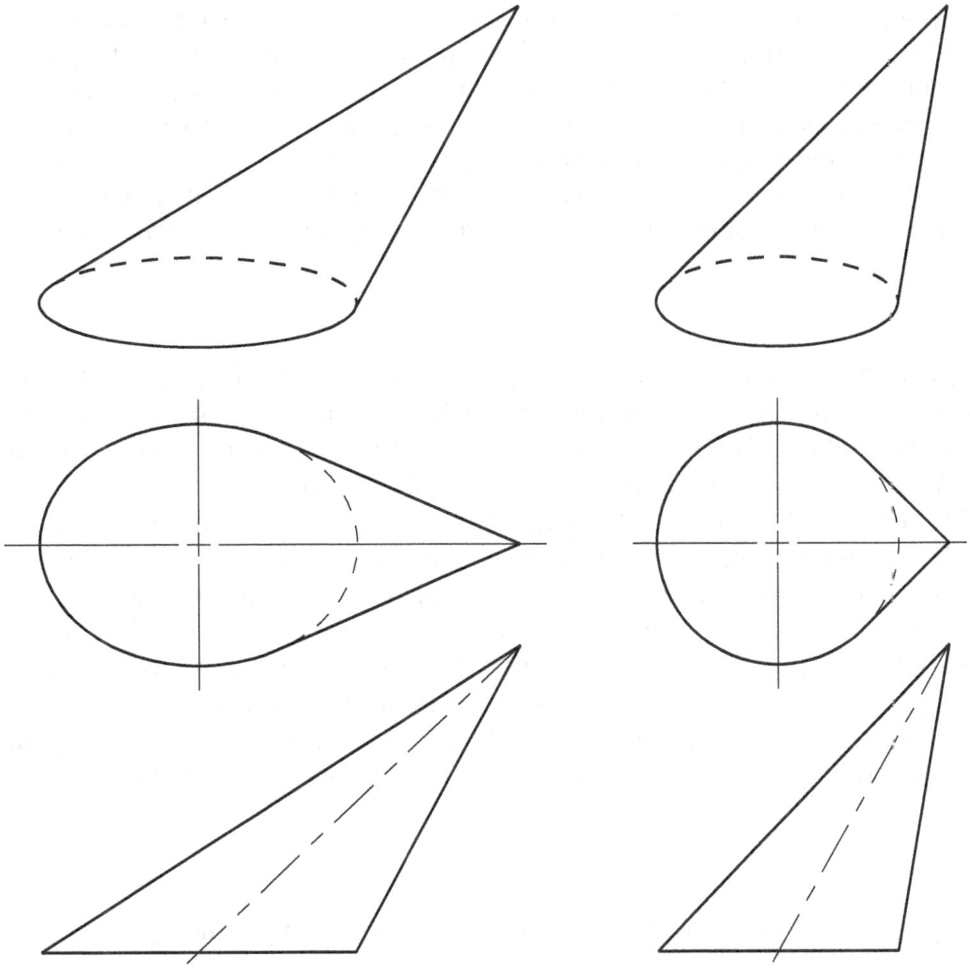

Elliptical base **Circular base**

Figure 9-20. Oblique cones can have an elliptical or a circular base.

Developing an Oblique Cone

The following steps outline how to fabricate an oblique cone. Refer to **Figure 9-21** while working through this process.

1. Draw the top and front views.
2. Divide the circular view into a number of equal parts.
3. Project these to the base in the front view.
4. Draw the elements from each division on the base to the top of the cone. This divides the surface into a series of triangles.
5. Find the true length of each element. Only 0-7 and 0-1 are true length. Refer again to **Figure 9-21**. Find the true length of the others by rotation.
6. On the development, lay out each triangle in the order in which it is found on the views of the cone. The first triangle is 0-1-2 in **Figure 9-21**. First lay out true length side 0-1. Next, measure the straight-line distance between 1 and 2 on the top view. Using 1 on the development as the center, swing an arc with this radius. Next swing an arc with a radius equal to the true length of element 0-2, using 0 as the center. Where the arcs intersect is one point on the pattern: point 2. Repeat these steps for each element. Connect the points with an irregular or a French curve.

Figure 9-22 shows some examples of sheet metal developments that utilize the triangle in their production. These transition pieces are often used in commercial buildings for heat, air-conditioning and ventilation systems. As you can see, the triangle is used often in sheet metal work.

These examples are meant to provide basic information about triangulation. If you need more information about finding the true length of a line or pattern development, refer to a quality drafting book. Today, most of this work is done using computers that have programs designed to make calculations and drawings. However, it is still a good idea to understand how these drawings are made, using traditional drafting tools and methods.

Refer to the Reference Section in this text to see an additional triangulation problem. It shows the steps involved in transitioning from one shape to another.

Using the Computer for Pattern Development

Computer-aided drafting (CAD) packages allow the user to draw an object on a computer and then print it. If the printer is large enough, the drawing can be printed full sized and used as a paper template. No drafting instruments are required, just a computer and a printer.

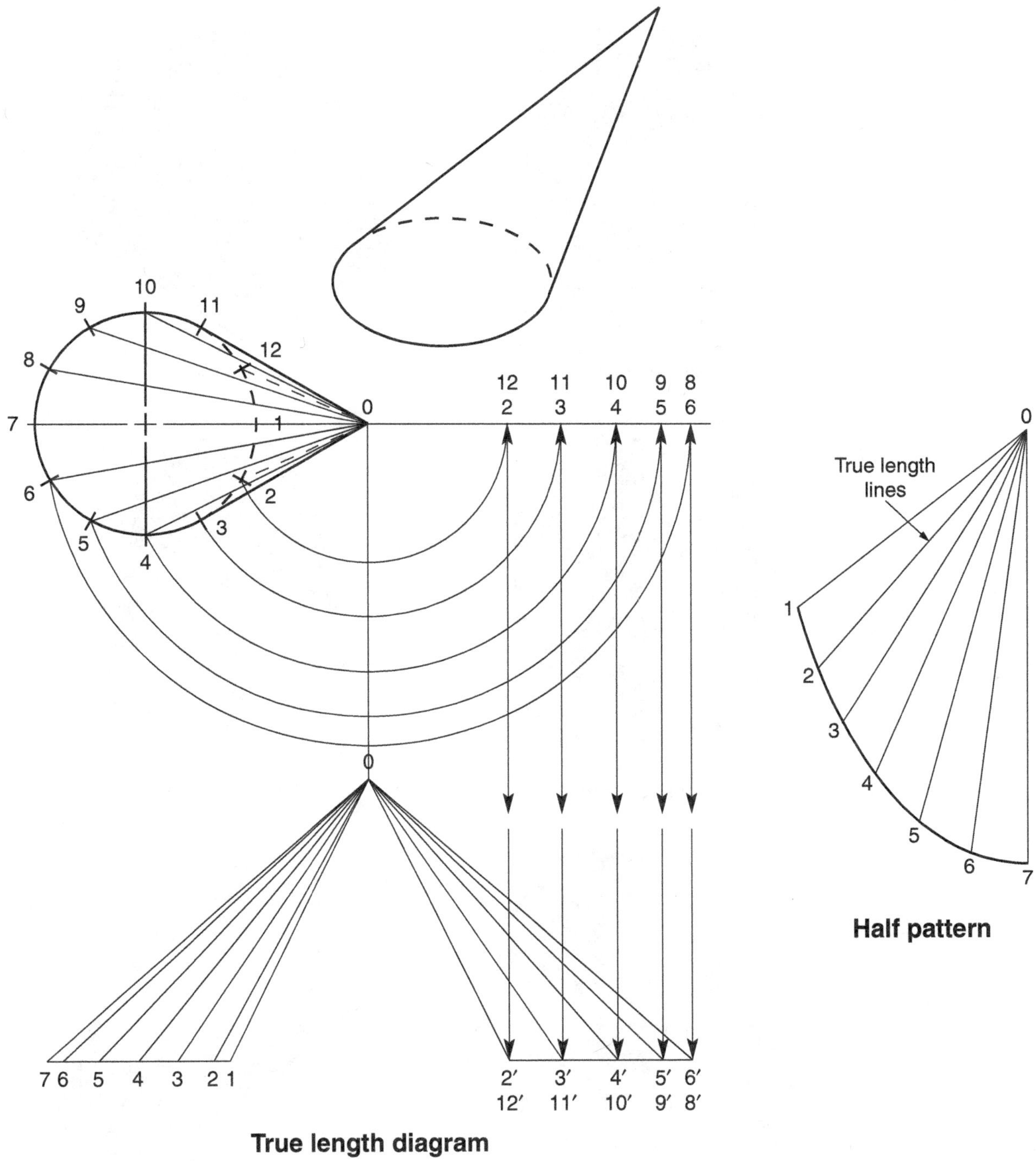

Figure 9-21. Development of an oblique cone.

There are many advantages to using CAD packages. They include:

- A drawing, and all its variations, can be saved and never has to be drawn again.
- Multiple copies of the same pattern can be printed.
- A drawing can easily be rescaled to make similar smaller and larger objects. For example, a 4″ (100 mm) diameter elbow can be doubled in scale to make an 8″ (203 mm) diameter elbow.

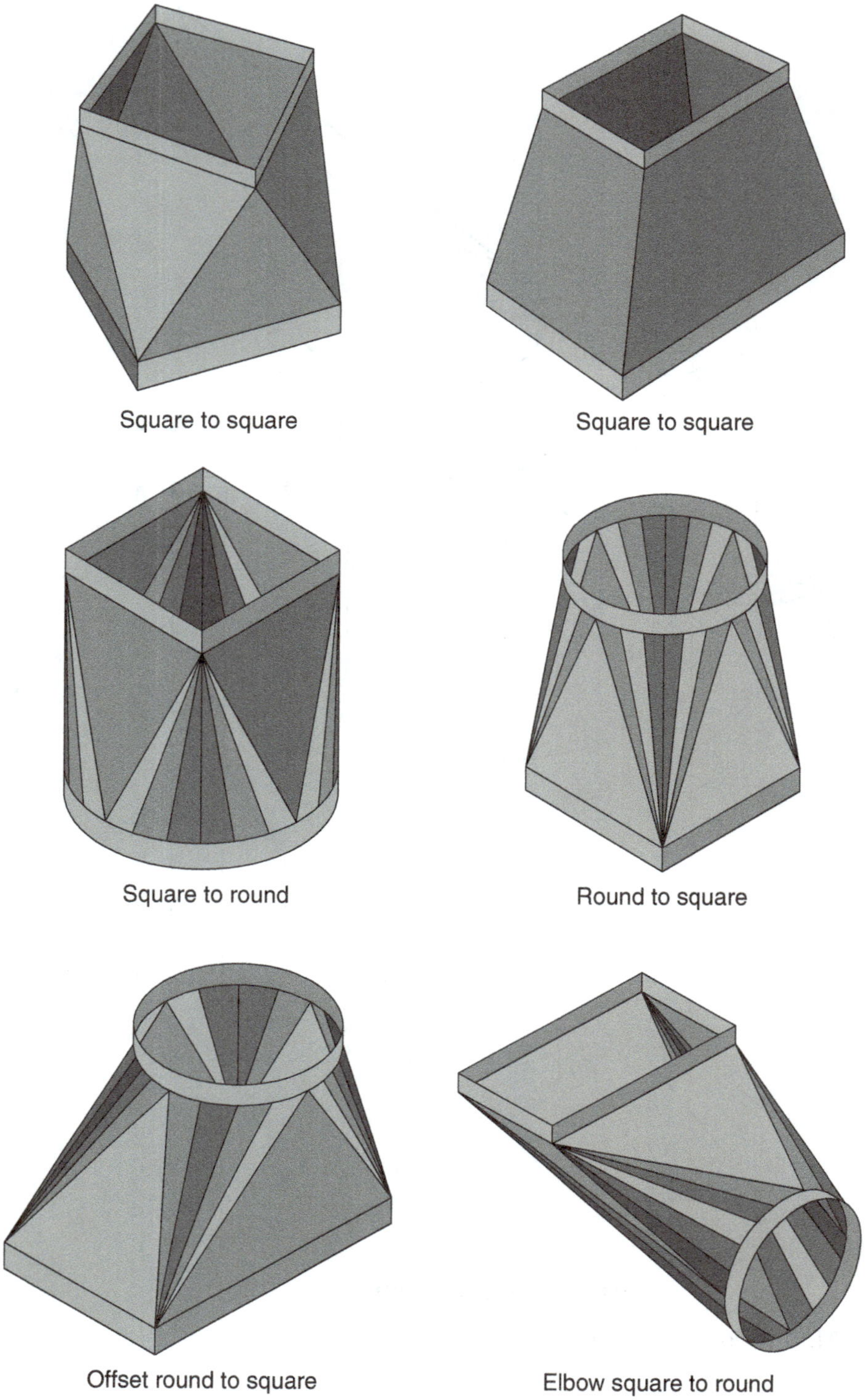

Figure 9-22. Examples of triangulation applied to sheet metal developments.

fabrication software: A computer program that contains ready-made patterns for common sheet metal pieces.

Shops that make a great deal of complicated, custom sheet metal duct often invest in ***fabrication software***. Fabrication software contains ready-made patterns for common sheet metal pieces. The user inputs the length, diameter and opening size of duct ends, and the type of seam needed. The software calculates all the angles and lengths required for bends. Users can print the pattern in full scale if a large plotter is available. Fabrication software allows inexperienced sheet metal workers to create the most complex parts in a matter of seconds. Mistakes made due to human error are virtually eliminated and production time is drastically reduced.

Summary

Finished sheet metal objects are developed from patterns. The method used to develop an object will depend on the object's shape. Three pattern development methods commonly used in sheet metal work include parallel line development, radial line development, and triangulation. Understanding which development method to use is an important task for the sheet metal worker.

Patterns can be developed directly onto the sheet metal to be used or on paper and transferred to the sheet metal. These patterns can be developed by hand or with computers. Computer-aided design packages help you develop your patterns accurately and efficiently. Computer software developed specifically for making sheet metal parts speeds the process even further.

Internet Resources

Sheet Metal Guy
www.sheetmetalguy.com/transition.htm

Outside Activities

1. Obtain a blueprint of an HVAC system. Make a list of the various lines used in the blueprint. Are there any lines you are not familiar with? Conduct research to learn about these lines.
2. Collect several pieces of HVAC ductwork. Determine which development method would be used to create a pattern. Working with at least one other classmate, develop a pattern for one piece.

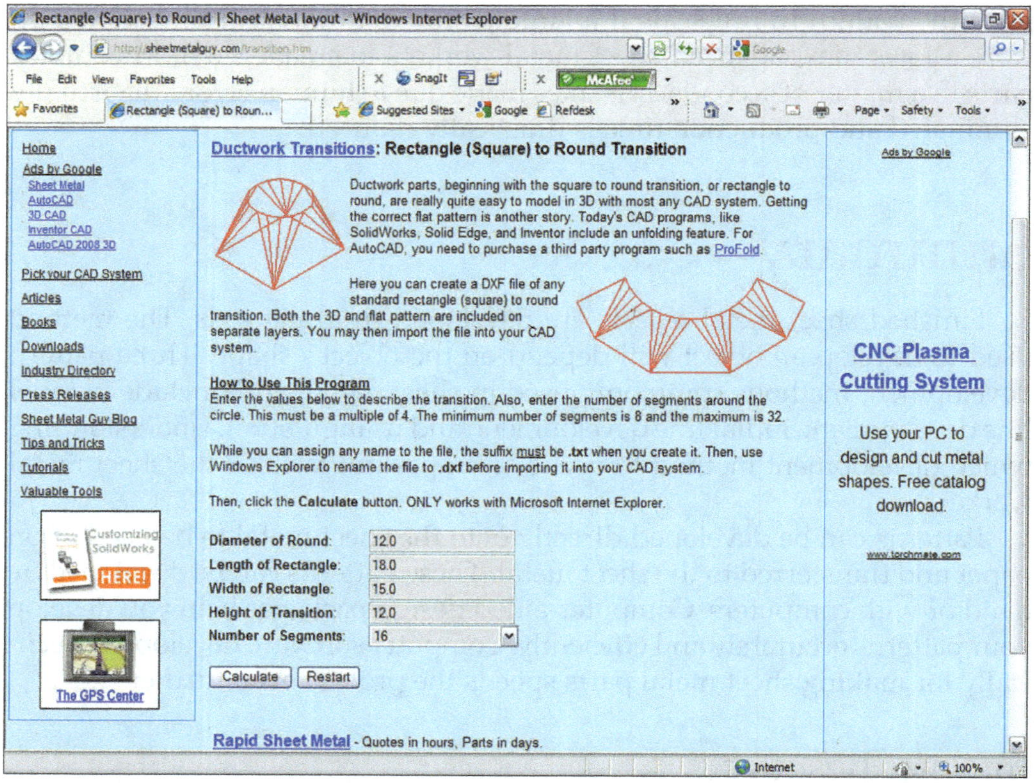

The Sheet Metal Guy Web site has a great deal of information regarding ductwork transistions.

Name ______________________________ Class ________________ Score __________

Chapter 9
Drafting and Pattern Development

Review Questions

Write your answers in the spaces provided.

1. What is a pattern? ____

____

____

____

2. Parallel line development, radial line development, and triangulation are all types of _____ method. 2. ________________

3. The ability to see an object as a collection of shapes and the way the shapes fit together is called _____. 3. ________________
   A. clairvoyance
   B. visualization
   C. abstraction
   D. None of the above.

4. A paper pattern is called a(n) _____. 4. ________________

Match each of the following descriptions with the correct term.

5. Used to draw circles and arcs.	A. Triangle.	5. ________________
6. Used to join plotted dots that indicate the location of a curve.	B. Rule. C. Compass.	6. ________________
7. Placed against the top edge of a T-square to draw vertical or angled lines.	D. T-square. E. Irregular curve. F. Protractor.	7. ________________
8. Used to locate and draw parallel lines.		8. ________________
9. Used to measure or lay out angles.		9. ________________
10. Used to measure length.		10. ________________

11. Which of the following statements about angles in *not* true? 11. ________________
    A. The vertex is the point at which two straight lines meet.
    B. Obtuse angles are less than 90°.
    C. Right angles are exactly 90°.
    D. Acute angles are less than 90°.

12. True or False? A hexagon is an eight-sided figure. 12. ____________________

13. The length of a straight line through the center of a circle that ends at each side is called a _____. 13. ____________________

Match each of the following terms with the correct drawing.

14. Centerline. 14. ____________________

15. Hidden line. 15. ____________________

16. Extension line. 16. ____________________

17. Dimension line. 17. ____________________

18. Object line. 18. ____________________

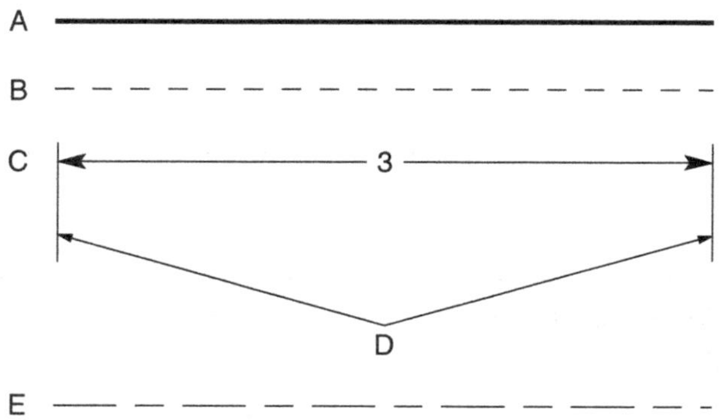

19. A(n) _____ drawing often includes a top view and front view. 19. ____________________

20. The best development method for ducts that change from square to round or some other shape is the _____ method. 20. ____________________

Chapter 10 Materials and Materials Testing

Objectives

After studying this chapter, you will be able to:

- Identify the classes of materials.
- Define the terms *ferrous* and *nonferrous.*
- List various types of nonferrous metals and alloys and how they are used.
- Summarize those properties that influence the selection of materials.
- Explain the different types of hardness testing.

Technical Terms

alloying elements
antimony
bismuth
Brinell hardness test
brittleness
bronze
compressive strength
ductility
elastic limit
elasticity
fatigue strength
flex strength
hardness
impact strength
materials
plasticity
properties
Rockwell hardness test
shear strength
shore scleroscope hardness test
strain
strength
stress
tensile strength
torsion strength
zinc

Knowledge of various metal materials and their properties is useful for sheet metal workers. Most workers will never be asked to purchase or test the materials needed for a job, but understanding how these materials are created, and knowing their strengths and weaknesses, results in a more knowledgeable worker.

Materials

materials: The physical substances used in production or manufacturing.

Materials are physical substances used in production or manufacturing. Materials can be raw, semifinished, or finished. Sheet metal is a finished material. It is produced from raw materials like iron and carbon that are mined and processed into steel. This semifinished material is then processed further into the finished material: sheet metal.

Materials are generally classified based on their properties and characteristics. A basic classification method divides materials into metals and nonmetals. Metals are further divided into nonferrous (contains no iron) and ferrous (iron, steel) categories. Nonmetals can be further classified into polymer, ceramic, and composite categories, **Figure 10-1.**

Metals

Metals are a substance that have the ability to conduct electricity and heat. Metals are strong and malleable. For these reasons, metals have been used for hundreds of centuries. In the past, they were made into tools and other goods that eased and improved people's lives. They played an important role in advancing civilizations. Metals, both ferrous and nonferrous, have many useful properties. This ensures that they will continue to be used now and in the future.

Materials Classification System	
Materials Family	**Subcategories and Examples**
Metals	Ferrous–iron, steel
	Nonferrous–aluminum, copper, copper alloys
Polymers	Natural–wood, rubber
	Synthetic–plastics, glues
Ceramics	Natural–glass, diamond
	Synthetic–zirconia, aluminum oxide
Composites	Polymer matrix–fiberglass, plywood
	Metal matrix–titanium, silicon-carbide
	Ceramic matrix–concrete, lithium-alumina-silicate

Figure 10-1. Material classification helps designers, engineers, and scientists choose the proper materials for specific applications.

Ferrous Metals and Alloys

Ferrous metals contain iron and are magnetic. Iron and steel are the most common ferrous metals. They are an economical material. They are widely used in machinery, construction, and other applications where the ability to carry a load is important.

Pure iron (ferrite) is a relatively soft element. In order to be useful, it is alloyed with other elements. Cast iron contains 2–4 percent carbon, 1–6 percent silicon, and traces of manganese. Types of cast iron include white cast iron and gray iron. Cast iron properties vary, depending on the form of the carbon within the iron. Wrought iron contains less than 0.25 percent carbon and is malleable and tough. *Steel* is a general term used to describe alloys in which iron is the base metal and carbon is the most important added element. Steel is usually classified as either plain carbon steel or alloy steel.

Plain carbon steel is made up of iron, carbon, and trace amounts of ***alloying elements***. The carbon content in plain carbon steel does not exceed 1.7 percent. The higher the carbon content, the harder and stronger the steel. The alloying elements in plain carbon steel are limited to manganese, silicon, copper, and traces of other elements. The properties of the steel depend on carbon content and heat treatment.

alloying element: A substance added to a base substance to improve specific properties.

Plain carbon steel is divided into three groups: low-carbon steel, medium-carbon steel, and high-carbon steels. Low-carbon steel, or mild steel, contains a maximum of 0.30 percent carbon. It has less strength and hardness and so is easier to machine and work. Low-carbon steel is used to make galvanized sheets, storage tanks, and is used in bridge, ship, and building parts. It is inexpensive to produce and purchase.

Medium-carbon steel contains between 0.30 and 0.60 percent carbon. Heat treating increases its hardness and strength. It is versatile and is used in a variety of applications. Common uses for medium-carbon steel include automobile parts, forgings, and high-strength castings. It is more expensive than low-carbon steel.

High-carbon steel contains between 0.60 and 1.0 percent carbon. With heat treating, it becomes very strong and very hard. However, it is also brittle. It is generally used where high strength is needed, such as lawn mower blades, knives, railroad rails, hand tools, and dies.

Alloy steel is carbon steel that contains 1–4 percent alloying elements. These alloying elements are added for several reasons:

- To improve mechanical properties.
- To improve resistance to corrosion and oxidization.
- To improve hardenability of the steel.
- The table in **Figure 10-2** lists the effect of alloying elements on steel.

Like plain carbon steel, alloy steel is also divided into groups: low-alloy steel and high-alloy steel. Low-alloy steel contains less than 10 percent alloying elements. The amount of alloy is sufficient to change the steel properties. High-alloy steel contains more than 10 percent alloying elements. Stainless steel and tool steel are two types of high-alloy steel.

Stainless steel resists oxidation and corrosion when correctly heat treated and finished. It is also easy to clean. Stainless steel sheets are used to make industrial kitchen equipment such as sinks, food preparation surfaces, and

Effects of Alloying Elements on Steel	
Element	**Effect**
Phosphorus	Prevents sticking in light-gage sheets, slightly strengthens low-carbon steel, improves machineability in some steels.
Sulfur	Improves machineability in some steels but is generally regarded as a contaminant.
Manganese	Increases hardness and strength, although too much can create brittleness and hardness that makes cutting difficult. Has a neutralizing effect on sulfur.
Nickel	Increases both strength and toughness.
Aluminum	Improves strength of ingots and castings.
Vanadium	Improves susceptibility to fatigue and to shock. Improves ductility, tensile strength, and elasticity.
Molybdenum	Deepens hardening, counteracts tendency to brittleness, improves corrosion-resistance in stainless steel, forms abrasion-resisting particles.

Figure 10-2. Each alloying element has a different effect on the properties of steel.

range hoods. They are also used to make hospital equipment. Stainless steel is a widely used material in the sheet metal industry.

Tool steel is carbon or alloy steel with properties suitable for use in tools. High-speed steel is one type of tool steel. It is used in cutting tools found on lathes and milling machines.

Nonferrous Metals and Alloys

Nonferrous metals and alloys contain no iron. Nonferrous metals include copper, zinc, tin, antimony, lead, and aluminum. Nonferrous alloys using those metals as a base element include brass, bronze, and aluminum alloys.

Copper is ductile and malleable. It is an excellent conductor of heat and electricity. It also resists corrosion so it is often used as roofing material. Alloys of copper include brass and bronze.

Brass is an alloy of copper and zinc. Small percentages of tin, lead, and other metals are also found in some types of brass. The amount of zinc in brass can be determined by its color, **Figure 10-3.** Tensile strength varies, according to composition and treatment. Brass is easily polished to a bright finish that is useful in decorative applications.

bronze: A nonferrous metal that is an alloy of copper and tin, or copper and alloying elements other than zinc.

In the past, ***bronze*** was defined as an alloy of copper and tin. Today, however, alloys of copper and any alloying element other than zinc are also considered bronzes. Bronze is used in a variety of applications, from decorative bells to marine parts. In tin bronzes, the amount of tin alloyed with the bronze determines its use, **Figure 10-4.** Special bronzes, such as phosphor, manganese, gun, and Tobin bronze, each have different properties that are adapted for special uses. Bronze is used instead of brass in some instances for its unique appearance and greater strength.

Brass Composition	
Brass color	**Amount of zinc**
Red	5 percent
Bronze	10 percent
Light orange	15 percent
Greenish yellow	20 percent
Yellow	30 percent
Yellowish white	60 percent

Figure 10-3. The amount of zinc in brass determines its color.

Bronze Composition	
Treatment Method/Use	**Amount of Tin**
Cold-rolled into wire	1–6 percent
Art bronze	3–10 percent
Worked at red heat	6–15 percent
Machine parts	9–20 percent
Bell bronze	20–30 percent

Figure 10-4. Bronze becomes more brittle when the amount of tin increases above 5 percent.

Zinc is brittle at room temperature, but malleable and ductile between 212°F (100°C) and 300°F (149°C). It is commonly used as a coating to prevent corrosion. Steel sheets are dipped in liquid zinc, resulting in galvanized steel sheets. These are used in car bodies to reduce corrosion resulting from road salt. Zinc is also used for eaves, gutters, and roofs because it is light and easy to handle.

zinc: A nonferrous metal used to galvanize iron and steel against corrosion.

Tin is ductile and malleable but has low tensile strength. It can easily be drawn into wire at 212°F (100°C) and rolled into very thin sheets. Tin is used as a protective coating for iron and copper because it resists corrosion.

Antimony is a hard, brittle metal that resembles tin. It is an elementary substance that resembles metal in its appearance and physical properties, but is chemically related to the class of nonmetallic substances. It is commonly mixed with lead in battery plates, and is used in manufacturing semiconductors, paint, and flameproofing compounds.

antimony: A hard, brittle nonferrous metal that resembles tin.

Lead is the heaviest of the common metals. It melts at 621°F (327°C) and is soft enough to be cut with a knife. Lead is also malleable, ductile, and has low tensile strength. Compared with other metals, it is not a good conductor of heat or electricity. It is used to alloy with other metals for bearings and solders. Lead has largely been replaced in the plumbing industry.

Bismuth has two remarkable properties: its specific gravity decreases under pressure and it expands on cooling. The melting point is about 520°F (271°C). Bismuth is often alloyed with antimony to fill molds completely after solidification.

bismuth: A nonferrous metal whose specific gravity decreases under pressure and expands on cooling.

Aluminum is lightweight but can be made extremely strong through the use of alloying elements or heat treatment. Common aluminum alloying elements include copper, silicon, magnesium, nickel, iron, zinc, and manganese. Aluminum is corrosion resistant, attractive, malleable, ductile, and a good conductor of heat and electricity. Thermal expansion of aluminum is slightly more than twice that of steel and cast iron.

Aluminum alloy castings are made in sand molds, permanent metal molds, or pressure die-casting molds. Permanent molds are practical only when a large number of identical castings are needed. The minimum number that will justify the production of a metal mold, or die, varies greatly with the nature of the casting.

Aluminum is often rolled into sheets to be stamped, punched, or shaped into many projects. Sheet aluminum is also used for siding on houses and commercial buildings. It can be used for various purposes and a number of thicknesses are available.

Babbitt metal is an antifriction alloy of tin, antimony, and copper. It was invented in 1839 by Isaac Babbitt, a Boston goldsmith. At that time, it was used in steam engines. Today, Babbitt metal is used as bearing material in automobile crankshafts and axles, and in marine applications. The alloy is made of a combination of hard crystals and softer alloys, which allows it to conform to bearing shafts.

White metal is a term often applied to various alloys containing mainly zinc and tin, or zinc, tin, and lead. White metal is used for bearings.

Miscellaneous Metals

Monel metal is an alloy of copper and nickel and a small percentage of iron. Its melting point is 2480°F (1360°C), and it may be forged from 165°F (74°C) to 1100°F (593°C). An important use of monel metal is in ship propellers.

Muntz metal is an alloy containing 60 percent copper and 40 percent tin. It is used when a hard sheet brass is desirable.

Tobin bronze is an alloy containing 58–60 percent copper, about 40 percent zinc, and a small percentage of iron, tin, and lead. Its tensile strength is about 60,000 pounds per square inch. Delta metal is similar in composition and properties to Tobin bronze.

Tantung® is a trade name for a series of alloys that have great hardness, strength, and toughness, and resistance to wear, heat, impact, corrosion, and erosion, even at extremely high temperatures. These alloys are composed chiefly of cobalt, chromium, and tungsten, with either tantalum or columbium carbide and other components added.

Tantung G® is widely used in tipped lathe tools, milling cutters, and in solid bits. It is available in rods and bars that can be used as tool bits or converted into punches, rollers, drills, and other special tools or wear-resistant parts.

Material Selection

Materials are selected for use in products based on their properties, cost, source, and environmental impact. The most important of these considerations is the properties of the materials.

The material chosen must have the properties needed for the product. For example, steel is strong, tough, and elastic. These properties make it useful in the construction of bridges, buildings, and certain types of machines. These properties can also be manipulated through the use of alloying elements and heat treatment. Steel is made even more desirable because its cost is reasonable and it is easy to obtain.

Properties

Sheet metals are made from various metals, including iron, aluminum, steel, and copper. Each material has ***properties*** that define its character or behavior under various conditions. Knowing the properties of all types of metals helps in selecting the right type of sheet metal for a particular job. In some instances, sheet metal sections are part of a larger project. Knowing the properties of the sheet metal and how it interacts with the other materials is important.

properties: The characteristics or qualities of a material when that material is placed in certain conditions or combined with other materials.

Materials have three basic properties: mechanical, physical, and chemical. Mechanical properties determine the ability of a material to resist or carry mechanical stresses or forces. Physical properties are the characteristics of a material as it interacts with various forms of energy (light, heat, electricity, and magnetism). Chemical properties are the characteristics of a material as it reacts with its environment, such as gases, liquids, and solids.

In sheet metal work, engineers are most concerned with a material's mechanical properties. These properties include strength, hardness, elasticity, plasticity, ductility, brittleness, and toughness.

strength: A material's ability to resist stress.

stress: The result of a force or load applied to a material.

tensile strength: A material's ability to resist pulling forces.

compressive strength: A material's ability to resist the force created by opposite objects pushing against each other.

flex strength: The ability of a material to withstand bending forces.

shear strength: A material's ability to resist sliding forces.

torsion strength: Related to shear strength; the ability of a material to resist a rotating force.

Strength

Strength is the ability of a material to resist stress. ***Stress*** is the result of a force or load applied to a material. For example, steel beams used to frame a building are under stress simply from their own weight. Those beams require incredible strength to stay in place and not collapse under that weight. All metals need strength to withstand the stresses encountered in different applications. If material is to be useable and practical, it requires some type of strength to withstand the stress it will encounter. Some types of forces and stresses are shown in **Figure 10-5.** Refer back to this figure as you read the following paragraphs.

Tensile strength is a material's ability to resist a pulling force. This force is tensile stress. Imagine a rope hammock. It is attached tightly to either end of the frame to keep it taut. The tensile force pulls against the tensile strength of rope used in the hammock.

Compressive strength resists the force created by materials pressing together or compacting. Compressive force is created by opposite forces pushing against each other.

Flex strength is the ability to withstand tensile and compressive forces. When a material is flexed, it experiences tension on one side and compression on the opposite side.

Shear strength is a material's ability to resist sliding action. The sliding action or force is parallel to the resisting area. Shear force is usually not as strong in most materials as tensile or compressive force. A related strength is ***torsion strength,*** the ability to resist a rotating force.

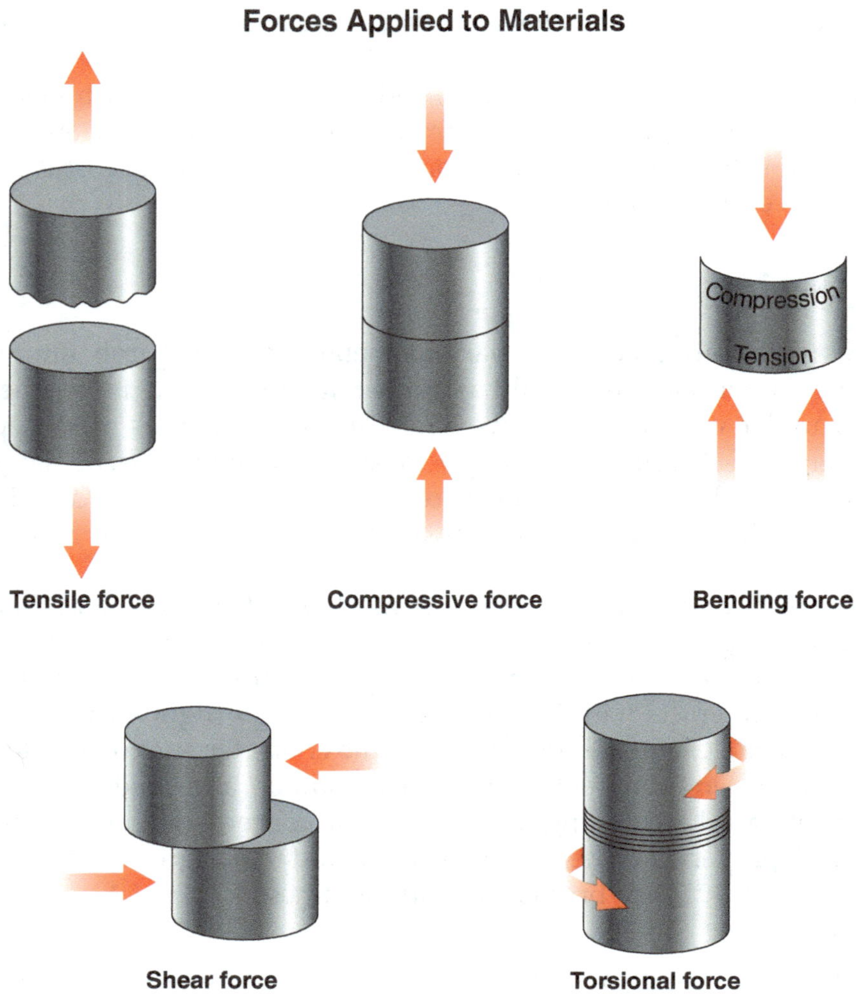

Figure 10-5. Study the difference in the types of force applied to materials.

impact strength: The ability of a material to withstand a sudden shock or load.

Impact strength is a material's ability to withstand a sudden shock or load. When a rock hits the windshield of a car, the impact strength of the windshield will be tested. If it is strong enough, the rock will not shatter the windshield. However, if the rock is too large or heavy, it will fracture the windshield.

fatigue strength: The ability of a material to withstand repeated stress.

Fatigue strength is the ability to withstand repeated stress. Fatigue stress is cyclic, occurring over and over. Bending a metal rod back and forth repeatedly at the same point is an example of fatigue stress. Eventually, the rod will break.

Hardness

hardness: The ability of a material to resist penetration or scratching by another material.

Hardness is the ability of material to resist penetration or scratching by another material. It is one of the most important properties of metal as it relates directly to several other properties. Knowing the hardness of a material also indicates the strength, brittleness, and ductility of that material. Hardness is measured using several methods that will be explained later in this chapter.

Elasticity and Plasticity

Elasticity is the ability of a material to return to its original shape after a load is removed. For example, a rubber band returns to its original shape after it is stretched. The ***elastic limit*** is the maximum load this material can hold without being permanently deformed.

Plasticity is the ability of a material to be permanently deformed without breaking. Nearly all metals have this property. It is a useful property for materials that are forged or extruded.

Ductility and Brittleness

Ductility applies to material that can be drawn out, stretched, twisted, or bent when cold, without breaking. Ductile materials are flexible and include iron, nickel, and lead.

Material that breaks easily, with a somewhat smooth fracture is called brittle. ***Brittleness*** usually increases as material becomes harder.

Toughness

Toughness is a property related to strength and ductility. Tough materials have the ability to absorb shock and to be deformed without breaking.

elasticity: A material's ability to return to its original shape after a load is removed.

elastic limit: The maximum load a material can hold without being permanently damaged.

plasticity: The ability of material to permanently deform but not break.

ductility: The ability of material to be drawn out, stretched, twisted, or bent when cold, without breaking.

brittleness: A characteristic of a material that allows it to break suddenly.

Material Testing

Materials are tested to determine what properties they possess and to what degree. This information is then used when choosing materials for a specific application. The materials chosen must be able to support the expected load. For example, bridge-building materials need completely different properties than materials used for gutters. Safety is the most important reason material testing is done.

Testing is conducted using a material sample. The results of testing are assumed to apply to all material from which the sample is taken. In other words, the test results from a bolt taken from a group of 500 bolts will apply to all the bolts in that group.

Testing can be destructive or nondestructive. In destructive testing, the test piece is destroyed and is no longer usable. In nondestructive testing the test piece is not destroyed and can, therefore, be used in a finished product.

Stress and Strain

Recall that stress is defined as a naturally occurring force or load working against the material. Stress is measured as force per unit area, such as pounds per square inch or newton per square meter. This stress causes the tested material to deform. The deformation is known as ***strain*** and causes elongation, compression, or distortion of the material.

strain: Deformation caused by stress that causes elongation, compression, or distortion.

Elasticity and Plasticity

Elasticity and plasticity determine the ability of a material to return to its previous shape after stress is released. Elastic material will return to its original shape and plastic material will retain some permanent deformation after the load is removed, **Figure 10-6.**

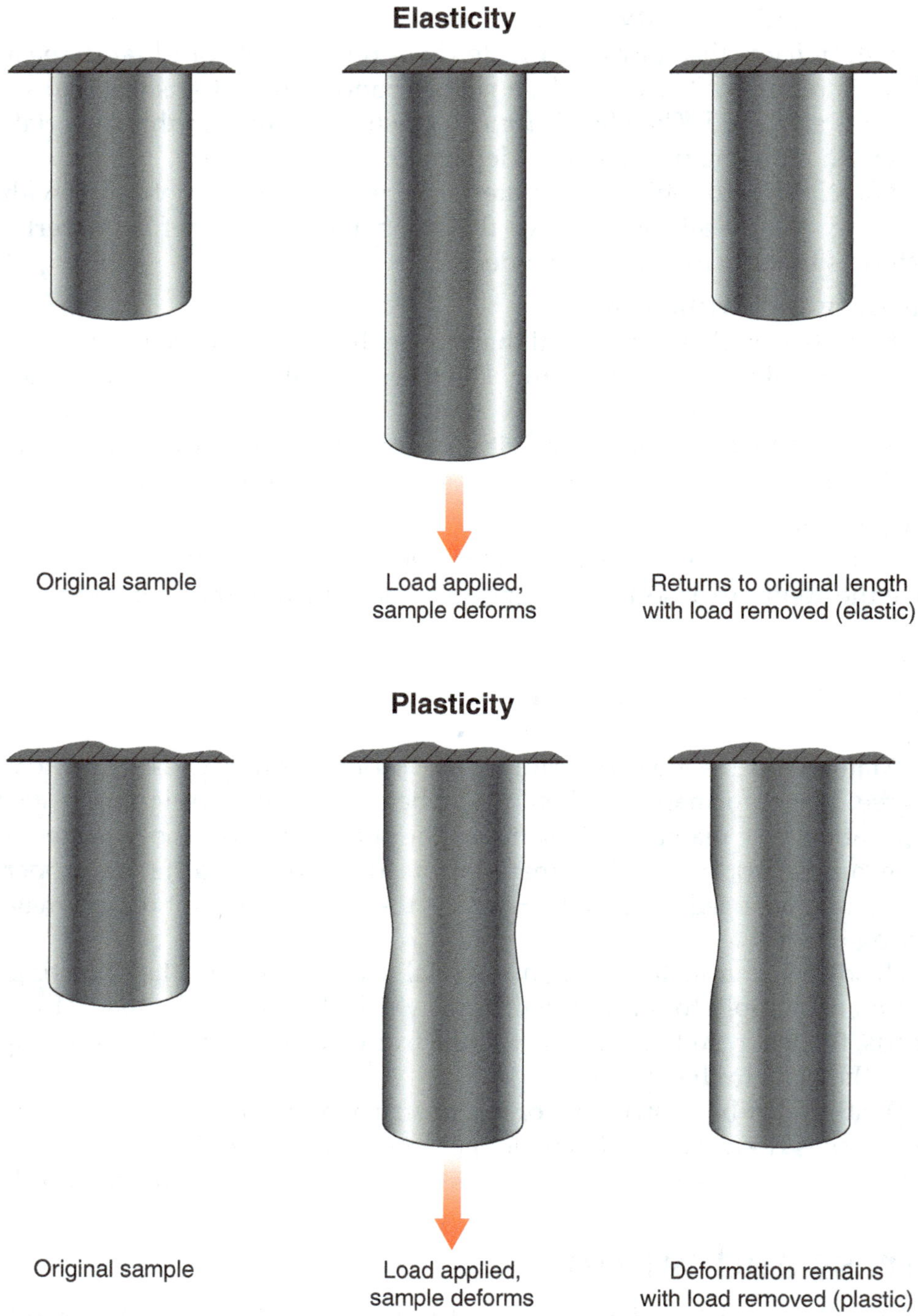

Figure 10-6. Note the differences between elasticity and plasticity.

Tensile Strength and Ductility

Tensile strength and ductility are both measured by applying a pulling force to a sample. The tensile strength machine has a set of jaws. The material sample is placed in the machine and then pulled apart by the jaws until the material breaks. The machine also provides a profile, outlining how the material reacted to the force being applied. Tensile strength is measured by the elongation of the material. Ductility is also tested using a tensile test and is measured by the elongation of the material and the reduction in cross-sectional area.

Toughness

Toughness is measured using an impact test. Impact tests are designed to test the ability of a notched sample to absorb sudden force. A striker is mounted on the end of a pendulum. When it is released, it breaks the sample. The height of the swinging pendulum after it breaks the sample is the amount of energy absorbed by the sample piece.

Hardness

Hardness is measured using one of a variety of hardness tests. Hardness can be determined using a scratch test, an indentation test, or a rebound test.

In a scratch test, a sharp file is drawn slowly and firmly across the surface of the sample material. If the file does not bite into the surface, it is considered file hard. If the file cuts quickly and easily into the surface, the material is soft. There are several obvious disadvantages to this test. There are differences in the files used for testing. There are differences in the way operators conduct the test. The hardness cannot be recorded as numerical data. The advantages include that the test is inexpensive, rapid, and nondestructive. A skilled inspector may be able to use the test to discard unsatisfactory pieces.

shore scleroscope hardness test: A rebound test used to determine the hardness of a material.

The ***shore scleroscope hardness test*** is a type of rebound test. A small diamond-pointed hammer is dropped from a predetermined height onto the surface of the test piece. The rebound height of the hammer is measured and this number translates to a scale. The corresponding number on the scale is the scleroscope hardness number. Softer material has greater deformation and less energy available for rebound.

Indentation testing methods are the most commonly used method. In these tests, an indenter of a known force is pressed into the material. The depth or area of the resulting indentation is measured and given a hardness value. Values differ based on which test is used.

There are four different hardness tests:

- Rockwell.
- Brinell.
- Knoop.
- Vickers.

Each test has its own scale and these scales do not correlate with each other. So, a five on a Rockwell test is not the same as a five on a Knoop test. However, all values increase as hardness increases.

Rockwell hardness test: A type of indentation test used to determine the hardness of a material. An indenter is forced into the material being tested. The resulting dent is measured and given a hardness value.

The ***Rockwell hardness test*** uses either a hardened steel ball or a diamond-tipped cone as the indenter. A minor load is applied first and then a major load. The minor load produces an initial indentation. The dial is then set at zero, and the major load is applied for a set time interval. Hardness numbers are read directly from the indicating dial. A Rockwell hardness tester is shown in **Figure 10-7.**

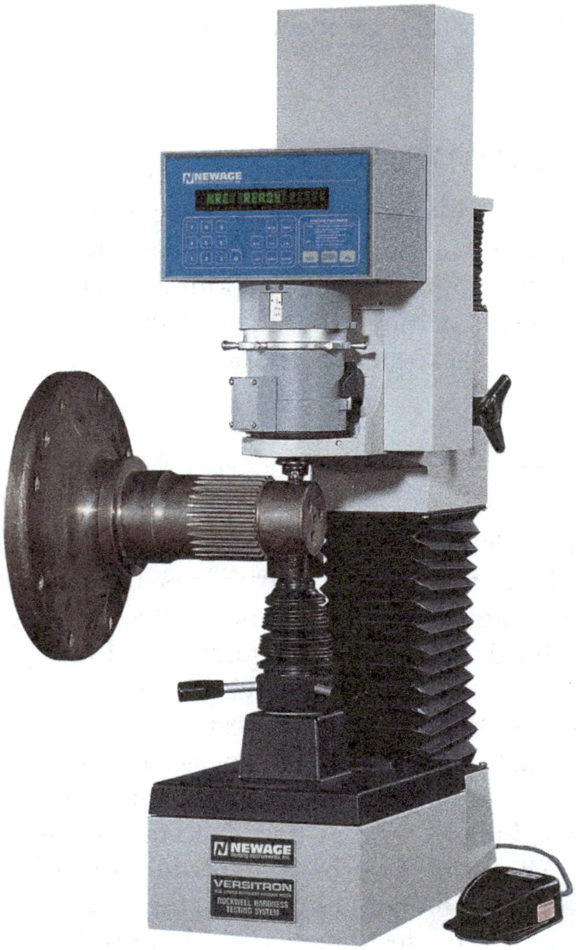

Figure 10-7. One type of Rockwell hardness tester. (Newage Testing Instruments)

Rockwell tests are considered nondestructive because the indentations are so small they do not affect the integrity of the test piece. Rockwell tests are rapid and accurate. In the higher hardness range, it is considered more accurate than the Brinell test.

Brinell hardness test: A type of indentation test used to determine the hardness of a material. A load is applied and held for a predetermined time to an indenter. The resulting indentation is measured in two directions and averaged.

The ***Brinell hardness test*** applies a predetermined load onto a hardened steel ball that then presses into the material under that load for a short time interval (not less than 15 seconds). See **Figure 10-8.** The diameter of the indentation is measured in two directions and averaged. This average is converted to a Brinell hardness number using a chart or a formula. Generally, the softer the metal, the larger the indentation, and the lower the Brinell hardness number. The Brinell test is considered destructive under some conditions.

The Knoop hardness test is a microhardness test. It is used to test the hardness of very brittle or thin materials that can have only a small indentation made for testing purposes. It is considered a destructive test. Precautions that must be observed when conducting this test make it impractical in many situations.

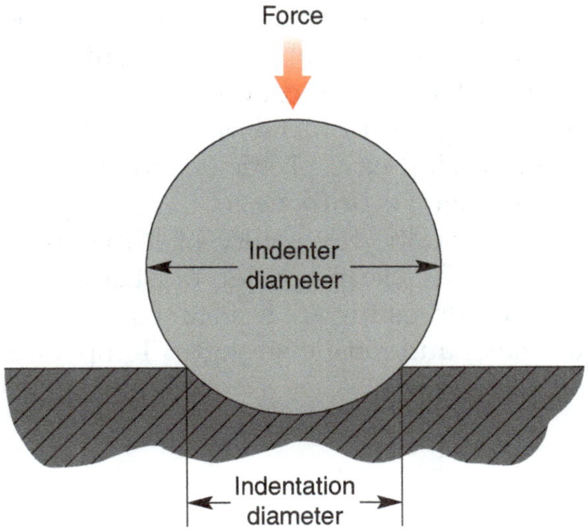

Figure 10-8. An illustration of the Brinell hardness test.

The Vickers hardness test is done in the same way as the Brinell test, but using a differently shaped indenter. The impression left by the Vickers indenter is clearer than that left by the Brinell indenter.

Relationship between Mechanical Properties and Hardness

All mechanical properties are directly affected by hardness.

- Toughness increases as hardness decreases. In contrast, toughness decreases as plasticity and ductility decrease.
- Plasticity increases as hardness decreases. When hardness of a metal becomes great enough, the metal ruptures before plastic deformation takes place.
- Ductility decreases as hardness increases. Decrease in ductility can make fabrication of a metal more difficult.
- Tensile strength increases as hardness increases.

Effects of Temperature

All mechanical properties are affected by temperature. As temperature increases, tensile strength and hardness decrease. As temperature increases, plasticity and deformation also increase. The effects of temperature on the mechanical properties of metals are numerous and complicated. A thorough study of the heat treatment of metals, however, is beyond the scope of this book.

Summary

Materials are physical substances used in production or manufacturing. They can be raw, semifinished, or finished. All materials are generally classified based on their properties and characteristics. A basic division of materials is metals and nonmetals. Metals include nonferrous (contains no iron) and ferrous (iron, steel) categories. Nonmetals include polymers, ceramics, and composites.

Materials are selected for use based on several considerations, the most important of these being the properties of the materials. Properties of a material define its character or behavior under various conditions. Mechanical properties are the most important to consider in metal working. Desirable properties include strength, hardness, elasticity, plasticity, ductility, and toughness. An undesirable property of metal is brittleness.

Material testing is necessary to determine the ability of a material to handle a particular job. The metals chosen for use in a project must be durable, dependable, and safe. Included in materials testing are tests for elasticity and plasticity, hardness, ductility, toughness, and tensile strength. Temperature also has an effect on metal.

Internet Resources

Instron

www.instron.us/wa/home/default_en.aspx

MatWeb

www.matweb.com/search/MaterialGroupSearch.aspx

Outside Activities

1. Collect a variety of materials from around your classroom or school. Break into small groups and categorize each material as either a metal, polymer, ceramic, or composite. Further classify the materials within their groups, using **Figure 10-1** as a guide.
2. Imagine that you must choose the proper material to make the supporting structure of a bridge. Write a short explanation of what materials you would choose and why. Explain what strengths your bridge would need in order to withstand the forces it would experience.
3. Observe as your instructor demonstrates the use of a hardness testing machine. Make a list of dos and don'ts to follow when using this machine. Review the user's manual for additional information on operating and caring for the hardness testing machine.

Name ______________________ Class ______________ Score ________

Chapter 10
Materials and Materials Testing

Review Questions

Write your answers in the spaces provided.

1. True or False. Materials are classified based on their properties. 1. ________

Match each of the following material subcategories with the correct materials family.

2. Ceramic matrix materials such as concrete. 2. ________
3. Nonferrous materials such as copper and copper alloys. 3. ________
4. Natural materials such as glass and diamond. 4. ________
5. Synthetic materials such as plastics and glues. 5. ________

A. Metals.
B. Polymers.
C. Ceramics.
D. Composites.

6. Define the terms *ferrous metal* and *nonferrous metal.* ________________

7. Plain carbon steel contains iron, carbon, and trace amounts of _____ elements. 7. ________
   A. fusion
   B. matrix
   C. compounded
   D. alloying

8. List three reasons that alloying elements are added to alloy steel. ________________

9. The _____ of a material determine its character and the way in which it behaves under certain conditions. 9. ________

10. Name the three basic properties of materials. ______________________________

____

____

____

Match each of the following definitions with the correct term.

11. The ability to resist pulling force. 11. ______
12. The ability to withstand tensile and compressive forces. 12. ______
13. The ability to withstand sudden shock or load. 13. ______
14. The ability to withstand repeated stress. 14. ______
15. The ability to withstand a rotating force. 15. ______

A. Fatigue strength.
B. Tensile strength.
C. Impact strength.
D. Flex strength.
E. Torsion strength.

16. What type of load is depicted in the following drawing? ______________________________

____

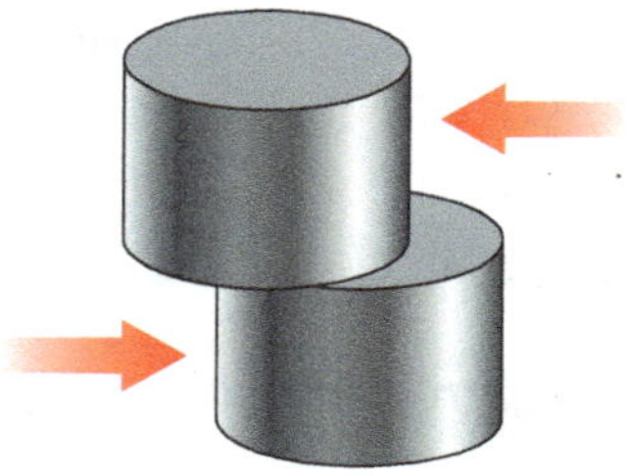

17. The _____ is the maximum load a material can hold without being permanently deformed. 17. ______
18. The ability of material to resist penetration or scratching by another material is called _____. 18. ______
    A. elasticity
    B. hardness
    C. plasticity
    D. flex strength
19. True or False? All hardness tests use the same rating scale. 19. ______
20. The _____ hardness test applies a single load onto a steel ball for a specific length of time. 20. ______

Chapter 11
Sheet Metal Forming

Objectives

After studying this chapter, you will be able to:

- Explain the process of sheet metal forming.
- List various methods of sheet metal forming.
- Summarize the sheet metal spinning process.
- Explain and compare the stretch forming, deep drawing, rubber forming, drop hammer, and HERF processes.
- List five types of HERF processes.

Technical Terms

annealing
blank
conventional spinning
deep drawing
high-energy-rate forming (HERF)
lathe
mandrel
pickling
shear forming
sheet metal forming
springback
stretch forming

sheet metal forming: The process of applying force to a metal sheet to change its surface shape. No material is removed. The metal sheet is deformed but not weakened by the applied force.

Sheet metal forming is the process of applying force to a metal sheet to change its surface shape. No material is removed. The metal sheet is deformed but not weakened by the applied force. Using sheet metal forming processes, the sheet can be formed into many shapes. Forming processes discussed in this chapter include conventional spinning, shear forming, stretch forming, deep drawing, rubber pressure forming, drop hammer forming, and high-energy-rate forming.

Sheet Metal Spinning

blank: A flat piece of metal, used in various machining processes.

mandrel: A cylindrical piece of metal used to support work pieces being formed.

conventional spinning: A type of sheet metal spinning in which the blank is pressed against the mandrel, forcing it into the shape of the form.

The art of sheet metal spinning has been practiced for many centuries and is still done, as both a hobby and as a commercial method of shaping sheets of metal. Sheet metal spinning is used to shape hollow-centered, cylindrical parts. It is usually done on a manual or a CNC lathe. A sheet metal ***blank*** is held in place between the headstock and tailstock of a lathe. The blank rotates at high speed, while a roller bends the blank against a ***mandrel,*** forming the piece, **Figure 11-1.**

Conventional spinning and shear forming (or shear spinning) are two types of metal spinning processes used today. In ***conventional spinning,*** the roller bends the blank against the mandrel, forcing it into the shape of the form. The finished piece has a smaller diameter than the blank, but the thickness remains the same, **Figure 11-2.** Several passes must be made over the blank to make the finished piece. During this process, a gradual hardening can occur, making the metal brittle and unsuitable for further spinning. In this situation, the work must be removed from the lathe and annealed before continuing the process.

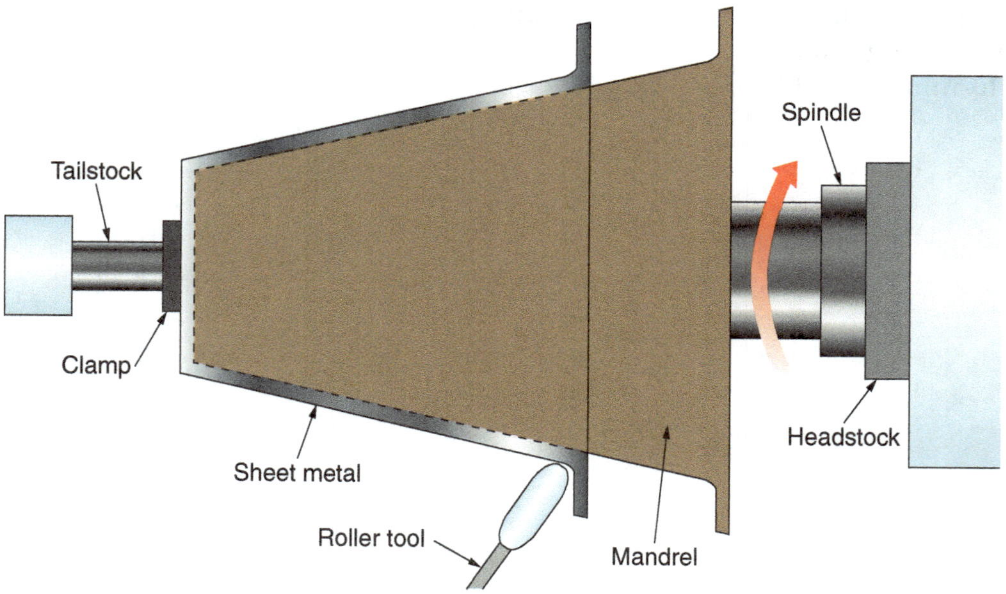

Figure 11-1. Basic sheet metal spinning process.

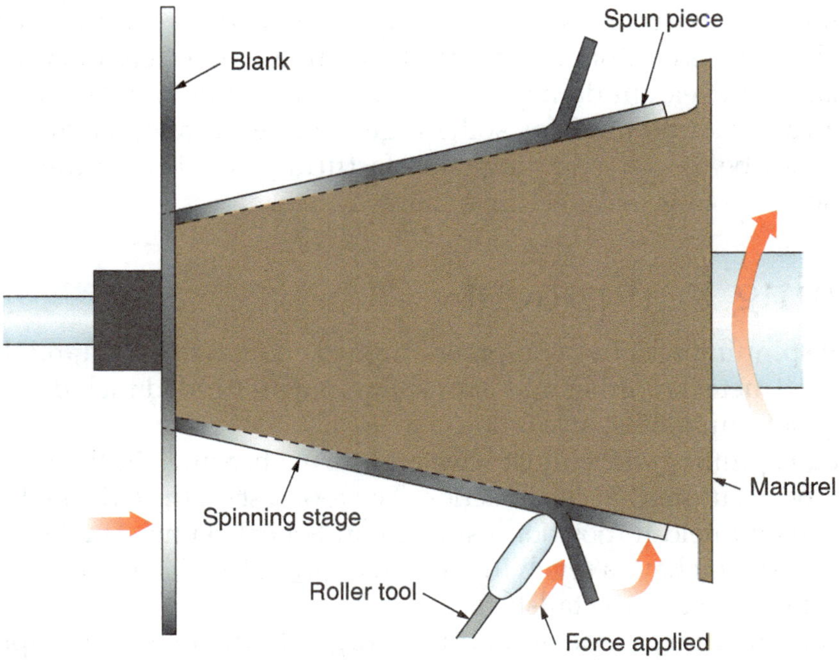

Figure 11-2. The stages of conventional sheet metal spinning.

In ***shear forming,*** the roller bends the blank and also applies downward force. This stretches the blank over the mandrel. The finished piece has the same diameter as the blank, but a different thickness, **Figure 11-3.** Usually only one pass is required in shear forming, eliminating the need for annealing.

shear forming: A simplified variation of the spinning process in which the roller bends the blank and also applies downward force.

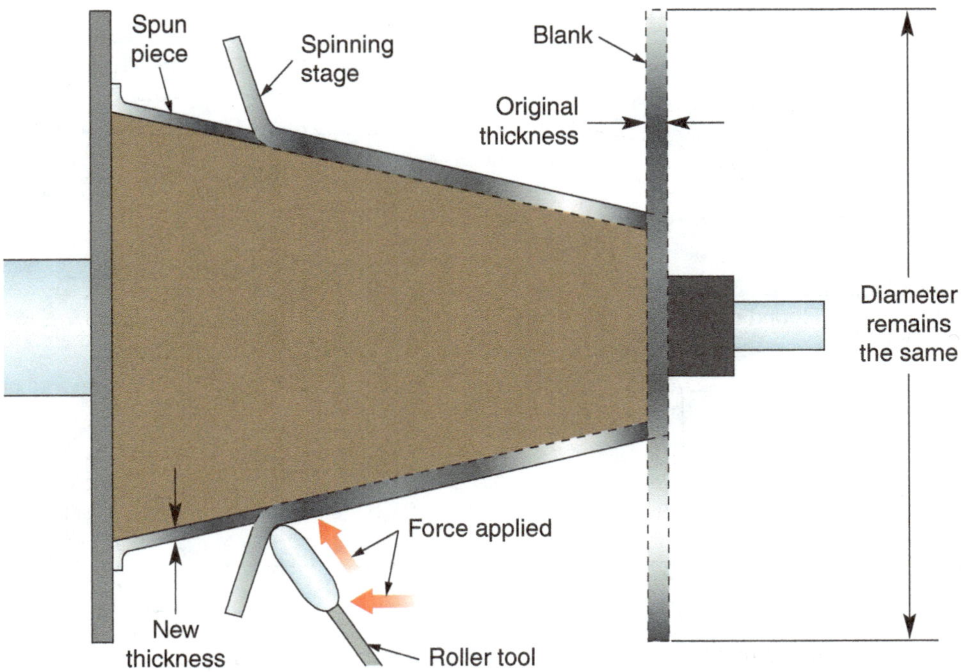

Figure 11-3. A cone being formed in the shear forming process.

Tooling costs for spinning are low compared with other forming methods, making the process economical for small quantities. Mandrels and rollers are easily made and seldom damaged. With the integration of CNC machines, spinning can even be used for some high-volume production runs. Materials that may be used in spinning manufacturing include aluminum, brass, copper, hot- or cold-rolled steel, and stainless steel.

Spinning Equipment

Metal spinning requires equipment rugged enough to withstand the end thrust of the metal spinning operations. Equipment needed includes a lathe, spinning tools, mandrels, and followers.

lathe: A machine that holds and rotates a work piece, which is machined by a stationary tool.

A metal spinning ***lathe*** must have end-thrust bearings in the headstock and must be of substantial construction. The tool rest on the lathe bed should be movable and held in position using a handwheel clamp. The lathe's tailstock center should have a revolving center, **Figure 11-4,** that turns in a socket fitted to a taper hole in the tailstock.

Spinning tools are made of tool steel, forged to the required shape from round steel. Sizes range from 1/2″ to 1 1/2″ (13-38 mm) in diameter and measure about 3′ (914 mm) in length. The length allows the operator to exert pressure on the work. Various spinning tools are shown in **Figure 11-5.**

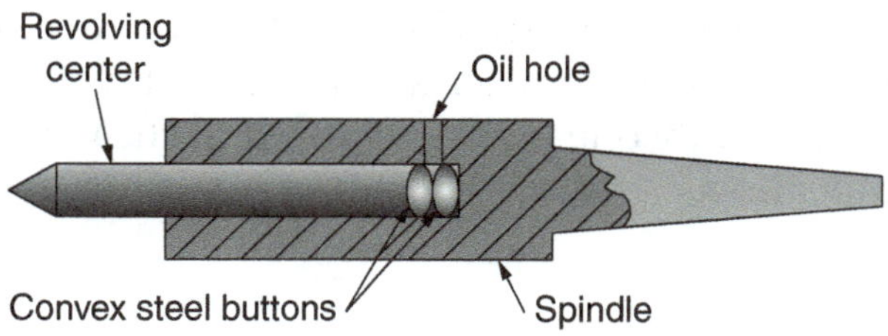

Figure 11-4. A revolving tailstock center for a metal spinning lathe.

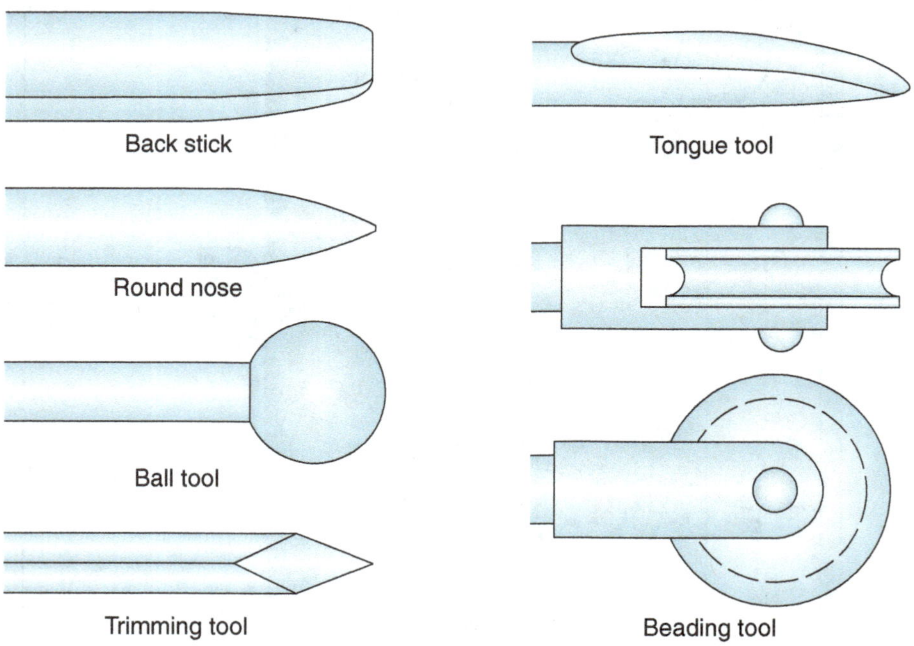

Figure 11-5. These metal spinning hand tools are used to shape the work piece.

The shape of the mandrel should be in the desired shape of the finished article. A solid mandrel is a plain form used for simple metal spinning, **Figure 11-6.** Solid mandrels are used for work that can be removed in one piece. A sectional mandrel is used when the work cannot be removed from the mandrel in one piece. Instead, the piece is removed from the mandrel by removing one section of the mandrel at a time after the work is completed, **Figure 11-7.**

A plug is a type of sectional mandrel that is used to create a shape that has projections or shoulders at both ends, but no bottom. An hour glass, **Figure 11-8,** is an example of such a shape.

A follower is the block of wood used to hold a sheet metal blank against the mandrel. When the lathe is in motion, friction between the work and the mandrel causes the work to turn. The follower presses the work firmly against the mandrel. A hole in the center of the follower engages the revolving center. The follower should be made with the largest possible surface bearing on the work.

Spinning Metal

The cover shown in **Figure 11-9** can be made using the spinning process. More than one mandrel and follower are used to create the desired shape, **Figure 11-10.**

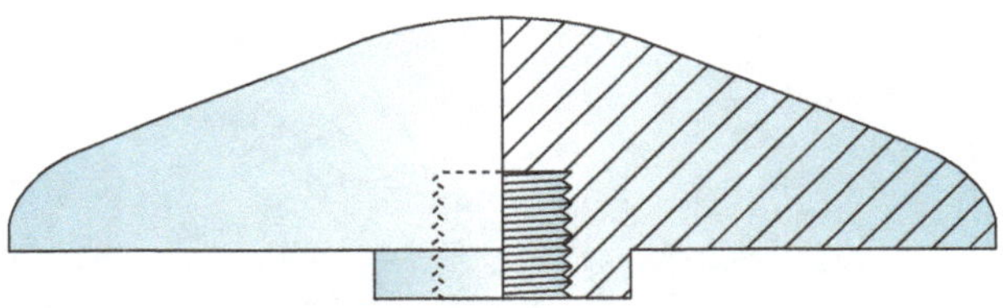

Figure 11-6. This solid, plain mandrel is used for one-piece work.

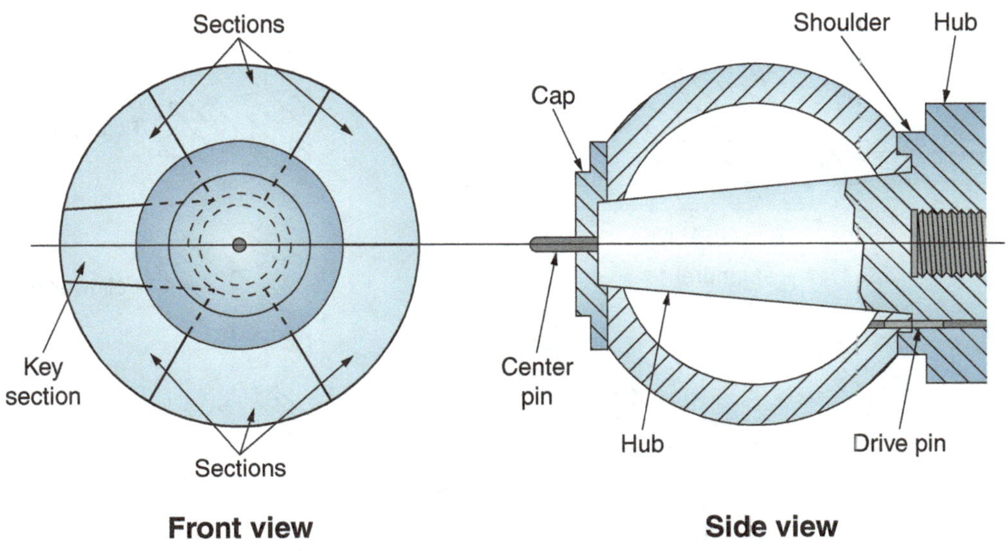

Figure 11-7. A sectional mandrel is used on work that requires the mandrel to be removed in sections after spinning is completed.

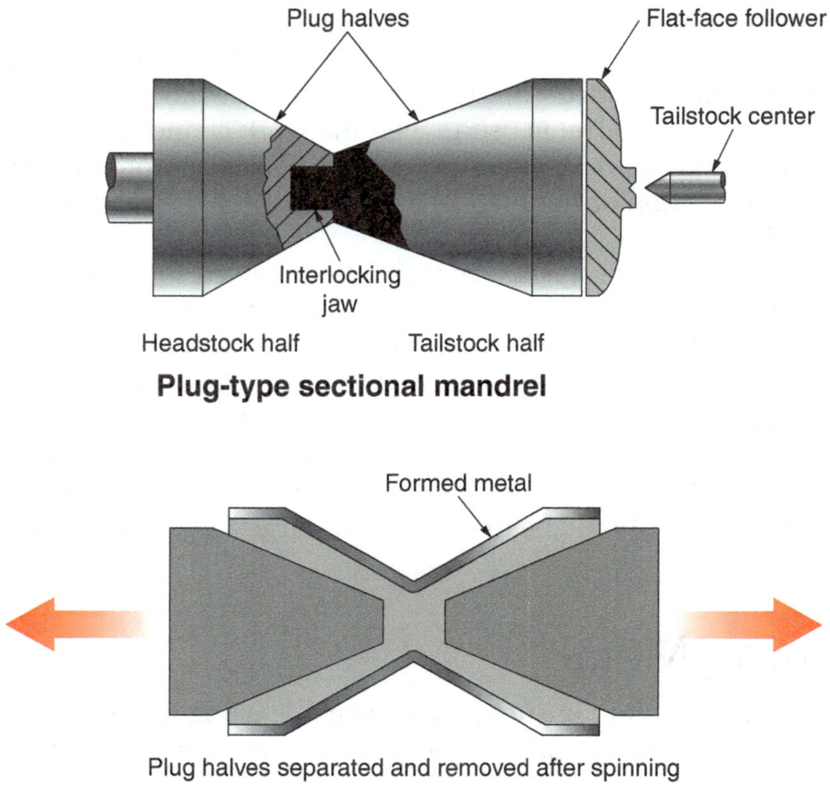

Figure 11-8. A plug mandrel and a flat-face follower.

Figure 11-9. More than one mandrel and one follower will be needed to create this metal cover.

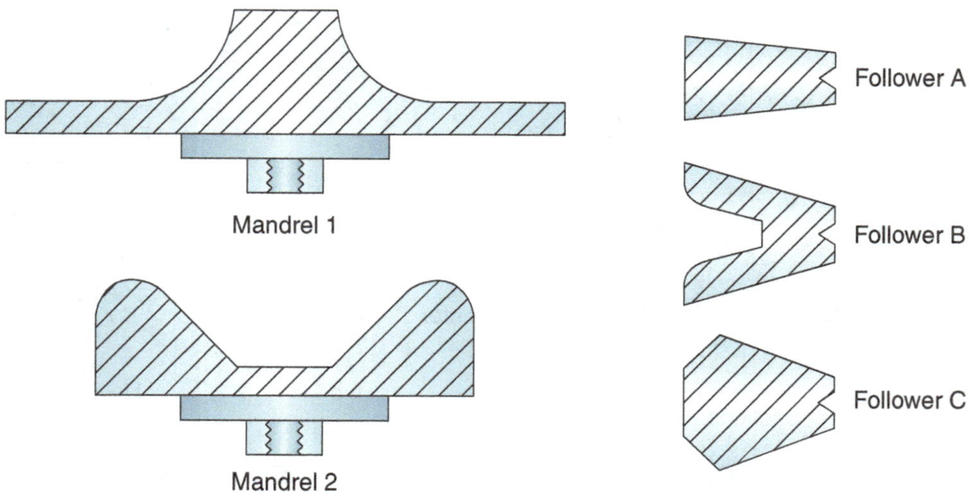

Figure 11-10. These mandrels and followers will be used to create the metal cover shown in Figure 11-11.

Place Mandrel 1 on the lathe spindle. Using your hand, press the blank against the mandrel and turn the tailstock center against the center of Follower A with just enough pressure to hold the blank in place. Start the lathe. Center the blank by pressing a hardwood stick against the edge of the blank. When the blank is centered, gradually increase the pressure of the center against the blank and clamp into place.

Safety Note

Never place a blank in the lathe while the lathe is in motion. The blank can become airborne, flying at a high rate of speed and turning at the same time. This can cause serious injury to anyone standing nearby.

Lightly lubricate the surface of the work with soap or beeswax before starting to spin. Petroleum jelly mixed with graphite is another suitable lubricant.

Adjust the prop pin on the tool rest so it points near the blank. This angle will allow the proper leverage for spinning. Hold the handle of the tool under the right armpit, using the spinning tool as a lever and the prop pin on the tool rest as a fulcrum, **Figure 11-11.** As pressure is applied, the metal disk is forced slowly against the mandrel.

Move the spinning tool in and out constantly. Do not permit it to rest in one spot. Apply pressure only when moving toward the edge of the blank, not when moving toward the center. Hold the back stick in your left hand.

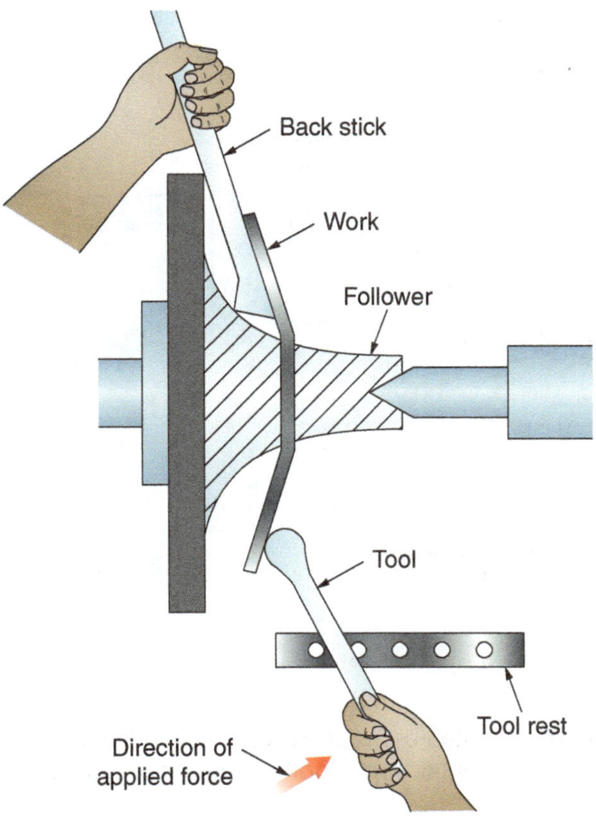

Figure 11-11. Note the position of the back stick and shaping tool. Force should be applied in the directions shown.

Press firmly against the reverse side of the metal at a constantly changing point opposite the tool. The back stick prevents wrinkling of the piece as it is worked toward the edge of the disk.

Work the metal back a short distance, **Figure 11-12.** Stop the lathe and substitute Follower B, **Figure 11-13.** Follower B provides a better grip on the work. Continue spinning until the work is in the shape of Mandrel 1. Stop the lathe and substitute Mandrel 2 and Follower 3, **Figure 11-14.** In this final spinning stage, shape the outer part of the work to Mandrel 2, **Figure 11-15.**

Annealing and Pickling

During the spinning process, the metal may gradually harden. It may fracture and become difficult to work. ***Annealing*** will soften the metal and make it workable.

annealing: A heat treatment process used to reduce the hardness caused by working a material.

Metal is annealed by placing it in a heat-treating furnace. The metal is then cooled slowly in an insulating material. The temperature of the furnace and the amount of time spent in the furnace is based on the type and size of the metal being worked.

The annealing process forms an oxide scale on the metal. This coating will mar the surface of the piece if it is not removed. To remove it, the annealed piece is bathed in a ***pickling*** solution. Once the piece is clean, it is removed, rinsed thoroughly in running water, and dried.

pickling: A cleaning process that removes oxide scale formed during the annealing process. The process involves submerging the work piece into a diluted sulfuric acid bath.

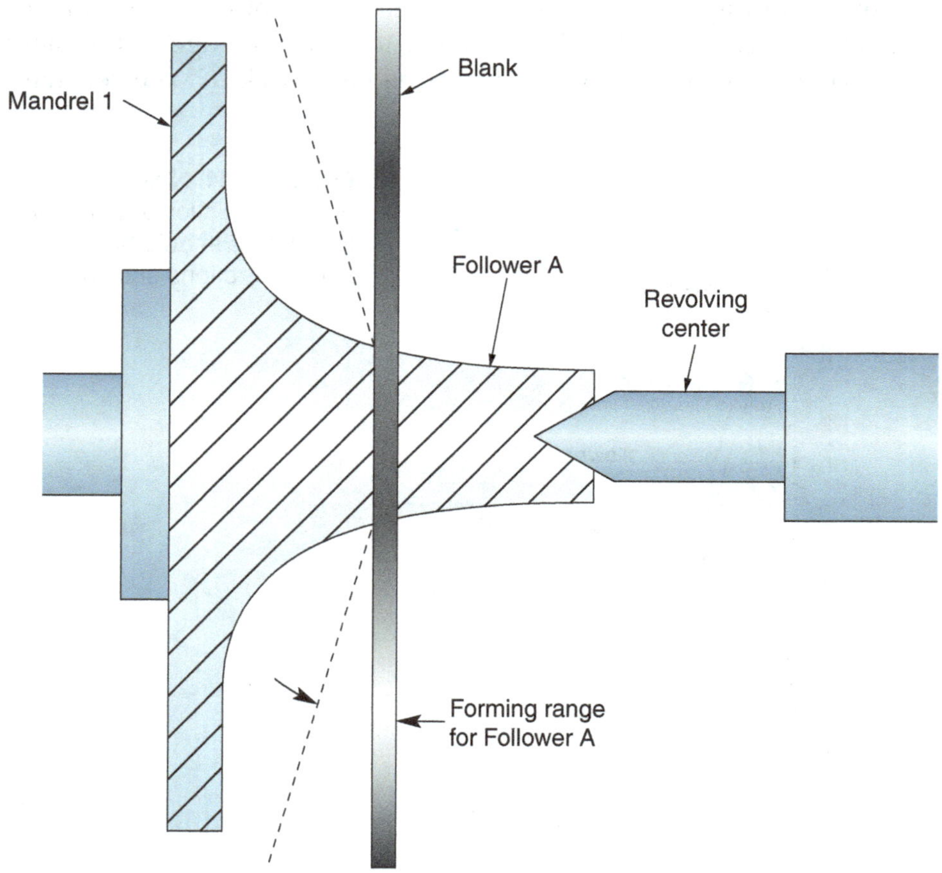

Figure 11-12. The metal blank is properly mounted between Mandrel 1 and Follower A. It should be worked toward the mandrel only as far as the dotted line before substituting Follower B.

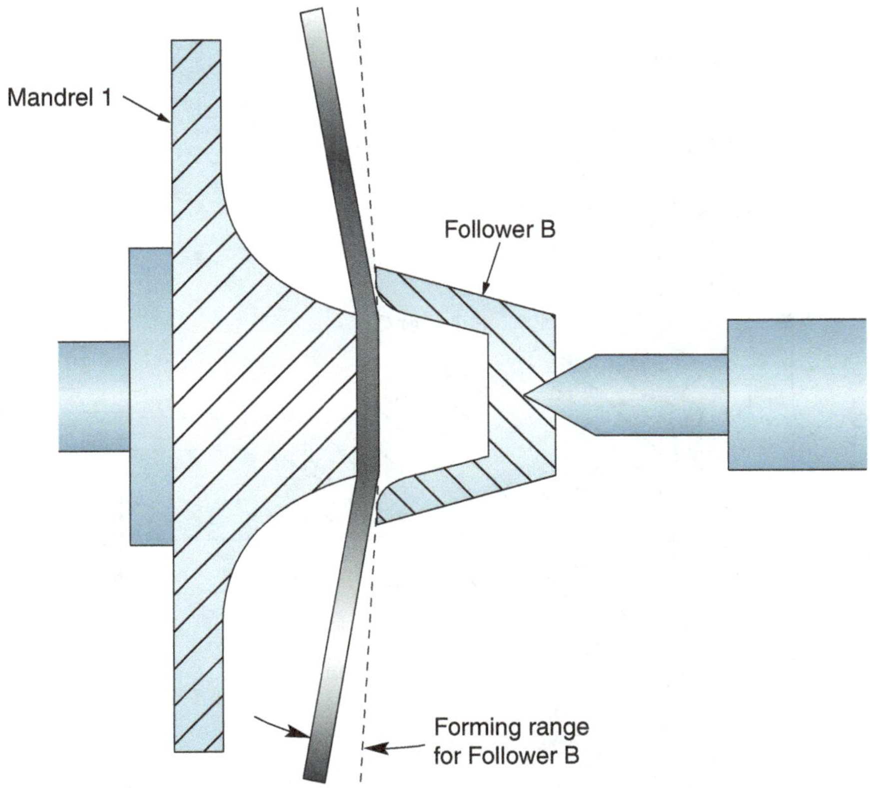

Figure 11-13. After substituting with Follower B, continue spinning until the work has the same form as Mandrel 1.

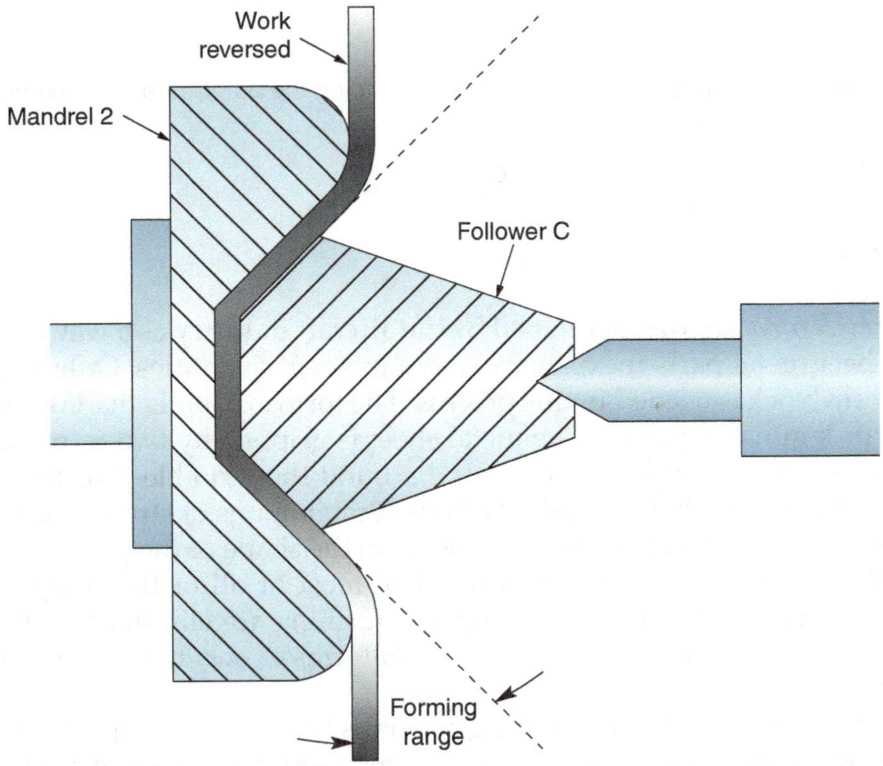

Figure 11-14. After the shape of Mandrel 1 has been created, substitute Mandrel 2 and Follower C. The dotted line indicates the position of the work at the beginning of this operation.

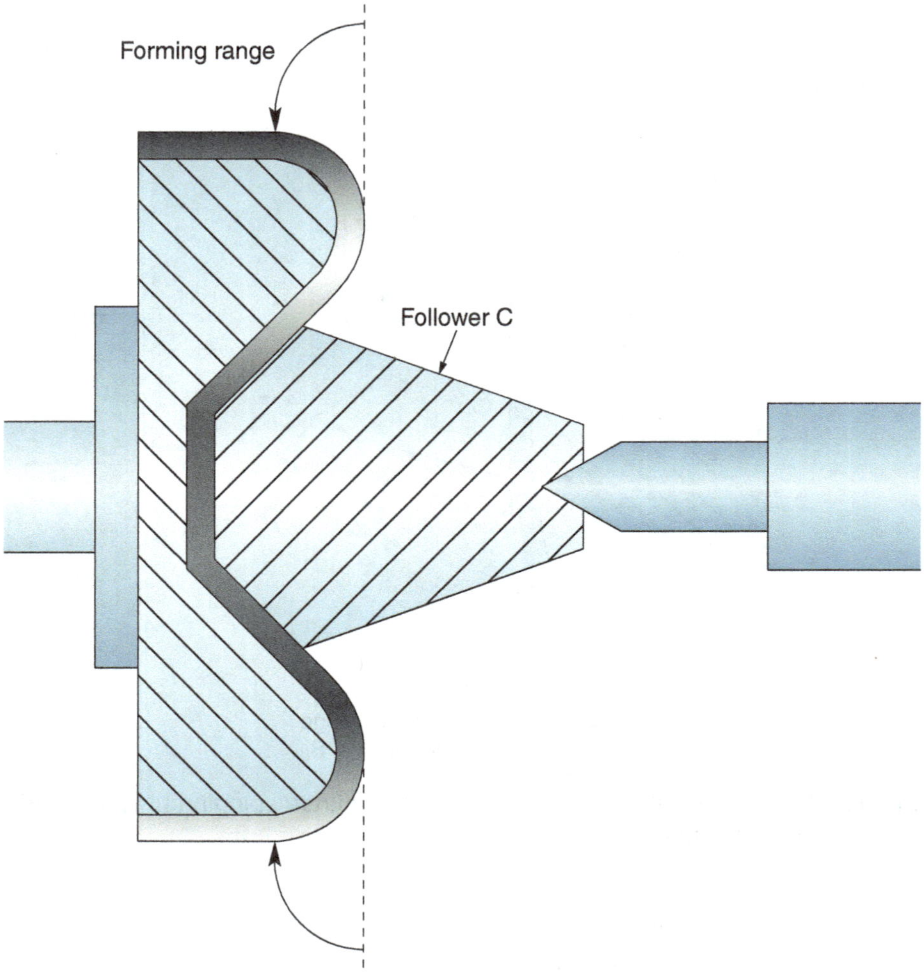

Figure 11-15. The finished piece is created by shaping the outer portion of the blank over the mandrel.

Stretch Forming

stretch forming: Shaping a metal sheet by first applying suitable tension or stretch and then wrapping it around a form of the desired shape.

springback: Partial rebounding of formed material caused by its elasticity.

Stretch forming was developed by the aircraft industry as a way to make large sheet metal parts inexpensively and in small quantities. Only a single male form block is needed in this process. The form is usually made of wood, metal, or laminated plastic. The metal sheet is gripped by two or more sets of jaws that stretch and wrap the metal around the form block as the block moves upward. See **Figure 11-16.** Various combinations of stretching, wrapping, and upward motion are used, based on the shape of the part.

If the stretching is properly controlled, most or all of the compressive stresses that go with normal bending and forming are eliminated. There is very little ***springback,*** so the work piece conforms closely to the shape of the form block.

Stretch forming is used in making cowlings for aircraft engines, wing tips, scoops, and large aircraft panels. These parts are usually made of aluminum or stainless steel. Large automobile and truck panels can also be produced using this process.

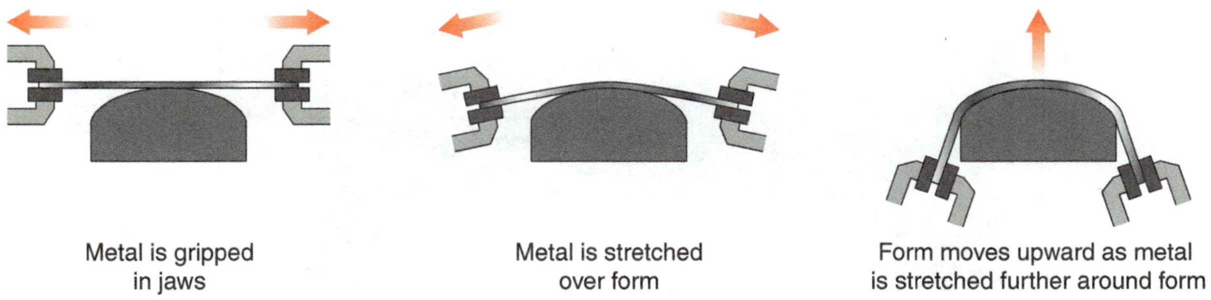

Figure 11-16. The steps in the stretch forming process.

Stretch-draw forming is a similar process to stretch forming. However, in stretch-draw forming, mating male and female dies are used to shape the stretched metal.

Deep Drawing

Deep drawing is used to form cup-shaped containers that have a depth greater than the diameter of their openings. The metal blank is held tight at the edges, while the center of the blank is forced into a die, **Figure 11-17.** This process is also known as shell drawing because its earliest use was in the manufacture of artillery shells and cartridge cases. On pieces on which the depth of the drawn part is less than the diameter or the minimum surface dimension of the blank, the process is called shallow drawing.

deep drawing: A stretching process in which a blank is placed in a draw-press die. The blank holder places pressure on the blank, holding it tight enough that it slides into the die, while keeping the metal flat and free from wrinkles.

Drawing causes materials to wrinkle and vary in thickness. But by controlling the flow of metal, wrinkling and thinning can be controlled. This is usually done using a pressure ring or band. In single-action presses, **Figure 11-18,** there is only one movement of the slide. A ring or air pressure controls the flow of metal between the upper die and the pressure ring. Double-action presses have two or more independent rams, **Figure 11-19.** The force applied to the pressure ring is controlled independently of the position of the main slide. This permits the pressure to vary as needed during the drawing operation. That is also the reason that double-action presses are most often used for drawing more complex parts. Single-action presses are best for simpler operations.

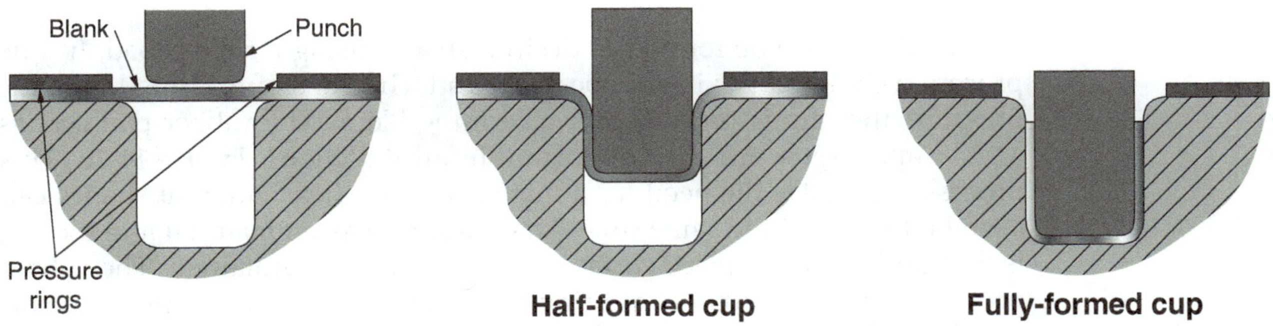

Figure 11-17. The steps in the deep drawing process.

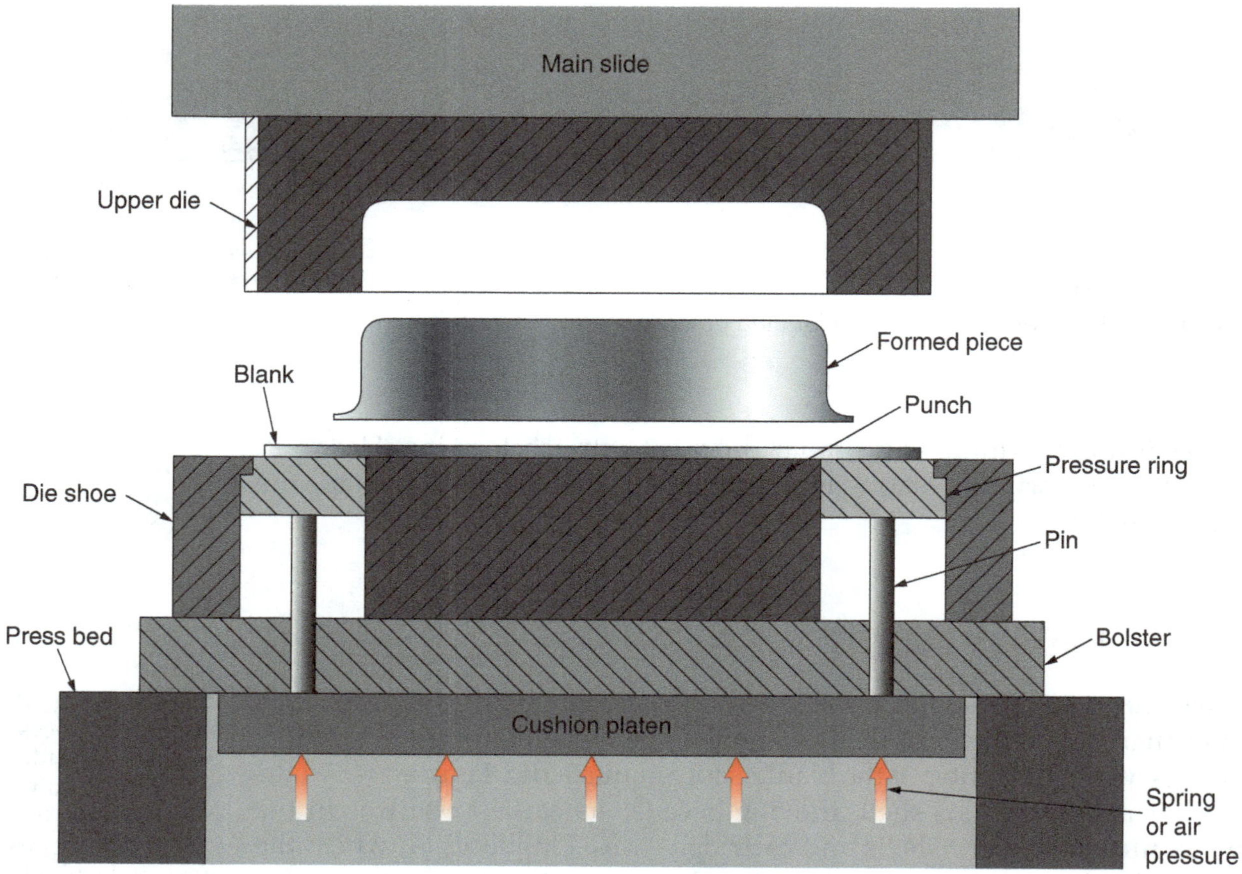

Figure 11-18. A single-action press is used to create simple sheet metal pieces.

Because metal does not always flow uniformly throughout a work piece, many drawn parts must be trimmed after forming to remove excess or undesired metal. The trimming is usually be done either by hand or by using a specially-made trimming die. This adds to the production costs. To avoid the need for trimming, a blank that has been cut to a special shape before it is drawn is often used. The choice whether to use a special die or cut a blank to a special shape often depends on the complexity of the part and the number being made.

Rubber Pressure Forming

Metal can also be formed to desired shapes using a rubber pad. In this process, a confined, flexible rubber pad is attached to a ram. Using hydraulic pressure, the ram is pressed over a forming block. The rubber pad acts as a fluid and exerts pressure uniformly in all directions, **Figure 11-20.** This process eliminates the need for a matched set of dies. Different shapes can even be formed at the same time. That means wear on material and tooling is low, and the work piece surface quality is easily maintained. The rubber pressure forming method was originally known as Guerin forming and was developed by the aircraft industry.

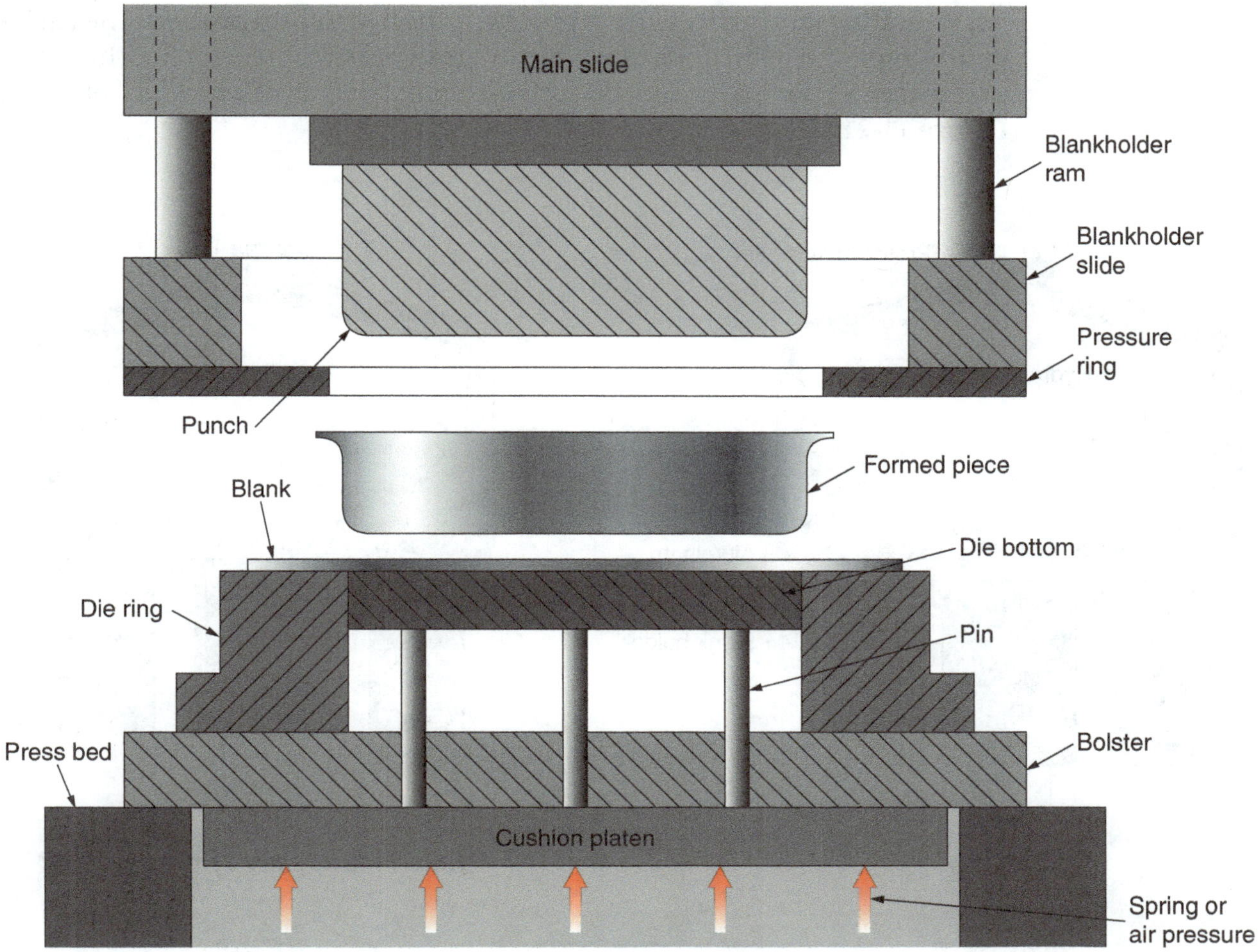

Figure 11-19. The pressure in a double-action press is controlled independently of the position of the main slide.

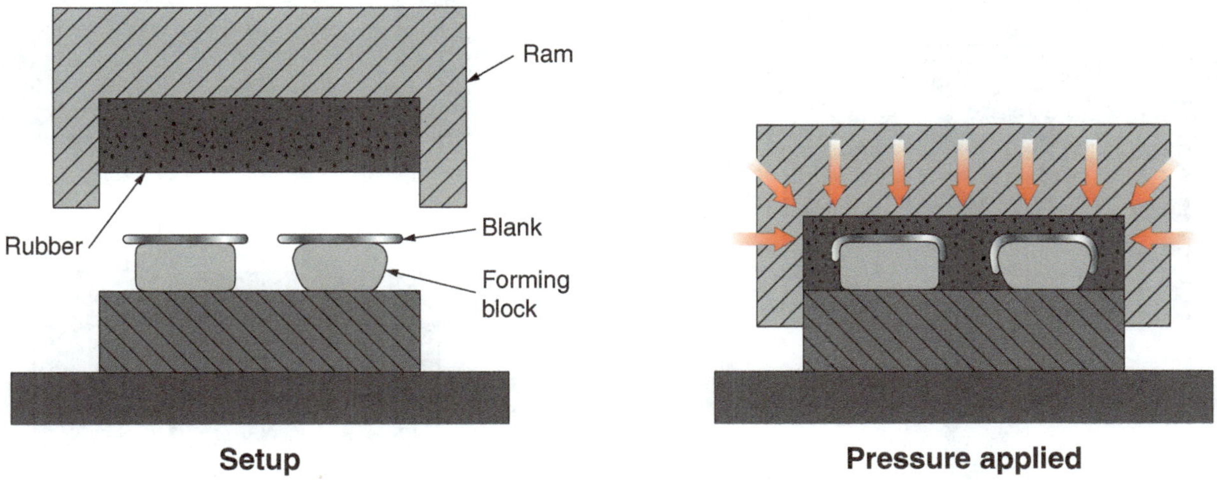

Figure 11-20. Forming sheet metal using the rubber forming process.

A modification of the Guerin process is the hydroform process. The rubber pad is replaced with a flexible diaphragm backed by controlled hydraulic pressure. Deeper parts can be formed with truly uniform fluid pressure, **Figure 11-21.**

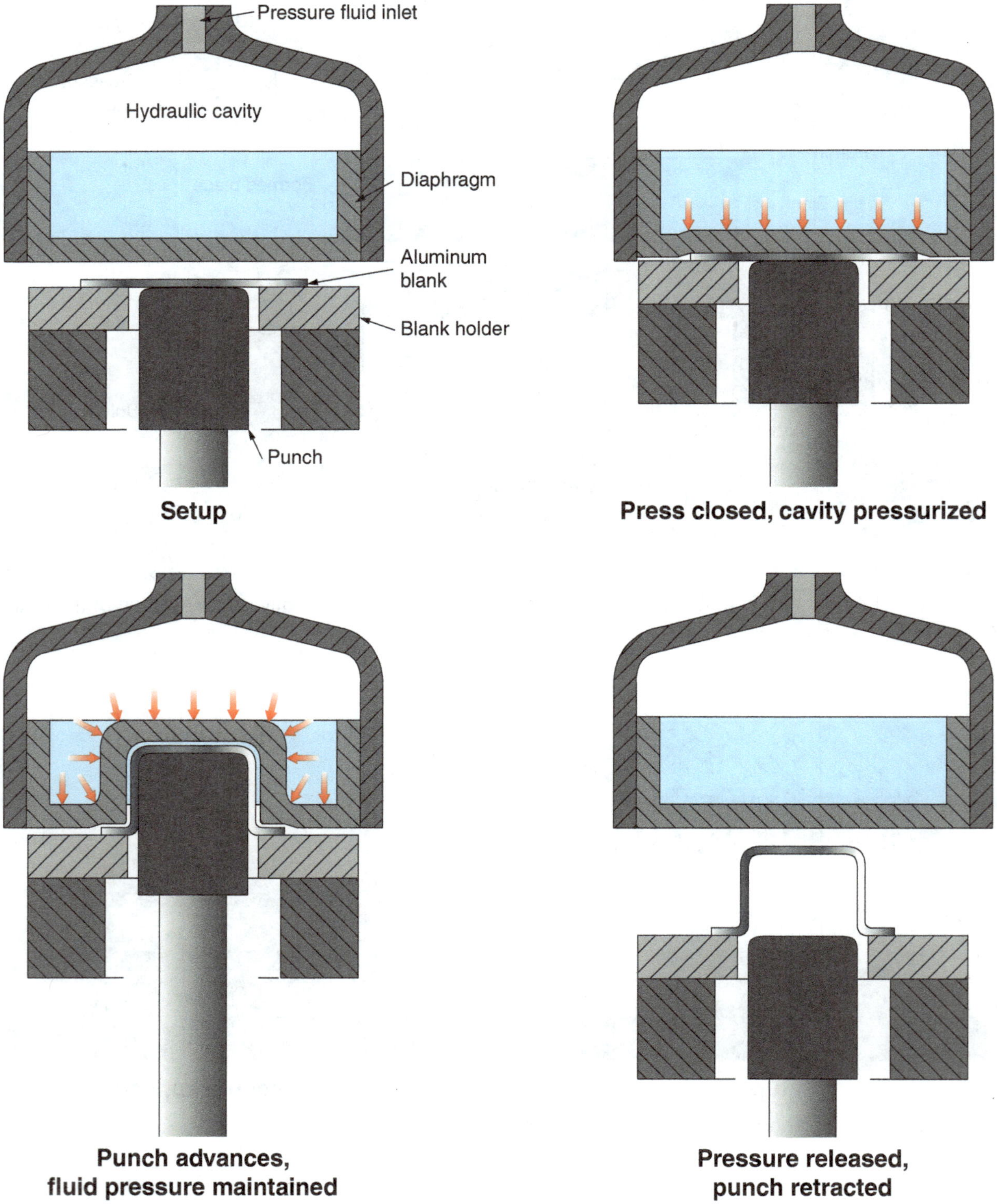

Figure 11-21. The steps in the hydroform process.

In the bulging process, an oil or rubber pad applies internal pressure to expand a metal blank or tube outward against a female die. This eliminates the necessity for a complicated, multiple-piece male die member. Split female dies are often used to make removal of the finished part easier. Complex equipment is used in this technique to provide the necessary seals. For less complicated shapes, rubber can replace the internal fluid.

Drop-Hammer Forming

Drop-hammer forming is a process in which a ram, or hammer, is dropped repeatedly from a predetermined height onto a metal blank placed over a die. This method is ideal for forming small quantities of shallow-drawn parts and is used in the aircraft and automobile industry. Kirksite dies are often used for this technique, avoiding the expense of costly machined steel dies. Kirksite dies can also be melted down and cast into other shapes when no longer needed.

During drop-hammer forming, the hammer is raised and lowered using a rope passed several times around a rotating shaft. This allows for good control of the hammer. Wrinkles may form on the first drop of the hammer, but those wrinkles can be hammered out between strokes using a hand mallet. In some cases, a stack of shims, made of thin plywood, is placed on top of the sheet of metal, **Figure 11-22.** When the ram is dropped, it descends through only a portion of the complete draw. At the same time, the shims act as a pressure pad to restrict and control the flow of metal, preventing wrinkling. One shim is withdrawn after each stroke of the ram. This allows the ram to fall farther and thus deepen the draw on each successive stroke.

Drawing on a drop hammer is considerably cruder than drawing with steel dies. However, it is often the most economical method for small quantities. It is most suitable for aluminum alloys, but can also be used successfully on thin carbon and stainless sheets.

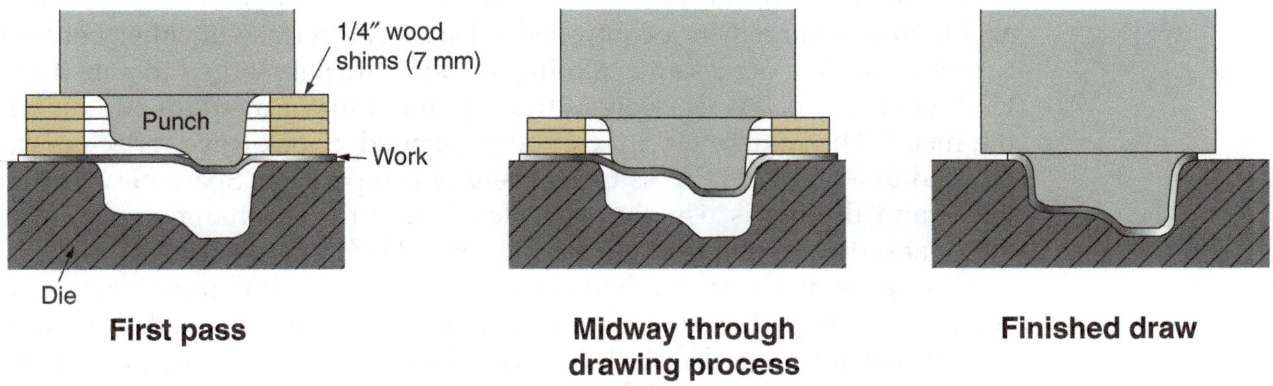

Figure 11-22. Deep drawing using a drop hammer, kirksite die, and a book of shims.

High-Energy-Rate Forming

high-energy-rate forming (HERF): A group of sheet metal forming methods that shape a work piece using the force created by a sudden blast of extremely intense energy.

High-energy-rate forming (***HERF***) is a group of forming processes. In these processes, metal is deformed by applying a high rate of strain to the metal through the application of high rates of energy transfer. The major advantage of HERF is that many metals tend to deform more readily using these ultrarapid rates of load application. As a result, HERF makes it possible to form large work pieces and difficult-to-form metals with less-expensive equipment and tooling than would otherwise be needed.

Another advantage of HERF is that there is little springback. This is because:

- High compressive stresses are set up in the metal when it is forced against the die.
- Some minor elastic deformation of the die may occur under the high pressure, resulting in the work piece being slightly overformed, giving the appearance of no springback.

The intense energy used in HERF processes is created using one of five methods:

- Explosive charge.
- Spark discharge.
- Pneumatic-mechanical method.
- Internal combustion method.
- Electromagnetic method.

The explosive-charge method creates a high-energy rate using blasting material placed in a tank of water or other liquid. The work piece and a female die are also placed in the tank, along with the explosive charge. When the explosives are detonated, the energy flows through the liquid, contacts the metal piece, and deforms it. The explosive-charge method is best for use on large parts or thick material. It can be used to form a variety of shapes, **Figure 11-23.**

The spark-discharge method uses the energy of an electrical discharge to shape metal, **Figure 11-24.** Electrical energy is stored in banks of capacitors and then discharged in a controlled release. The release is either between two electrodes or across an exploding bridge wire submerged in a medium. High-energy shock waves spread through the transfer medium and deform the metal. The initiating wire can be preshaped. Shock-wave reflectors can be used to adapt the process to a variety of components. Space between the blank and the die is usually evacuated before the discharge occurs. This eliminates the possibility of puckering caused by trapped air.

The spark-discharge method is most often used for bulging operations in small parts, but parts up to 50″ (1.27 m) in diameter can be formed. Compared to explosive forming, the discharge technique is easier and safer. It does not require as large a tank, and does not have to be performed in remote areas.

Pneumatic-mechanical and internal combustion HERF techniques were developed for use inside buildings. In pneumatic-mechanical forming, one segment of the forming die is attached to the stationary bolster of the press bed and the other to a piston rod. Low-pressure gas then acts on the entire

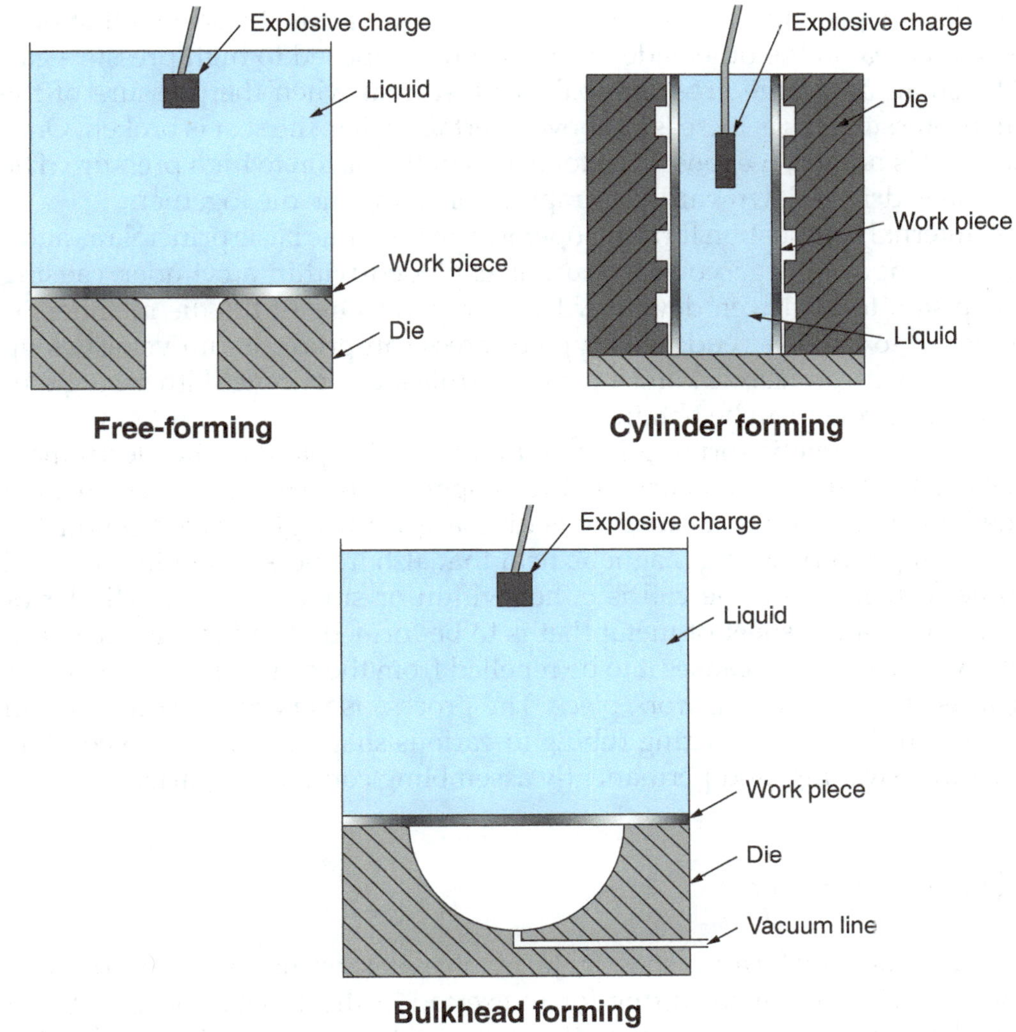

Figure 11-23. Three methods of high-energy-rate forming with explosive charges.

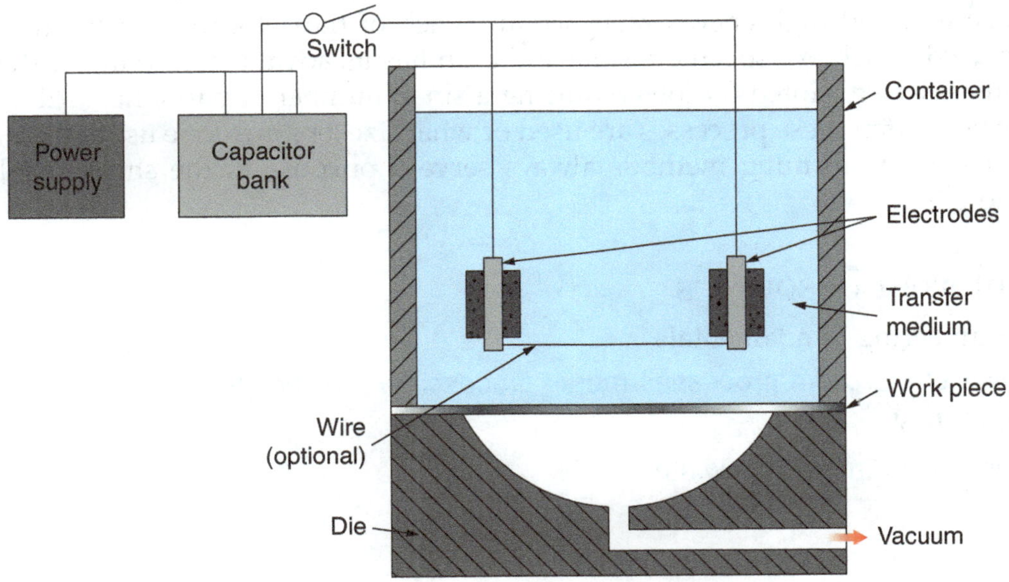

Figure 11-24. The components in the spark-discharge forming method.

area of the piston in a cylinder. This holds it up against a seal so that only a small area on the other side of the piston is exposed to high-pressure gas. The small and large areas are balanced, so that when the pressure of the high-pressure gas is increased above a certain value, the seal is broken. Once the seal is broken, it exposes the total area of the piston to high pressure. The piston is driven downward very rapidly, bringing the die together.

Internal combustion forming operates on the same basic principle as automobile engines. A gaseous mixture is exploded within a cylinder, causing the piston to be driven downward very rapidly. One of the die members is attached to its lower end. This type of press can produce ram velocities up to 50′ (15 m) per second and up to 60 strokes per minute. Either single or repeated strokes can be obtained.

Electromagnetic forming is based on the principle that the electromagnetic field of an induced current always opposes the electromagnetic field of the inducing current. A capacitor is discharged through a coiled conductor, producing a very strong magnetic field that also induces a field in the metal to be formed. Since the coil is either within or surrounding a cylinder or adjacent to a flat sheet of metal that is to be formed, it induces a current in the work piece. This causes it to be repelled from the coil and to be deformed against a die or mating work piece. The process is very rapid and is useful for expanding or contracting tubing to various shapes. It is used in coining, forming, swaging, and permanently assembling component parts.

Summary

The process of forming sheet metal involves deforming, but not destroying, the metal. There are no cutting forces exerted in sheet metal forming. Sheet metal forming processes change the shape and even the thickness of metal, but never weaken the metal.

Several sheet metal forming methods were reviewed in this chapter, including metal spinning, stretch forming, deep drawing, drop-hammer forming, and high-energy-rate forming. Each of these forming techniques is used to achieve specific results and each has its advantages. Many of the processes are suited for jobs requiring a small number of parts. Regardless of how often these processes are used or what size jobs are done using them, sheet metal forming methods always serve a purpose in the sheet metal industry.

Internet Resources

Engineering Fundamentals

www.efunda.com/processes/metal_processing/bending.cfm

Wise Tool

wisetool.com/deepdraw.htm

Outside Activities

1. Obtain and review literature from companies that use spinning, electromagnetic forming and the spark discharge method of metal forming. Ask the companies if they will lend your class samples of some of the items they make. If the companies are unable to do that, make a list of common, everyday objects made using these processes. Are you surprised by the number of items on your list? Why or why not?
2. Study the various manufacturing methods used to make sheet metal items. Many of these sheet metal items are manufactured further, to create another item. For example, the sheet metal work done to create the doors of an automobile is just one step in a many-stepped process. Create a timeline showing where the sheet metal manufacturing processes are done as part of a larger manufacturing process.

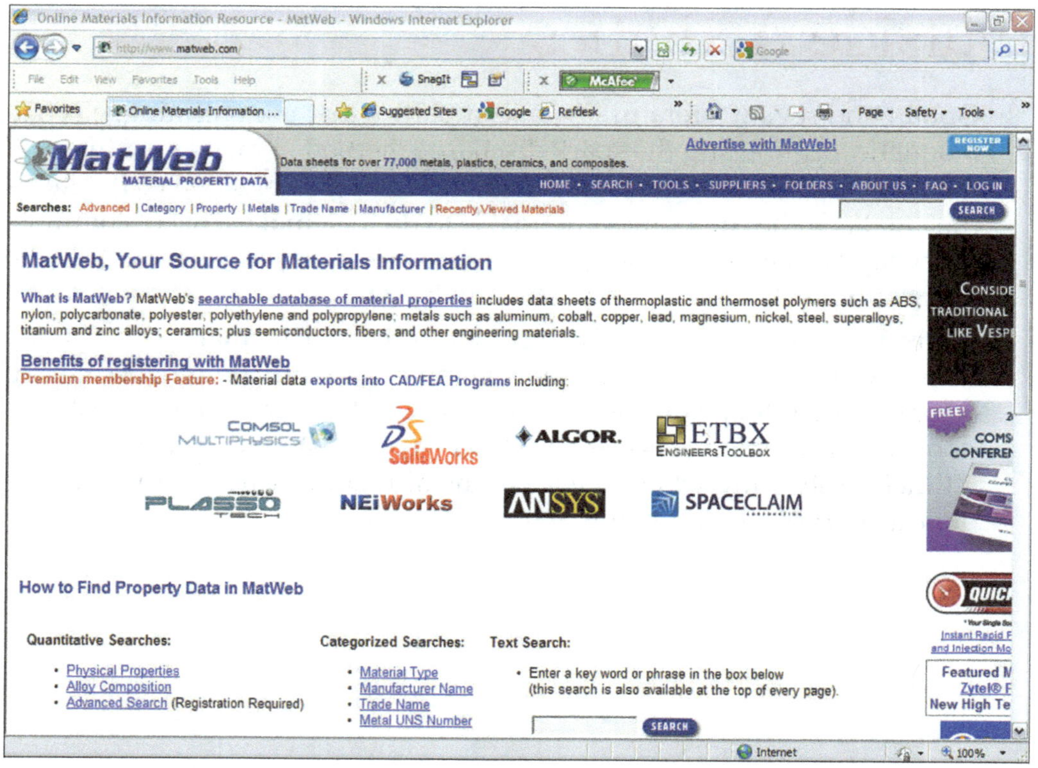

Learn more about different types of materials at the MatWeb site.

Name ______________________ Class ______________ Score __________

Chapter 11
Sheet Metal Forming

Review Questions

Write your answers in the spaces provided.

1. Briefly explain the sheet metal forming process.______________________________

____

____

____

____

2. Sheet metal spinning is used to form _____, cylindrical parts. 2. ______________
   A. machined
   B. hollow
   C. solid
   D. None of the above.

3. True or False? In conventional spinning, the roller bends the blank against the mandrel and also applies downward pressure on it. 3. ______________

4. In _____, the finished piece has the same diameter as the blank but a different thickness. 4. ______________

5. Which piece of equipment is needed to spin metal? 5. ______________
   A. Spinning tools.
   B. Mandrel.
   C. Followers.
   D. All of the above.

Match each of the following descriptions with the correct term.

6. Used to create a shape that has projections or shoulders at both ends, with no bottom pieces. 6. ______________

7. Used when the work cannot be removed from the mandrel in one piece. 7. ______________

8. Used for work that can be removed in one piece. 8. ______________

A. Sectional mandrel.
B. Solid mandrel.
C. Plug-type mandrel.

9. What is the purpose of annealing? ______________________________

____

____

10. To remove the oxide scale formed on metal during annealing, bathe the piece in a _____ solution. 10. ____________

11. What forming process is being shown in the following drawing? ____________

____

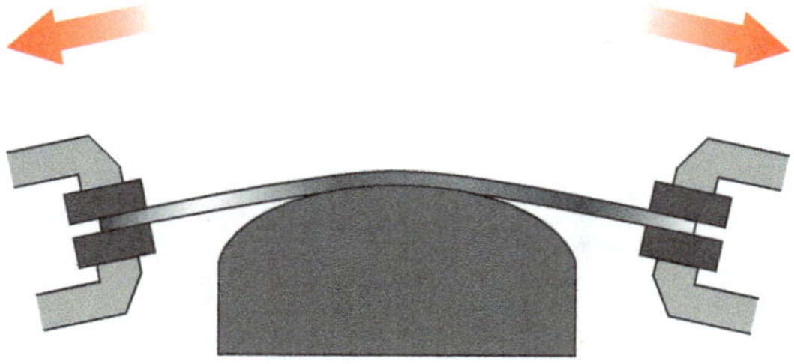

12. The partial rebounding of formed material caused by its elasticity is known as _____. 12. ____________

13. In order to control wrinkling during deep drawing, control the _____ of the metal. 13. ____________

14. True or False? Many drawn parts must be trimmed after forming to remove excess or undesired metal. 14. ____________

15. Which of the following statements about rubber pressure forming is *not* true? 15. ____________
    A. Was originally known as the Guerin method.
    B. Was developed by the aircraft industry.
    C. Requires a matched set of dies.
    D. Uses a confined, flexible rubber pad attached to a ram.

16. What does HERF stand for? ____________

____

17. What is the major advantage of HERF? ____________

____

____

18. The _____ method of HERF uses explosives detonated in water to deform work pieces. 18. ____________

19. List two reasons why using the spark-discharge method is easier and safer than using the explosive-charge method. ____________

____

____

____

____

____

20. True or False? Pneumatic-mechanical and internal combustion HERF techniques were developed for use inside buildings. 20. ____________

Chapter 12
Sheet Metal Careers

Objectives

After studying this chapter, you will be able to:

- Outline the tasks involved in sheet metal work.
- Compare and contrast different metalworking work environments.
- Explain the difference between apprenticeship training and on-the-job training.
- List the qualifications of a good sheet metal worker.
- Understand the types of jobs available in different occupational areas.
- Discuss the responsibilities of a good employee.

Technical Terms

bachelor's degree
initiative
manager
overtime
TAB technician

The Nature of Sheet Metal Work

People who work with sheet metal design, lay out, fabricate, and install sheet metal parts, equipment, and other products, **Figure 12-1**. Some are trained in one or two of these areas, while others are trained in all of them. During the design process, a sheet metal worker may read and interpret plans and specifications for a project. Using this information, the worker will determine the sequence and methods to use for fabrication, assembly, and installation. The worker may also need to select the gage and type of sheet metal to be used.

In the layout phase, sheet metal workers use their knowledge of math and layout techniques to develop patterns. The dimensions and reference lines from the patterns are then marked on the material. Patterns are developed and created either manually or with a computer. If they are developed by hand, a worker will use rulers, scribers, awls, and other measuring and marking tools to create a pattern.

During fabrication, parts are cut or stamped out and then assembled. All patterns are measured again before assembly to ensure accuracy. Each piece is then bent and connected. It may be connected using bolts, screws, welds, rivets, or using other joining techniques. Fabrication may be done in a shop or on the job site. If assembled in a shop, the parts will need to be moved to the job site for assembly.

At this point, installation can begin. Individual pieces are put in place and then assembled to create the finished product. For instance, the individual ducts in an HVAC system would be assembled on-site to create the finished system. For pieces that are already complete, they are moved into place and

Sheet Metal Processes and Tasks

Sheet Metal Process	Tasks Conducted during Process
Design	• Read and interpret plans and specifications for a project. • Determine sequence and methods for fabrications, assembly, and installation. • Select gage and type of sheet metal to use.
Lay out	• Develop patterns. • Mark dimension and reference lines on material.
Fabrication	• Remeasure pattern. • Cut out parts. • Assemble parts.
Installation	• Move assembled parts to job site. • Join individual pieces into completed product, if needed. • Move piece into place. • Finish attaching as needed.

Figure 12-1. Sheet metal work involves all of these processes. However, workers may or may not be trained to do all of them.

any additional assembly is then completed. Roofing systems, gutters, and downspouts are usually fabricated, assembled, and installed entirely at the job site.

Work Environment

Sheet metal work can be hazardous. Some of the dangers to workers include cuts from sharp metal, burns from soldering and welding, and falls from ladders and scaffolding. Workers must follow proper safety procedures and be alert and cautious at all times while on the job, **Figure 12-2**.

Sheet metal workers employed in shops enjoy proper lighting and ventilation, as well as protection from the elements. Those working on job sites may be working inside or outside. For example, installers of downspouts and gutters work outdoors, so they must deal with the weather. Those working inside do not have that worry. However, a building under construction will not have heating and cooling. Even in finished buildings, sheet metal workers often conduct their work in warehouses, attics, and on roofs.

Sheet metal work is physically demanding. You stand for long periods and lift heavy materials and finished pieces. You do considerable bending, lifting, standing, and climbing. Installers often work in close quarters or in awkward positions. Sheet metal workers in the manufacturing industry are subject to repetitive motions and a somewhat unchanging environment.

A 40-hour work week is standard in the sheet metal industry. ***Overtime*** may be required when meeting deadlines or working on a large job.

overtime: Time spent working beyond a typical 40-hour work week.

Training

Sheet metal workers can learn their trade either by working with a contractor who is willing to provide training or by entering an apprenticeship program that provides formal training.

On-the-job training provided by a contractor is an informal but effective and thorough way to learn all aspects of the sheet metal trade. On-the-job trainees will usually begin as helpers. They receive safety training and then learn to perform simple bending and cutting jobs. This allows them to begin work immediately. Helpers also perform other simple jobs like carrying parts and cleaning job sites. They observe and help experienced workers while learning about materials and tools. As their skills improve, they are allowed to work on more complicated tasks. Helpers continue to learn by observing

Figure 12-2. This sheet metal worker is properly clothed for the job.

their more experienced coworkers. Those who acquire their skill this way often take vocational school courses in mathematics or sheet metal fabrication to supplement their work experience. To become a journeyman, helpers must pass written examinations during their apprenticeships.

Many sheet metal contractors consider an apprenticeship the best way to learn the trade. Apprenticeship programs include approximately 4–5 years of comprehensive on-the-job training and classroom instruction. These programs are administered by local joint committees composed of both local and national union trade groups.

Sheet metal workers often choose to specialize in a specific type of sheet metal construction. Some of the more popular specialties include architectural sheet metal, sheet metal detailing, sign fabrication, and residential or commercial HVAC. Sheet metal workers also trained as welders are especially valuable in construction-related industries. They save time by completing work that might otherwise have to wait until a welder is available to complete the job. ***TAB technicians*** are sheet metal workers who specialize in testing, adjusting, and balancing existing HVAC systems. They ensure that systems are working properly and efficiently. This skill is very marketable as conservation of energy becomes increasingly important.

TAB technician: A sheet metal worker who specializes in the testing, adjusting, and balancing of existing HVAC systems to ensure they are working properly.

Qualifications

Sheet metal apprentices or helpers should be in good physical condition because the work is strenuous. They should enjoy working with their hands and have good eye-hand coordination. Sheet metal workers also need mechanical and mathematical aptitude.

Local apprenticeship committees require a high school education or its equivalent. Courses in algebra, trigonometry, geometry, mechanical drawing, and technical education provide a helpful background for learning the trade.

Learning does not have to end once a trainee or apprentice completes training. Workers often receive additional training to improve existing skills or to learn new ones. This is typically provided by a union or an employer. These workers can then move to supervisory positions.

Specialization in the areas mentioned earlier requires additional, formal training. On completion of the training, the worker receives certification in that sheet metal specialty.

Employment

Over 80 percent of sheet metal workers are employed in either construction-related industries or in manufacturing industries. Construction-related industries include residential and commercial building construction, HVAC, and kitchen equipment design. Examples of manufacturing industries include automobile and airline assembly lines.

Job Outlook

Average growth is projected for sheet metal work. Growth areas include residential and commercial HVAC, decorative sheet metal fabrication, and architectural restoration work. Employment will likely decrease in manufacturing areas as jobs continue to move to other countries and remaining plants operate more efficiently. Most job openings will result from the need to replace workers who retire or leave the occupation for other reasons.

Individuals trained through apprenticeship programs and those who complete advance training and certification will generally be in high demand. These workers will possess the skills needed to work in an ever-changing work environment.

Occupational Classifications

The sheet metal industry employs many people in a variety of occupations. While this text has focused on the job of sheet metal workers, many other jobs are made possible by the sheet metal industry, **Figure 12-3**. The sheet metal industry employs people in all the following areas:

- Professional and technical.
- Management.
- Administration.
- Sales.

Jobs in each group require different degrees of education and training in order to successfully perform the work. Individuals with specialized training and advanced education create an opportunity for themselves to find and keep secure employment. These types of jobs provide higher wages and additional benefits, such as retirement plans and insurance. Individuals with little or no education and training will have difficulty finding jobs that will provide long-term, secure employment. These individuals are often the last worker hired and the first worker laid off. Employers always prefer workers who have completed their high school education and have acquired a marketable skill.

Job Classifications

Classification Title	Qualifications	Responsibilities
Professional	• Bachelor's degree required	Design, develop, test, and evaluate processes, tools, systems, machines, and materials used in the metalworking industry.
Technical	• Some college preferred	Support those in the professional classification.
Management	• Bachelor's degree	Direct or plan the work of others.
Administration	• Some college preferred	Provide assistance to more senior employees in the department.
Sales	• Bachelor's degree preferred	Selling manufacturer's product to other businesses or industries.

Figure 12-3. The sheet metal industry employs a wide range of people besides sheet metal workers.

Professional and Technical

Metalworking industries employ many professionals and technical support staff, most notably in the engineering field. Types of engineers you might find in this industry include mechanical, industrial, manufacturing, and metallurgical engineers. These professionals use their skills to design, develop, test, and evaluate processes, tools, systems, machines, and materials used in metalworking industries. All of these occupations require a minimum of a ***bachelor's degree***. Many require advanced degrees. These occupations are highly specialized and technical in nature.

bachelor's degree: An academic degree awarded to individuals who complete an undergraduate course of study. It is awarded by a college or university.

Technical occupations work in support of professional occupations. These occupations are an outgrowth of today's complex society. Opportunities in this group include technical report writers, laboratory technicians, and quality-control workers. Advanced training, including some college, is required for these types of occupations.

Management

The metalworking industry employs all types of managers, from chief operating officers to human resource managers. The duties of a ***manager*** include planning, directing, coordinating, and budgeting the activities of workers for the benefit of the company as a whole. For example, an industrial production manager oversees the work activities of employees in a factory. The manager also makes sure that these workers have the materials they need to do their jobs properly. In general, managers direct or plan the work of others.

manager: An employee who is responsible for assigning and directing work for other employees.

Most managers have bachelor's degrees. It is still possible, however, for a person to be promoted "through the ranks" to become a manager in a specific occupational specialty. The number of individuals employed as managers has increased as the complexity of industrial organizations has increased.

Administration

Administrative support workers provide assistance to more senior employees in their department. This support work varies from position to position, but often includes duties such as data entry, customer service and support, scheduling and organizing meetings.

Administrative support workers usually have at least a high school education. Many administrative workers have additional business school preparation.

Sales

Sales-related occupations involve millions of workers in the labor force. At least half of this number are involved in retail sales. In the metalworking industries, this occupation is involved primarily with the sale of manufacturers' products, **Figure 12-4**. Salespeople do the actual selling of the product. They deal with the public every day as a part of their job.

College graduates are usually preferred for training as salespeople in the metalworking industries. Employers find that college-educated people are extremely comfortable dealing with company officials. Individuals must know the product they are selling to be successful. For example, an engineer familiar with a specific type of machinery may be hired by a company to sell that machinery.

Additional Occupational Information

An excellent source for occupational information is the Bureau of Labor Statistics Occupational Outlook Handbook. This handbook is updated every two years. The information is very reliable and includes training and education needed, earnings, expected job prospects, what workers do on the job, and working conditions.

Another excellent source for occupational information is at your school. Guidance counselors and teachers often have a great deal of useful information about employment. Trade union members and state employment service counselors are another excellent source of information for local employment opportunities. Finally, many community and junior colleges offer information on employment opportunities in metalworking occupations. Many of these colleges offer programs in a number of technical areas.

Figure 12-4. Salespeople must be knowledgeable about their products and able to talk easily with other people.

The World of Work

Eventually, you will become a member of the working world. In order to be successful, you must have some idea of what your employer will expect of you. As a metalworker, you will be expected to maintain and improve your skills. Many employers will expect you to keep up with the latest information in the field. You may also be asked to join a professional organization to stay informed. There are few skilled trades that do not require additional study and training. The best way for a company to stay competitive is to have well-trained, informed employees.

As an employee, your responsibilities will also include the following:

- Arrive on time, ready to start work. It is best to be a few minutes early.
- Avoid excessive absenteeism. If you routinely fail to come to work, you will probably be fired.
- Never come to work under the influence of alcohol or drugs. This is a fireable offense. It also puts you and others at risk.
- Be cordial and respectful to fellow employees, especially those you work with closely. The inability to get along with others is also a fireable offense.
- Take care of the tools and equipment assigned to you.
- Keep your work area neat. A neat work area is a safe work area.
- Be enthusiastic about your job. Have a plan for the best, most efficient way to complete your work. Employers appreciate an employee who shows ***initiative***.

initiative: The ability to start and finish a task or job.

- Know your rights as an employee. If you are entitled to health insurance or a retirement plan, educate yourself about these programs. Know what is available to you and how to use it. Well-informed employees are a benefit to their employers.

Summary

Sheet metal workers perform a variety of tasks in the process of doing their jobs. They design, lay out, fabricate, install, and maintain sheet metal products. In order to perform these tasks, they must be well trained and dependable. The need for sheet metal workers will continue to increase in the future due mostly to the retirement of current sheet metal workers. A career in sheet metal work provides an excellent living and job security.

The sheet metal industry also provides employment opportunities to others. Professional and technical workers, managers, administrative support workers, and salespeople can all find work in the sheet metal industry.

Being the employee of a company involves responsibilities. Employees must be well trained, knowledgeable about their jobs, and willing to learn new skills when necessary. Additional responsibilities include showing up for work every day, on time, with a good attitude. These actions let your employer know that you are a valuable employee.

Internet Resources

U.S. Department of Labor, Bureau of Labor Statistics

www.bls.gov/oco/ocos214.htm

International Training Institute for the Sheet Metal and Air Conditioning Industry

www.sheetmetal-iti.org

Sheet Metal Workers International Association

www.smwia.org

Outside Activities

1. Visit the Web site of the Bureau of Labor Statistics. Find the entry for sheet metal workers and read it. Write a short report based on this information. In your report, include the current salary range of sheet metal workers and the number employed in this field.
2. Invite a sheet metal apprentice to your classroom to discuss the apprenticeship process. Before the visit, assemble a list of questions you and your classmates would like to ask the apprentice. After the visit, discuss the pros and cons of an apprenticeship versus on-the-job training.

The International Training Institute for the Sheet Metal and Air Conditioning Industry maintains a Web site for both students and professionals in the sheet metal trades.

Name ______________________ Class ______________ Score __________

Chapter 12
Sheet Metal Careers

Review Questions

Write your answers in the spaces provided.

Match each of the following descriptions with the correct term.

1. Move assembled parts to job site.	A. Fabrication.	1. ______________
2. Mark dimension and reference lines on material.	B. Installation.	2. ______________
3. Read and interpret plans and specifications for a project.	C. Design.	3. ______________
4. Assemble parts.	D. Layout.	4. ______________

5. Which of the following is *not* true of the sheet metal work environment? 5. ______________
   A. The work is physically demanding.
   B. The work can be hazardous.
   C. Workers often work in warehouses, attics, and on roofs.
   D. Overtime is never required.
6. True or False? In order to become a journeyman, helpers are required to do a variety of jobs but are not required to pass any written examinations. 6. ______________
7. Apprenticeship programs include ____ years of both on-the-job training and classroom instruction. 7. ______________
8. What is a TAB technician? ______________________________

   ____
9. What qualities and skills does a successful sheet metal worker possess? ______________

   ____

   ____

   ____
10. In which two industries are the majority of construction workers employed? ______________

    ____

    ____

11. An academic degree awarded to an individual who completes an undergraduate course of study is called a(n) _____ degree. 11. __________
    A. associate's
    B. bachelor's
    C. graduate
    D. postgraduate

12. _____ occupations work in support of professional occupations. 12. __________

13. What are the duties of a manager? __________

____

14. True or False? The ability of employees to get along with each other is *not* important to employers. 14. __________

15. What is meant by *initiative?* __________

____

____

____

Experiments

There are five experiments in this section of the textbook. Each experiment starts with a Purpose statement. This statement is followed by Technical Information and a list of Materials needed to conduct the experiment. In the Procedure section, follow all instructions closely and answer the questions as you encounter them. Questions in the Summary section will give you the chance to sum up the main points of what you learned from the experiment.

The experiments in this textbook were designed to allow you the opportunity for hands-on work and to further your knowledge of sheet metal. Additional experience using materials and tools helps you to gain a larger understanding of metals in general and also provides you with valuable hands-on practice.

Experiment 1 Soldering and Fluxes
Experiment 2 Working Drawings
Experiment 3 Spot Welding
Experiment 4 Ferrous and Nonferrous Metals
Experiment 5 Cast Iron and Steel

Experiment 1
Soldering and Fluxes

Purpose

To observe the effects of flux in soldering.

Technical Information

Copper becomes coated after exposure to the atmosphere for a period of time. This coating prevents a good solder joint from being made. To make a solid joint, the metal must be cleaned and fluxed to prevent oxidation during the heating process.

Flux is a chemical applied to copper before soldering. The flux cleans the copper, prevents oxidation of the copper during the soldering process, and promotes adhesion of the solder to the copper. Soldering paste is a type of thick, rosin-based flux that can be applied to the copper with an acid brush. Heat is required to melt the solder to hold the pieces of metal together.

Safety Note

Safety is extremely important. The following guidelines will help you complete this experiment safely. Wear leather work gloves and eye protection at all times during this experiment. If you are using a propane, acetylene, or MAPP torch, use a spark striker to light it. Never light a torch with matches or a lighter! Make sure there are no people or combustible materials directly in front of the torch before lighting it. Always turn the torch gas valve(s) off when the torch is not in use.

Materials

- Leather gloves
- Safety glasses
- 2 pieces copper, 26 gage, 1″ × 4″
- 1 dust cloth
- 1 piece of emery cloth or sandpaper
- 2 pairs pliers, slip-joint or long nose
- 1 oz. rosin flux
- 1 acid brush
- 1 piece 50-50 wire solder
- 1 source of heat (open flame or soldering iron)

Name ______________________________ Score ______________

Procedure

1. Sand one strip of copper on one side.
2. Wipe the sanded area with the dust cloth until it is clean.
3. Apply flux to the cleaned side of the copper strip with an acid brush.
4. Do not sand or clean the other strip of copper.
5. Make sure there are no combustible materials in the immediate work area. Then, apply heat by holding the copper above the flame or soldering tip. The flux will melt and flow. Make sure there is nothing under the copper strip that can be damaged by dripping flux.
6. Apply solder to the fluxed area, **Figure EX 1-1.**

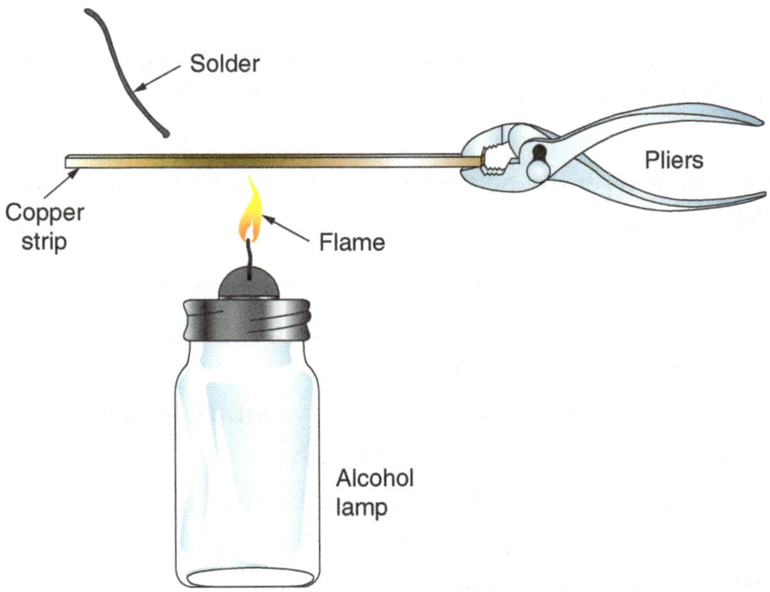

Figure EX 1-1. Apply solder to the fluxed area on the copper strip.

7. Solder will melt and coat the copper strip.
8. Set the copper aside to cool. Make sure the cooling area is free of combustible materials.
9. Place the other strip of copper (uncleaned) in contact with the heat source. Be sure the copper strip is held with the pliers, **Figure EX 1-2.**

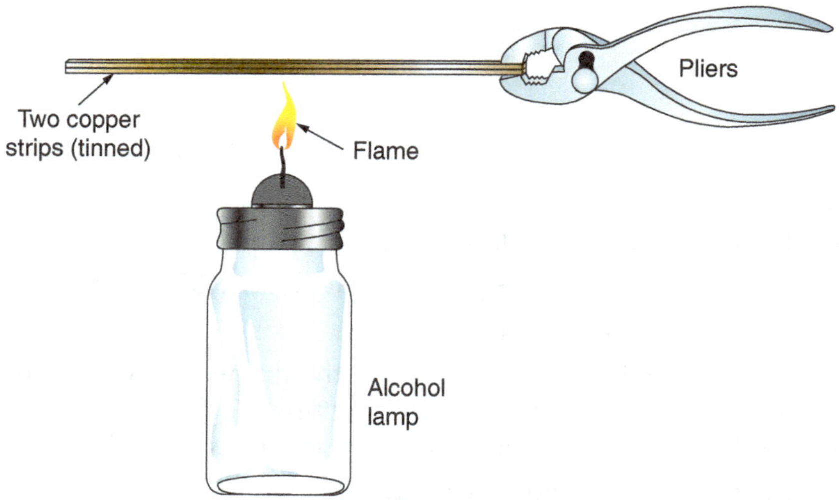

Figure EX 1-2. With the pliers, hold the two copper strips above the flame.

10. Hold the copper in contact with the heat source until the copper is hot enough to melt the solder. Make sure there is nothing under the copper strip that could be damaged by dripping solder.
11. Observe the solder flow over the heated area of the copper. Does it flow or does it form little balls of solder? ____

    ____

    ____
12. Permit the uncleaned piece of copper to cool.
13. Sand or clean the surface with sandpaper or emery cloth. Clean with the dust cloth.
14. Apply flux to the cleaned area using the acid brush.
15. Heat the cleaned and fluxed piece of copper. Make sure there is nothing under the copper strip that could be damaged by dripping flux or solder.
16. Apply solder to the heated fluxed area.
17. Permit the solder to melt and coat the piece of copper.
18. Set this piece aside, and permit it to cool. Make sure the cooling area is free of combustible materials.
19. Pick up the two pieces of solder coated copper, and place them with the solder-coated sides touching one another.
20. With the pliers, hold the two pieces in contact with the heat source. Refer again to **Figure EX 1-2.**
21. Keep the copper strips in contact with the heat source until the solder on the two pieces reaches melting temperature.
22. Remove the two pieces from the heat source, but continue clamping with the pliers until the solder cools.
23. Permit the copper pieces to cool until they can be picked up with your bare hands.

Name ______________________ Score ____________

24. Test the solder joint by pulling on each piece of copper with a pair of pliers to try to separate them. Can they be separated?

____

____

25. Soldering has been used to join two pieces of metal. Soldering not only is useful in preventing copper oxidation but also provides a good mechanical connection.

Summary

1. Why is a clean piece of copper needed for a strong solder joint? ______________________

____

____

____

____

____

____

2. Why was the flux needed? ______________________

____

____

____

____

____

3. What job did the flux perform? ______________________

____

____

____

____

4. Why did the uncleaned piece of copper react differently than the cleaned and fluxed piece?

____

____

____

____

5. What are the two most important reasons for soldering? ______________________________

6. Why is this type of soldering called sweat soldering? ______________________________

Experiment 2
Working Drawings

Purpose

To learn more about the basics of working drawings.

Technical Information

Working drawings are used to plan projects. The sheet metal worker visualizes the appearance of the project and then sketches working drawings to aid in assembly. With practice, the worker can quickly and accurately draw nearly any piece needed to complete the job.

Working drawings use several different types of lines. Each line has a definite form or weight (width). These lines provide meaning to the drawing. The Alphabet of Lines provides examples of the different lines and their uses.

Materials

2 pieces of drawing paper
1 2H pencil
1 30-60° triangle
1 engineer's scale

Procedure

1. Various lines used when drawing patterns are shown in **Figure EX 2-1.**

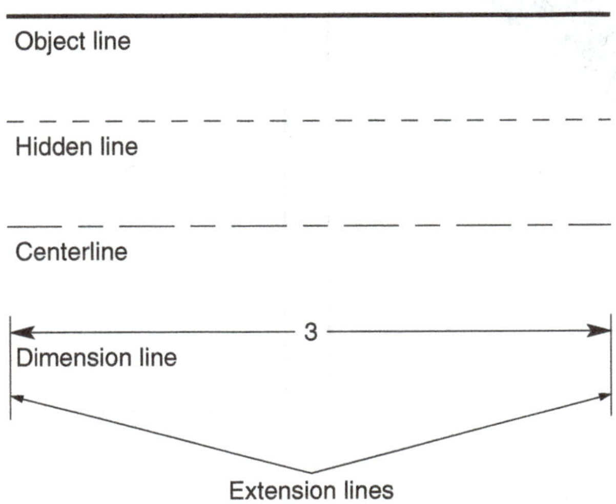

Figure EX 2-1. These lines are some of the most commonly used in working drawings.

Name ______________________________ Score ____________

2. Notice the varying line forms and weights of each line.
3. Following the lettering guide in **Figure EX 2-2,** write the name of each line shown in Step 1.

ABCDEFGHIJKLMNOPQRSTUVWXYZ..234567890 -- VERTICAL
ABCDEFGHIJKLMNOPQRSTUVWXYZ..234567890 -- INCLINED

Figure EX 2-2. Vertical and inclined single-stroke Gothic style lettering.

____

____

____

____

____

4. Sketch the screwdriver alongside the sample drawing.

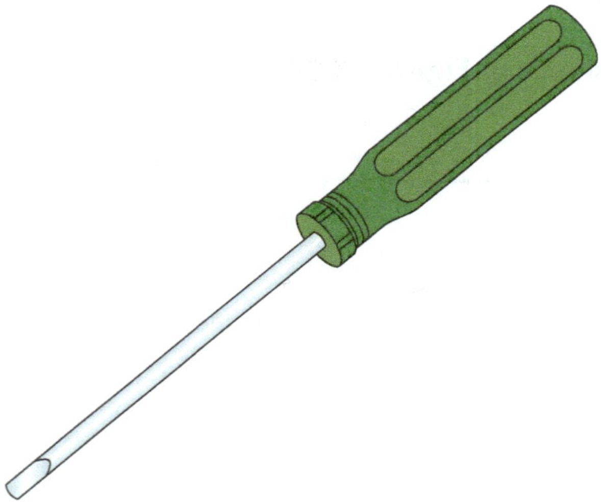

Name ______________________________ Score ____________

5. Sketch three views of each of the following objects. Label these dimensions, where appropriate:
   - Depth, designated with a *D*.
   - Width, designated with a *W*.
   - Height, designated with an *H*.

Label each view as shown in the following example.

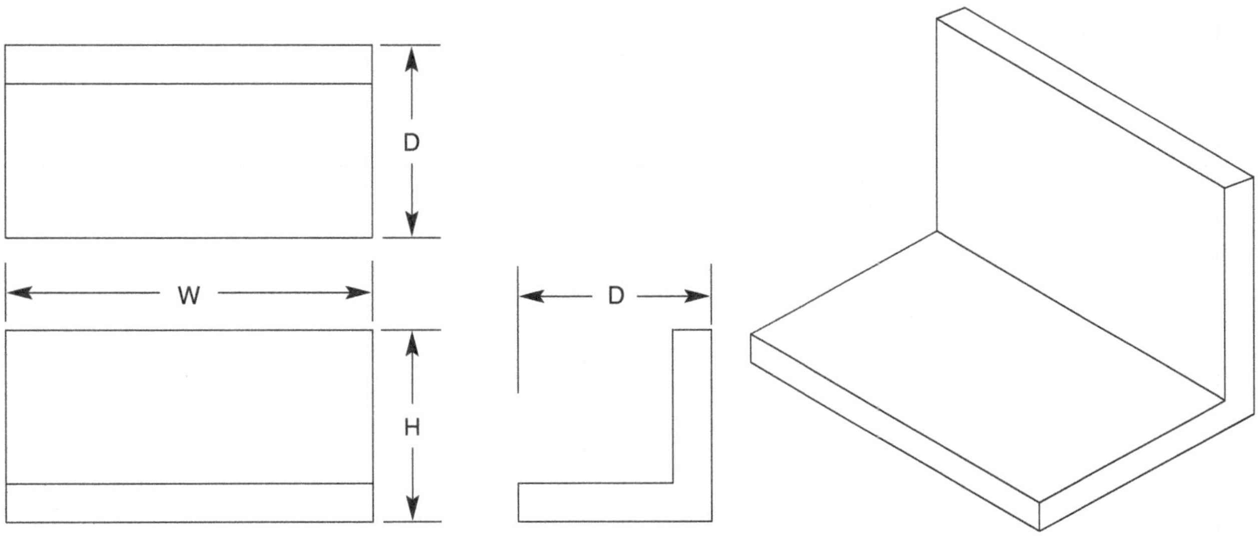

6.

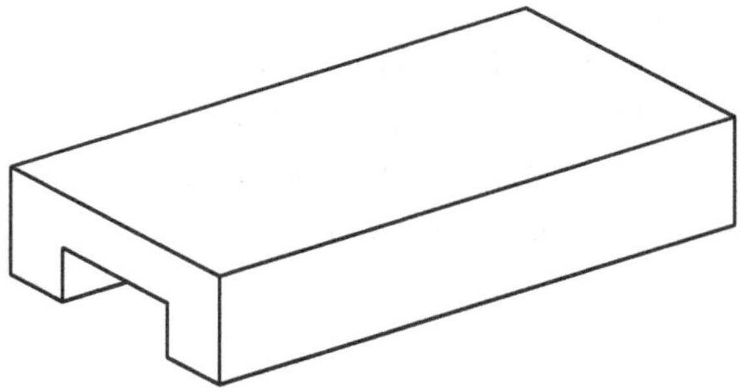

Name ______________________________ Score ______________

7.

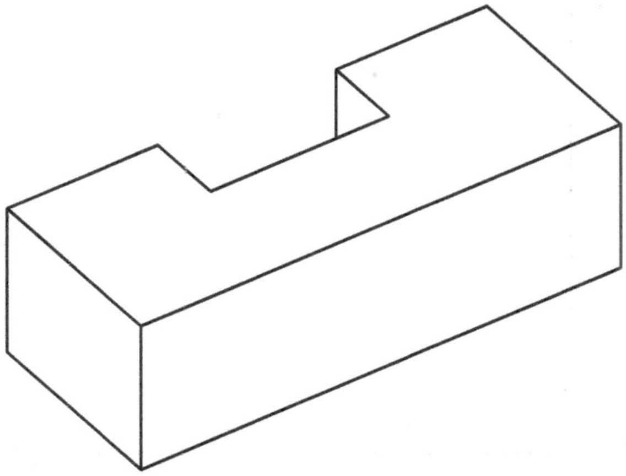

8.

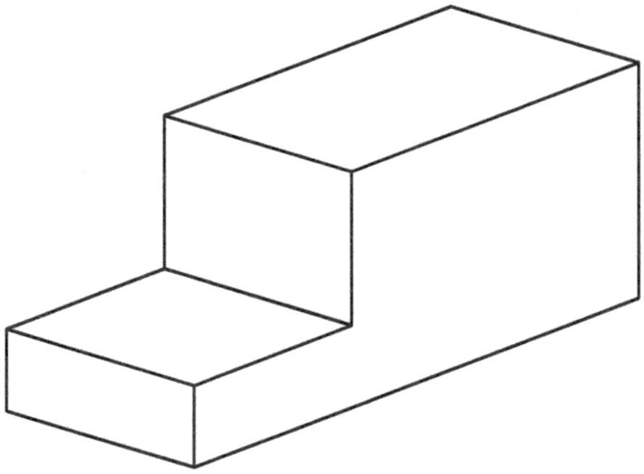

Name ______________________________ Score ____________

9.

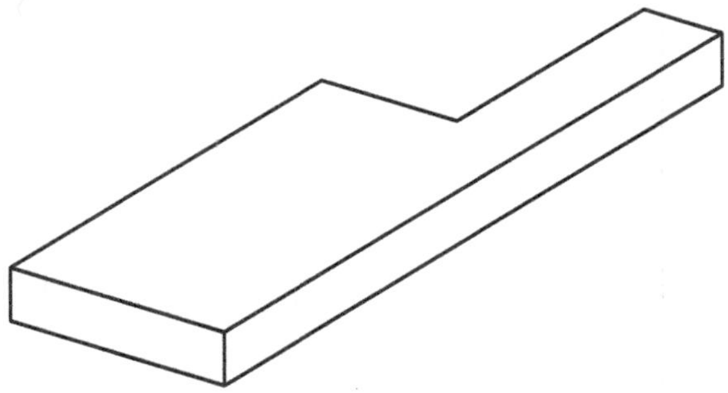

10.

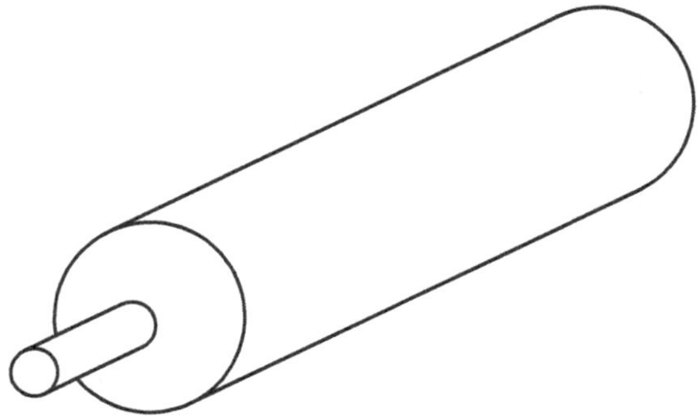

Name ______________________________ Score ______________

11.

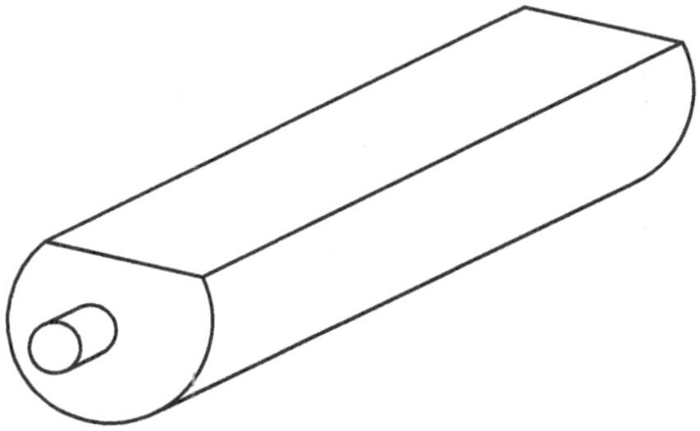

12.

Name ______________________________ Score ______________

13.

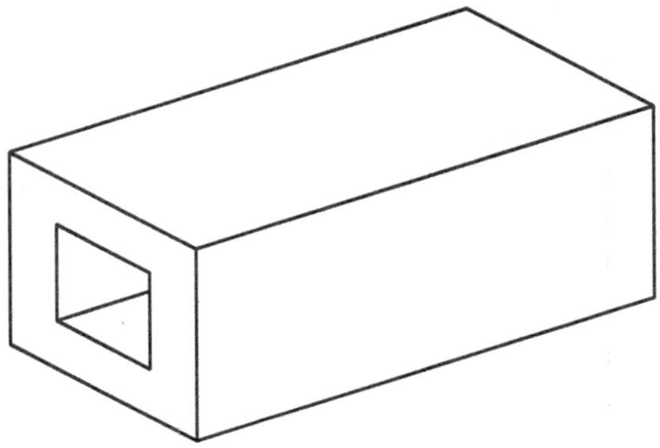

14.

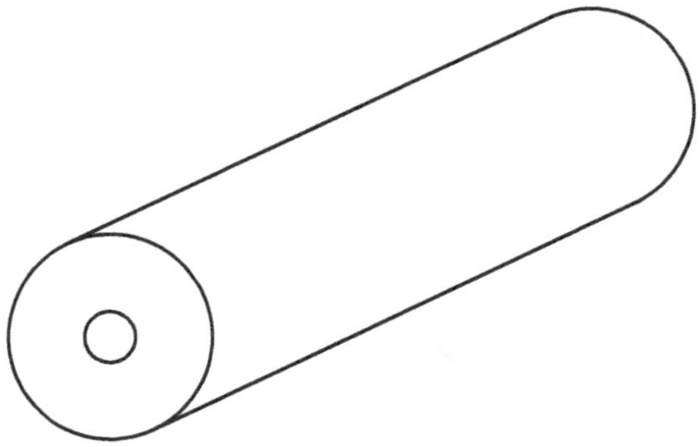

Experiment 3
Spot Welding

Purpose

To learn more about the basics of spot welding.

Technical Information

A spot welding machine is nothing more than an electrical transformer that steps down voltage, thereby increasing the amount of current available in the secondary circuit. The transformer in a spot welding machine produces enough current to melt metal. By confining current flow to a spot or small area, the machine can weld two pieces of metal together from the inside. The current flowing between a resistance welding machine's electrodes can very quickly raise the temperature of the sheet metal between the electrodes to its melting point.

Materials

- 2 pieces sheet metal, 2″ × 4″, 26 gage
- 1 portable spot welder
- Goggles
- Leather gloves
- 2 pairs pliers, slip-joint
- 1 clamp

Procedure

1. To begin this activity, have your instructor demonstrate the proper use of the portable spot welder and the heat settings you will need for welding the pieces of sheet metal. Wear eye protection and leather gloves at all times during this experiment.
2. Hold the two pieces of metal together with a clamp or pair of pliers.
3. Place the spot welder tips over the sheet metal. If the spot welder has a nonmovable jaw, place it in contact with the sheet metal. Make sure the space between the electrode tips and the sheet metal is free from obstruction.
4. Depending on the design of the welder, move the handle, pull the trigger, or push the switch on the welder so that the tips close on the sheet metal.
5. Release the handle, trigger, or switch and withdraw the spot welder from the assembly. Return the spot welder to its designated storage location.
6. Use the two pairs of pliers to try to pull the two pieces of sheet metal apart.

Name ______________________________ Score __________

7. Were you able to separate the two pieces?______________________________

____

8. Now try a hammer and chisel with the sheet metal clamped in a bench vise and see if the chisel can cause the two pieces to separate.
9. Did the chisel break the weld joint? ______________________________

____

10. What happened when the metal separated?______________________________

____

____

____

____

Summary

1. What happened when the two pieces of metal were clipped together and the switch on the welder was closed?______________________________

____

____

____

2. Did the welder burn a hole in the sheet metal? ______________________________

____

____

____

3. Why didn't the two pieces of metal separate when pulled by the pliers?______________________________

____

____

____

4. Why would you use spot welding to hold two pieces of sheet metal together?______________________________

____

____

____

Name ______________________________ Score ____________

5. How is the spot welder most frequently used? ______________________________

____

____

____

____

Experiment 4
Ferrous and Nonferrous Metals

Purpose

To learn how to tell the difference between ferrous and nonferrous metals.

Technical Information

Ferrous metals contain iron. Nonferrous metals do not contain iron. Iron is a magnetic material, meaning it is attracted to magnets. Nonferrous metals (aluminum, gold, silver, brass, copper, bronze, zinc, lead) are not attracted to magnets. Cobalt and nickel are also attracted to magnets, but are not considered ferrous metals.

Materials

- 1 bar magnet
- 1 aluminum strip, 1″ × 5″
- 1 lead strip, 1″ × 5″
- 1 tin strip, 1″ × 5″
- 1 zinc strip, 1″ × 5″
- 1 container of small carpet tacks
- 1 container of small brass tacks
- 1 container of small brads and staples

Procedure

1. Hold the bar magnet in one hand.
2. Spread the container of brass tacks onto a table top, **Figure EX 4-1.**

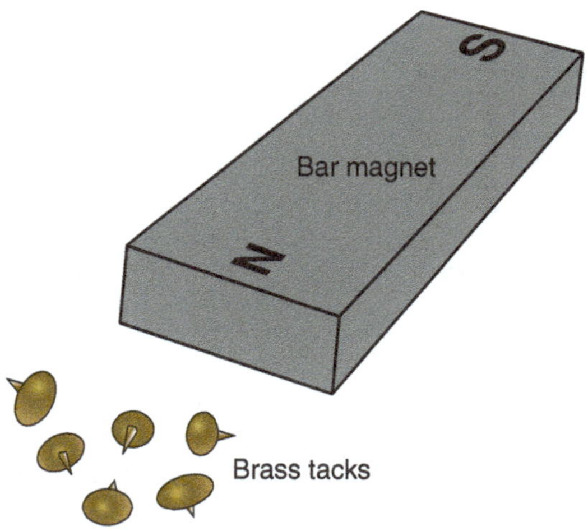

Figure EX 4-1. Brass tacks.

Name ______________________________ Score ______________

3. Are the tacks attracted to the magnet when the magnet is held near the tacks? ____________
4. Are the brass tacks ferrous or nonferrous? ______________________________
5. Spread the container of carpet tacks onto a table top, **Figure EX 4-2.**

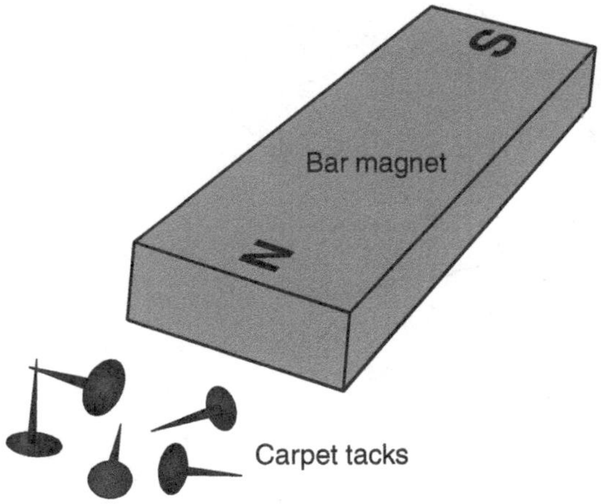

Figure EX 4-2. Carpet tacks.

6. Are the carpet tacks attracted to the magnet? ______________________________
7. Are the carpet tacks ferrous or nonferrous? ______________________________
8. Spread the container of small brads onto a table top, **Figure EX 4-3.**

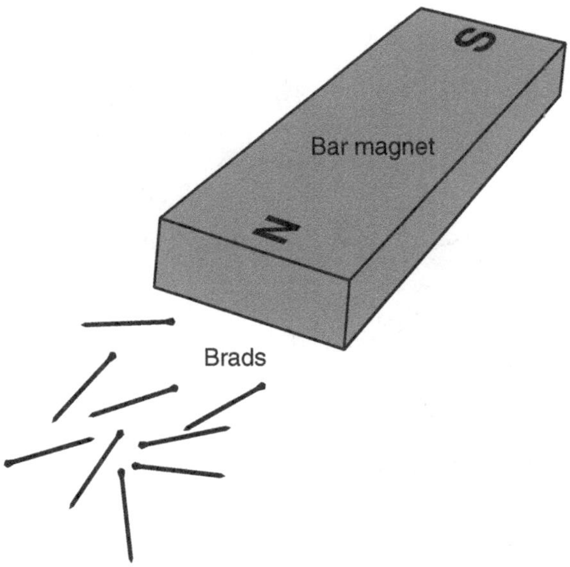

Figure EX 4-3. Brads.

9. Are the brads attracted to the magnet? ______________________________
10. Are the brads ferrous or nonferrous? ______________________________

11. Spread the container of nails onto a table top, **Figure EX 4-4.**

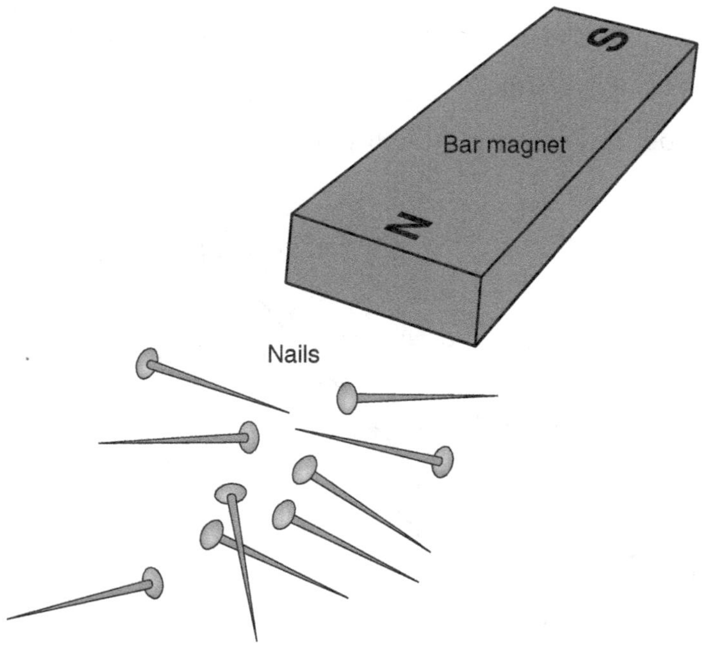

Figure EX 4-4. Nails.

12. Are the nails attracted to the magnet? ______________________________

13. Are the nails ferrous or nonferrous? ______________________________

14. Spread the container of staples onto a table top, **Figure EX 4-5.**

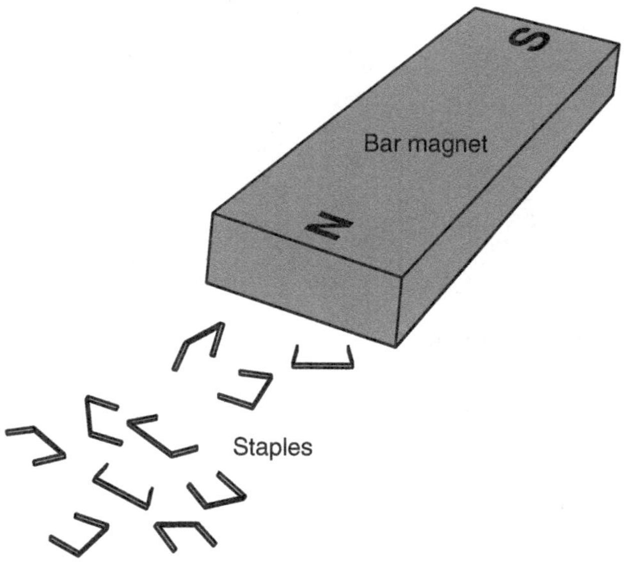

Figure EX 4-5. Staples.

15. Are the staples attracted by the magnet? ______________________________

16. Are the staples ferrous or nonferrous? ______________________________

Name ______________________ Score ____________

17. Place a strip of aluminum on the table top. Move the magnet near the piece of aluminum, **Figure EX 4-6.**

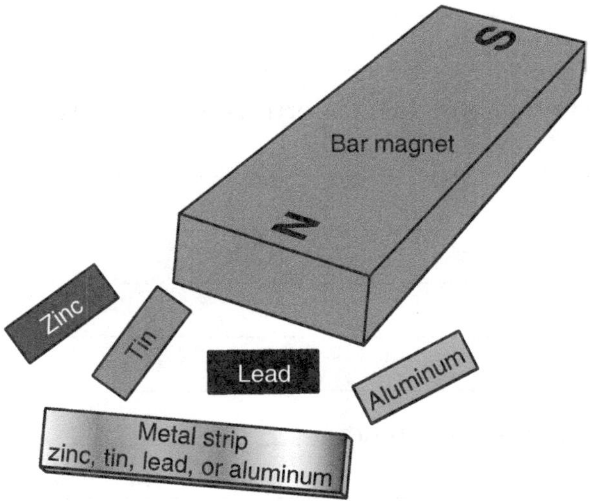

Figure EX 4-6. Aluminum, zinc, tin, or lead strip.

18. Is the aluminum attracted to the magnet? ____________
19. Is the aluminum ferrous or nonferrous? ____________
20. Place a piece of zinc on a table top. Move the magnet near the piece of zinc as you have done previously.
21. Is the zinc attracted by the magnet? ____________
22. Is the zinc ferrous or nonferrous? ____________
23. Place a piece of tin on a table top. Again, move the magnet near the piece of tin.
24. Is the tin attracted to the magnet? ____________
25. Is the tin ferrous or nonferrous? ____________
26. Finally, place the piece of lead on the table top as you have done before.
27. Is the piece of lead attracted by the magnet? ____________
28. Is the lead ferrous or nonferrous? ____________

Summary

1. Which metals are attracted to the magnet? ____________

____

____

2. Which metals were ferrous? ____

3. Which metals were nonferrous? ____

4. How can you determine whether a metal is a ferrous or a nonferrous metal? ____

5. What other metals are attracted by a magnet? ____

6. Does this mean they are ferrous metals? ____

7. How reliable is this test for ferrous and nonferrous metals? ____

8. In what type of situations would a sheet metal worker need to use this method to determine if something is ferrous or nonferrous? ____

Experiment 5
Cast Iron and Steel

Purpose

To distinguish between iron and steel.

Technical Information

Iron is a chemical element. Its symbol is Fe. Iron is silvery-white in color in its pure state and is lustrous and comparatively soft.

Pure iron is found in quantities sufficient only for laboratory experiments. In its natural state or in the state it is seen most commonly, it is found combined with other chemical elements. Pig iron is the type of iron most commonly found, with elements such as manganese, silicon, phosphorus, sulfur, and carbon. The cast iron made from melted pig iron usually is brittle and quite hard.

Steel contains iron and carbon. The quantity of carbon in the mixture determines the type of steel and its use. A commercial, plain-carbon steel may contain as much as 95.4 percent iron. It also contains about 1.7 percent carbon and about 0.6 percent manganese. Other elements are found in steel in smaller percentages.

The higher the carbon content of the steel, the harder and stronger the steel. Steels are classified as low-carbon, medium-carbon, and high-carbon steels. The low-carbon steels contain about 0.25 percent carbon; the medium-carbon steel contains 0.25 to 0.50 percent carbon; and the high-carbon steels contain 0.50 to 1.70 percent carbon.

The cast iron used commercially is remelted pig iron and scrap steel. It is cast into various shapes and works well with metalworking tools. The hardness of the metal is adjusted by changing the elements in the mixture. Heat treatment also can change the hardness. Cast iron is used in many parts of automobile engines, the heavy bases for various types of equipment, and the burner rings for gas ranges in the kitchen.

Soft steel, or low-carbon steel with less than 0.15 percent carbon, is neither as hard nor as brittle as cast iron. This type of steel is used in automobile bodies and fenders, washing machine tubs, wire, and nails, as well as in other sheet metal applications. When more carbon is added (up to 0.25 percent), the steel becomes harder and stronger. Some uses for this steel include beams in buildings and bridges, and axles of railway cars.

Materials

- Safety glasses
- 1 cast iron rod, 1/4″ × 1/4″ (cast iron welding rod can be used)
- 1 soft-steel nail, head and point removed
- 1 file
- 2 pairs of slip-joint pliers

Procedure

1. Put on safety glasses.
2. Use the file to cut one-third the distance through the midsection of both the rod and the nail, **Figure EX 5-1.**

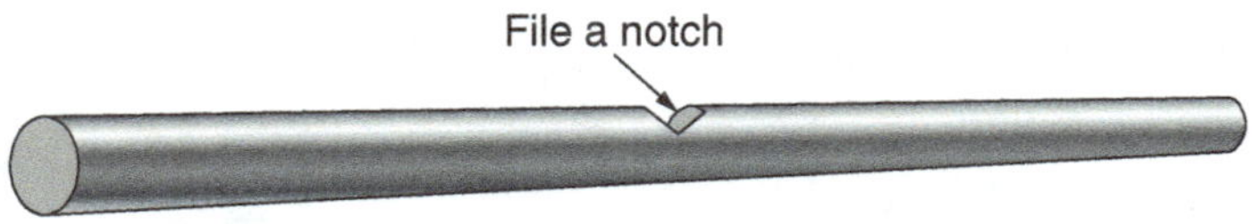

Figure EX 5-1. File a notch near the center section of the rod.

3. Grasp each end of the cast iron rod with the slip-joint pliers.
4. Bend the cast iron rod into a U shape as shown in **Figure EX 5-2.**

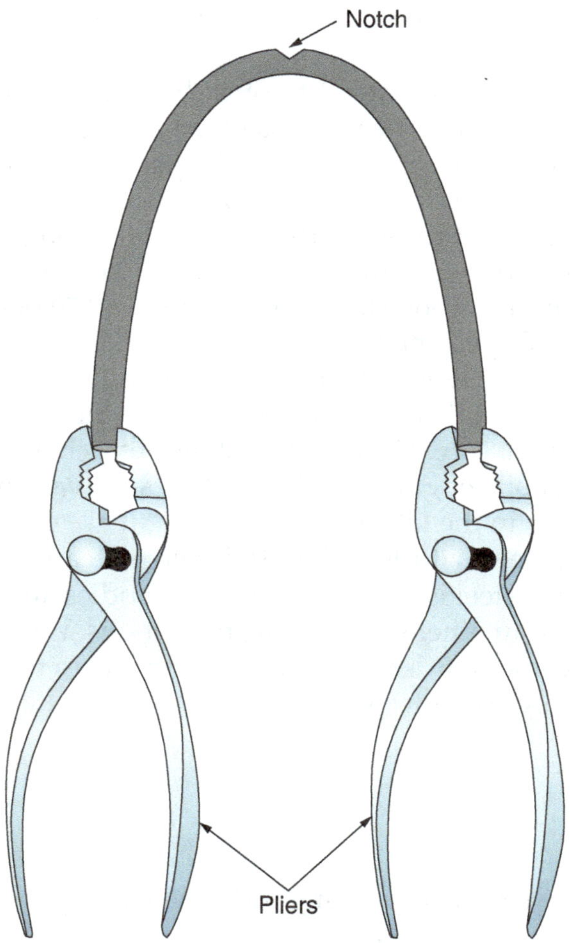

Figure EX 5-2. Using the two pliers, bend the rod into a U shape.

Name ______________________________ Score __________

> **Safety Note**
>
> Bend the rod away from your body or toward your feet. Wear safety glasses.

5. What changes occurred in the cast iron as the rod was being bent? ______________________________

____

____

____

6. Use the two pliers in the same manner as in Step 3, grasping each end of the soft-steel nail.
7. Bend the soft-steel nail into a U shape.

> **Safety Note**
>
> Bend the rod away from your body and toward your feet. Wear safety glasses.

8. Describe what happened to the soft-steel nail as it was being bent. ______________________________

____

____

____

____

Summary

1. Which material was easier to cut with the file? ______________________________

____

2. Was the soft-steel nail harder than the cast iron rod? ______________________________

____

3. Which piece was most brittle? ______________________________

____

____

4. What products are made using cast iron? ______________________________

____

____

____

5. What products are made using soft steel? ____

____

____

____

6. Which of these two is used in the form of sheets? ____

____

Projects

There are five projects in this section of the textbook. Each project begins with a brief introduction, followed by a bill of materials needed for construction of the project. Each project also includes a pattern for the item being made, as well as a detailed list of steps to follow when making the project.

The projects in this section were designed to give you the opportunity to work with tools and machines commonly used in the sheet metal trade. Working with these tools and machines helps you to increase your knowledge of sheet metal and its characteristics. It also improves the skills you need to work in the sheet metal profession.

Project 1
Basketball Hoop

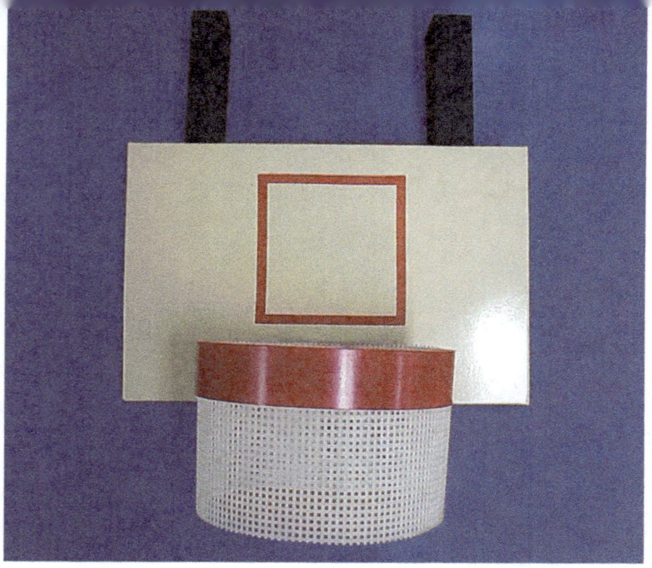

This basketball hoop can clip onto any door up to 1 3/4″ (45 mm) thick, without hindering the operation of the door. A 4″ (100 mm) foam ball can be purchased to use with this hoop.

Bill of Materials			
Item	**Quantity**	**Material Description**	**Size**
Backboard	1	28 gage (or thicker) galvanized sheet metal	8″ × 12″ (203 mm × 305 mm)
Door bracket	2	28 gage (or thicker) galvanized sheet metal	1″ × 12″ (25 mm × 305 mm)
Rim	1	28 gage (or thicker) galvanized sheet metal	1″ × 20″ (25 mm × 508 mm)
Red trim	4	1/4″ (7 mm) red automotive pinstriping	1/4″ × 4″ (7 mm × 100 mm)
Net	1	Plastic mesh	4″ × 19″ (100 mm × 482 mm)

Procedure

All of the information needed to build this basketball hoop can be found in either the bill of materials or the project pattern. Refer frequently to both of these resources while building this project.

Backboard

1. Cut out one backboard piece from the sheet metal in the size indicated in the bill of materials.
2. Apply blue layout die over the sheet metal piece.

Project Pattern

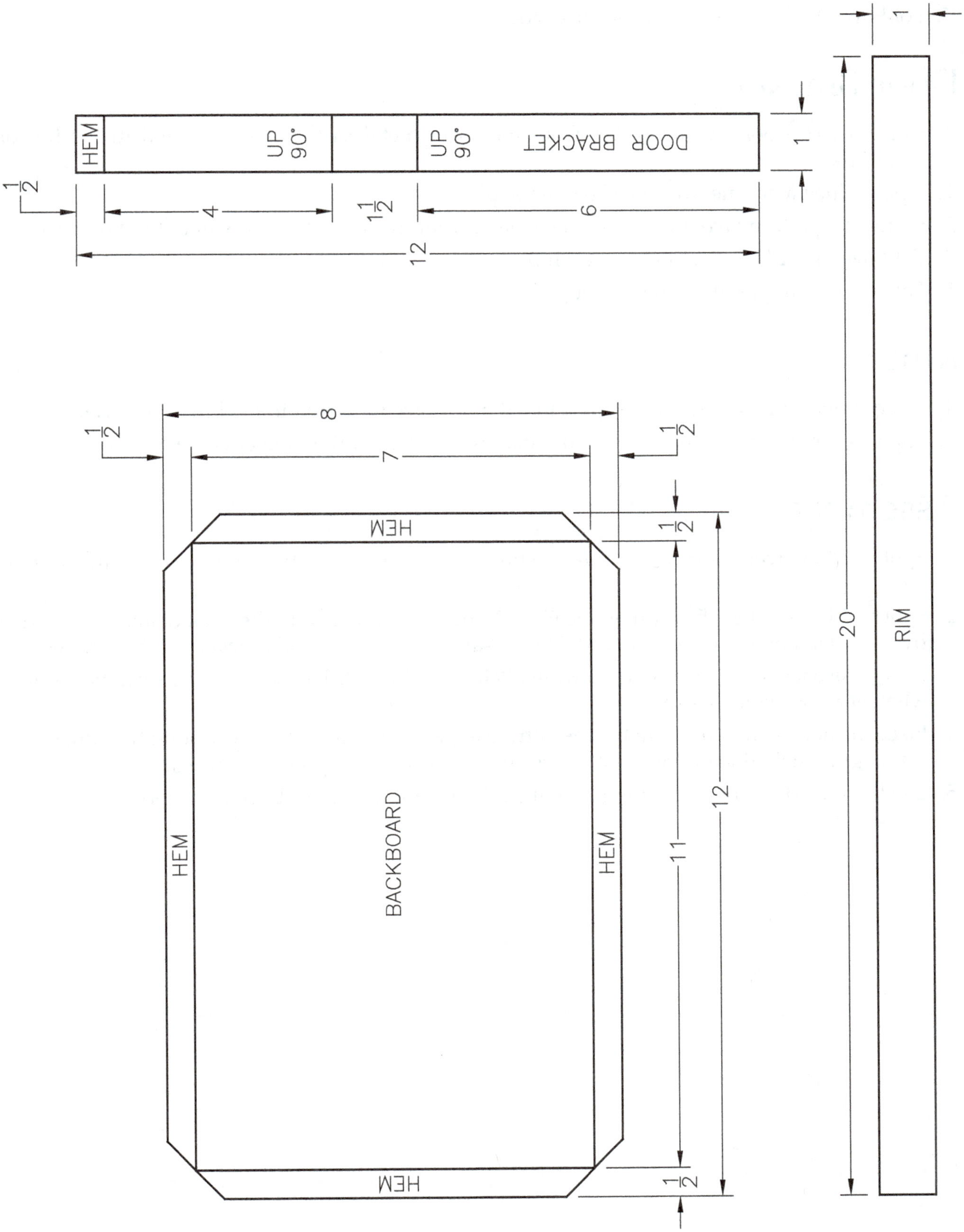

3. Scribe the pattern layout lines into the dried layout die. Cut out the pattern.
4. Notch all four corners at 45° angles.
5. Fold over 1/2″ (13 mm) hems on each side.

Door Bracket

1. Cut out two door bracket pieces from the sheet metal in the size indicated in the bill of materials.
2. Apply blue layout die over the sheet metal pieces.
3. Scribe the pattern layout lines into the dried layout die on each piece. Cut out the pattern.
4. Fold over a 1/2″ (13 mm) hem on each piece.
5. Fold two 90° angles in each bracket piece.

Rim

1. Cut out one rim piece from the sheet metal in the size indicated in the bill of materials.
2. Slip roll the 20″ (508 mm) piece until it makes a 6″ (152 mm) diameter circle.

Assembly

1. Spot weld or rivet the rim onto the backboard, 1 1/2″ (38 mm) from the center of the bottom edge.
2. Spot weld or rivet 2″ (51 mm) of one 6″ (152 mm) door bracket on the back of and 2″ (51 mm) from the left edge of the backboard. Do the same with the second bracket, on the right edge.
3. Prime the backboard, then paint it white. Paint the rim red. The door brackets can be painted whatever color you choose.
4. Place the netting into the rim. Let the netting expand until it fits tightly around the rim. Using a glue gun, melt glue on the plastic netting in three or four spots to secure it.
5. Use the four strips of pinstriping to make a 4″ (100 mm) square above the hoop.

Project 2
Dust Pan

This medium sized dust pan can be hung up in any closet. This practical project can be used by everyone. Tabs can be riveted or spot welded together, depending on equipment and materials that are available.

Bill of Materials			
Item	**Quantity**	**Material Description**	**Size**
Body	1	28 gage (or thicker) galvanized sheet metal	10″ × 12″ (254 mm × 305 mm)
Handle	1	28 gage (or thicker) galvanized sheet metal	3″ × 6″ (76 mm × 152 mm)

Procedure

All of the information needed to build this dust pan can be found in either the bill of materials or the project pattern. Refer frequently to both of these resources while building this project.

Body

1. Cut the body from the sheet metal material in the size indicated in the bill of materials.
2. Apply blue layout die over the sheet metal piece.
3. Scribe the pattern layout lines into the dried layout die. Cut out the pattern.
4. Cut out corners.
5. Fold over hems.
6. Fold up tabs and rear side approximately 100°.
7. Fold up sides 90°, until they touch the back. If necessary, push in the rear side a few degrees so the tabs will clear the back of it.

Project Pattern

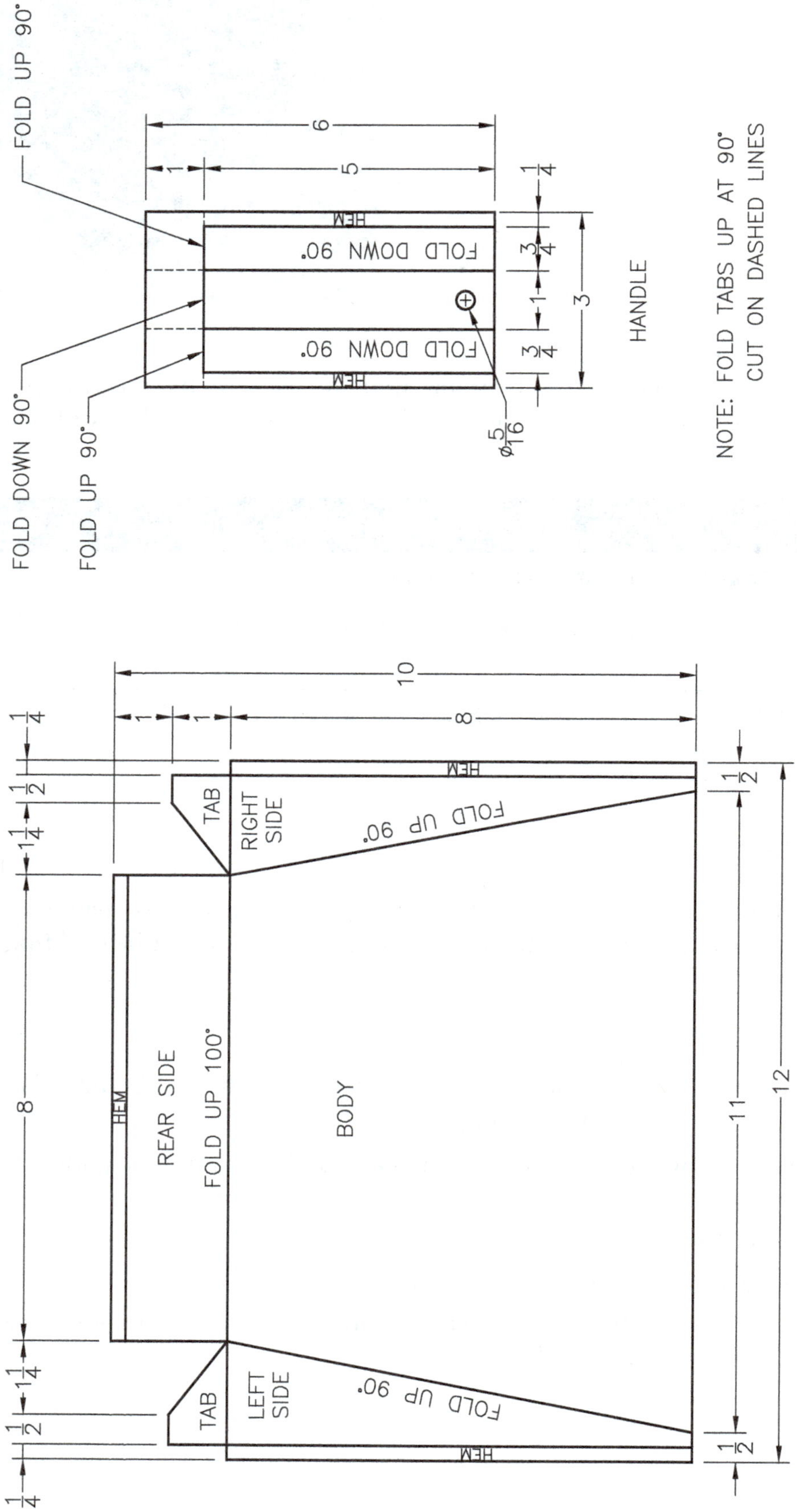

Handle

1. Cut the handle from the sheet metal material in the size indicated in the bill of materials.
2. Apply blue layout die over the sheet metal piece.
3. Scribe the pattern layout lines into the dried layout die. Cut out the pattern.
4. Drill or punch hole in the center of the bottom.
5. Cut along the dashed lines with snips.
6. Fold over hems.
7. Fold down long sides 90°.
8. Bend outside tabs out 90° from the sides.
9. Bend center tab down 90°.

Assembly

1. Center handle in the back of the rear side of the body.
2. Spot weld tabs to the body. If rivets are being used, punch or drill holes in tabs and body.
3. Prime the dust pan, then paint in the color you have chosen.

Project 3
Model Car

This model car is built on the same scale as most 1:25 scale plastic car models. Therefore, decals and stickers from these models can be used to dress up your vehicle, once it is painted. The car wheels can be ordered from a hobby or car model supply catalog. Axles for the project are 1/8″ (3 mm) steel wire.

Bill of Materials			
Item	**Quantity**	**Material Description**	**Size**
Body	1	28 gage (or thicker) galvanized sheet metal	5 3/4″ × 11″ (145 mm × 279 mm)
Top	1	28 gage (or thicker) galvanized sheet metal	3 1/2″ × 9 3/4″ (89 mm × 248 mm)
Front end	1	28 gage (or thicker) galvanized sheet metal	2 1/4″ × 3″ (57 mm × 76 mm)
Axle bracket	2	28 gage (or thicker) galvanized sheet metal	1″ × 5 1/2″ (25 mm × 140 mm)
Axle	2	1/8″ diameter steel rod or welding rod, 2″ long	1/8″ × 2″ (3 mm × 51 mm)
Wheel	4	Model car plastic wheels	1 1/2″ diameter × 1/2″ wide (38 mm × 13 mm)

Procedure

All of the information needed to build this car can be found in either the bill of materials or the project pattern. Refer frequently to both of these resources while building this project.

Body

1. Cut out the body material from the sheet metal in the size indicated in the bill of materials.
2. Apply blue layout die over the sheet metal piece.

Project Pattern

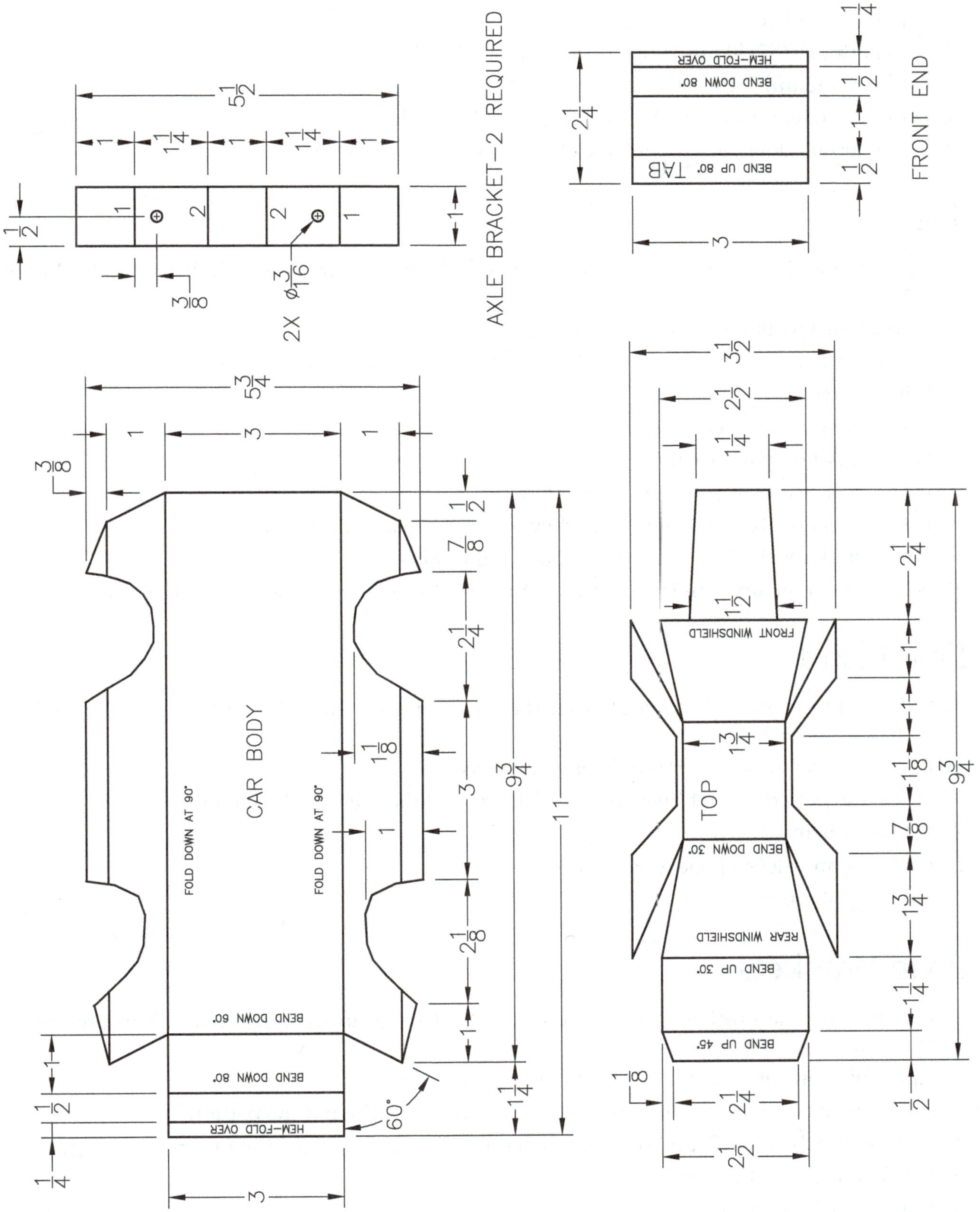

3. Scribe the pattern layout lines into the dried layout die. Cut out the pattern.
4. Notch corners.
5. Cut wheel wells with snips.
6. Fold rear end hem over.
7. Fold bottom 3/8″ (10 mm) edge in at 30° on both sides of the car.
8. Fold 1/2″ (13 mm) rear valence piece down at 80°.
9. Fold sides of car down at 90°, 1 3/8″ (35 mm) from edges.
10. Fold rear end down approximately 60°, until it touches the sides.

Top

1. Cut out material for the top of the car from the sheet metal in the size indicated in the bill of materials.
2. Apply blue layout die over the sheet metal piece.
3. Scribe the pattern layout lines into the dried layout die. Cut out the pattern.
4. Cut or notch piece.
5. Fold rear spoiler up 45°.
6. Fold trunk lid up approximately 30°.
7. Fold top of rear roof down approximately 30°.
8. Fold right and left sides down until they touch the rear windshield.
9. Fold front windshield down until it touches the sides.
10. Fold hood bulge up until it is parallel with the bottom edges of the sides of the roof.

Front End

1. Cut out the front end material from the sheet metal in the size indicated in the bill of materials.
2. Apply blue layout die over the sheet metal piece.
3. Scribe the pattern layout lines into the dried layout die. Cut out the pattern.
4. Fold over hem.
5. Fold bottom valence panel down 80°.
6. Fold top tab up 80°.

Axle Brackets

1. Cut out the axle bracket material from the sheet metal in the size indicated in the bill of materials.
2. Apply blue layout die over the sheet metal piece.
3. Scribe the pattern layout lines into the dried layout die. Cut out the pattern.
4. Drill or punch 3/16″ (5 mm) diameter holes 3/8″ (10 mm) up from the first bend of each end.
5. Fold bends labeled 1 up 90°.
6. Fold bends labeled 2 down 90°.

Assembly

1. Spot weld axle brackets to car while centering axle hole in the wheel well.
2. Spot weld the top to the body of the car on the trunk lid and hood bulge.
3. Spot weld the front end using the tab.
4. Prime the car.
5. When the primer has dried, paint the car in the color you have chosen.
6. Cut 1/8″ (3 mm) diameter axles to 2″ (51 mm) lengths.
7. After paint has completely dried, install axles into the axle brackets.
8. Install wheels on axles.

Project 4
CD Carrying Case

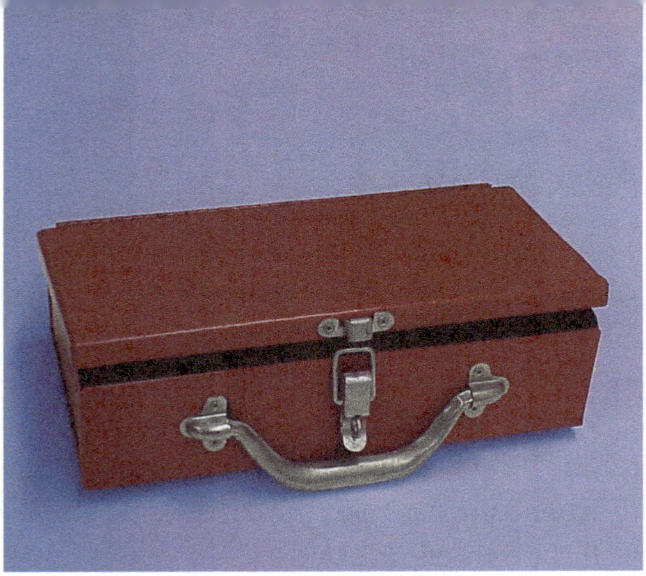

This metal carrying case can hold 14 CDs stacked in two groups of seven. A steel catch with a slot allows the case to be secured with a padlock. Chrome metal corners can also be attached to the bottom corners to prevent the case from scratching other items. Personalize your case by painting it bright colors or applying various recording artist stickers to it.

Bill of Materials			
Item	**Quantity**	**Material Description**	**Size**
Body	1	28 gage (or thicker) galvanized sheet metal	12 1/2″ × 12 1/2″ (317 mm × 317 mm)
Lid	1	28 gage (or thicker) galvanized sheet metal	6 1/2″ × 12 3/4″ (165 mm × 324 mm)
Side	2	28 gage (or thicker) galvanized sheet metal	3 1/2″ × 5 3/4″ (89 mm × 146 mm)
Continuous hinge	1	Steel: no holes	1 1/8″ × 12″ (28 mm × 305 mm)
Catch	1	Steel: type 2	1 5/8″ × 2 1/2″ (41 mm × 63 mm)
Handle	1	Steel: nickel plated	1 1/2″ × 7 3/8″ (38 mm × 187 mm)

Procedure

All of the information needed to build this case can be found in either the bill of materials or the project pattern. Refer frequently to both of these resources while building this project.

Body

1. Cut out the body material from the sheet metal in the size indicated in the bill of materials.
2. Apply blue layout die over the sheet metal piece.

Project Pattern

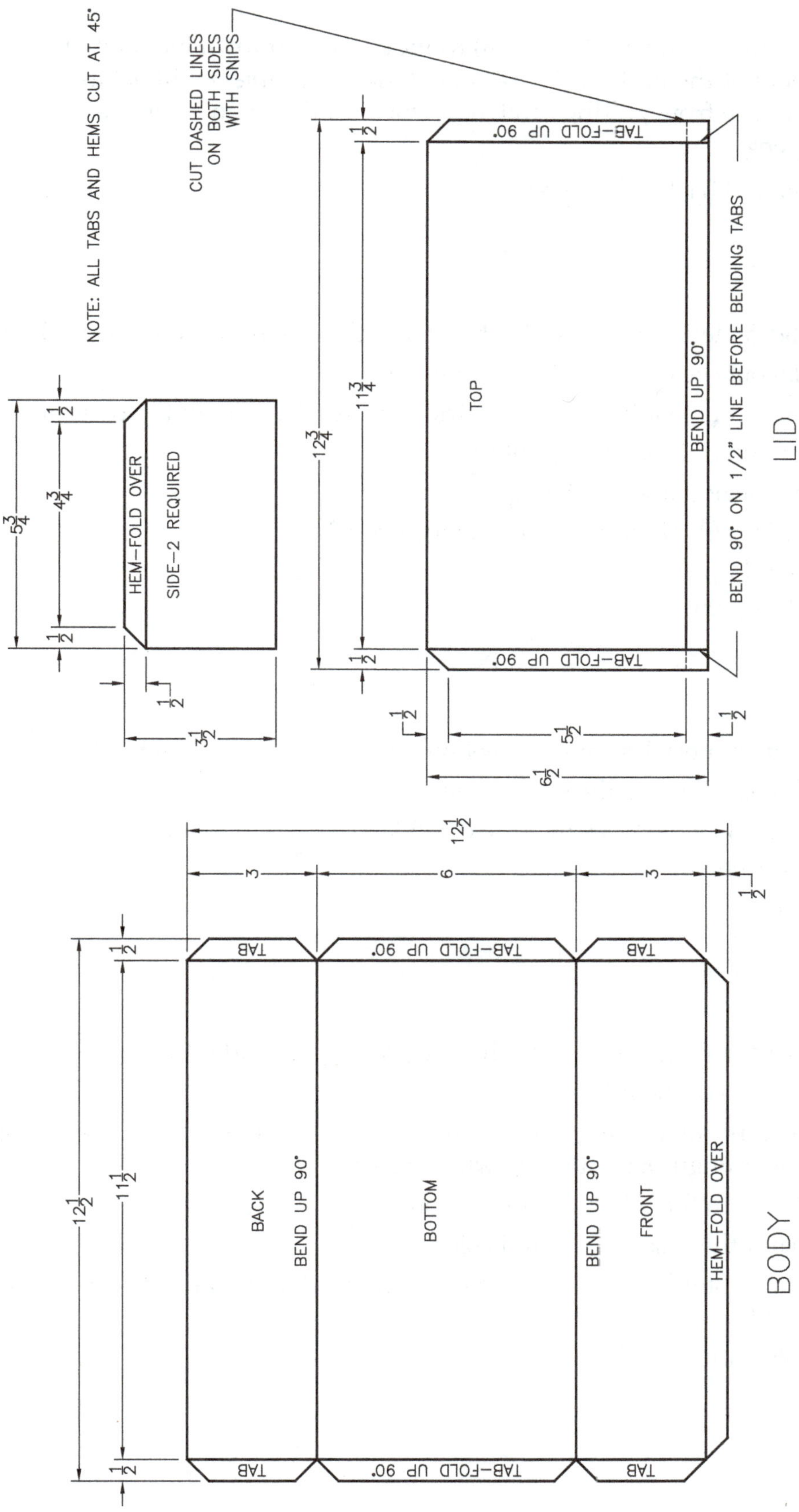

3. Scribe the pattern layout lines into the dried layout die. Cut out the pattern.
4. Notch 45° tabs and hems.
5. Fold over hem.
6. Using a steel punch with a 1/8″ (3 mm) rounded tip, stamp an indentation 1″ (25 mm) in from all four corners of the bottom. These indentations (dimples) will act as feet that prevent the bottom of the case from getting scratched when set down on a surface.
7. Fold tabs up 90°.
8. Fold the front and back sides up 90°.

Lid

1. Cut out the lid material from the sheet metal in the size indicated in the bill of materials.
2. Apply blue layout die over the sheet metal piece.
3. Scribe the pattern layout lines into the dried layout die. Cut out the pattern.
4. Make a 45° cut on the back corner of the tabs.
5. Cut along the dashed lines with snips.
6. Fold both 1/2″ × 1/2″ (13 mm × 13 mm) tabs up 90°.
7. Bend front of lid up 90°.
8. Bend side tabs up 90°.

Side

1. Cut out the side material from the sheet metal in the size indicated in the bill of materials.
2. Apply blue layout die over the sheet metal piece.
3. Scribe the pattern layout lines into the dried layout die. Cut out the pattern.
4. Cut 45° edges on hem.
5. Fold over hem.

Assembly

1. Place sides with hem facing up into the body and spot weld tabs.
2. Spot weld the tabs of the lid.
3. Spot weld the continuous hinge to the top edge of the back of the body. Vise grips may be used to ensure proper alignment of hinge while welding.
4. Spot weld the lid to the other side of the continuous hinge.
5. Prime and paint the case the desired color.
6. After paint has completely dried, center latch and handle and drill or punch 1/8″ (3 mm) holes for assembly.
7. Blind rivet latch and handle to case.

Project 5
Storage Unit/ Organizer

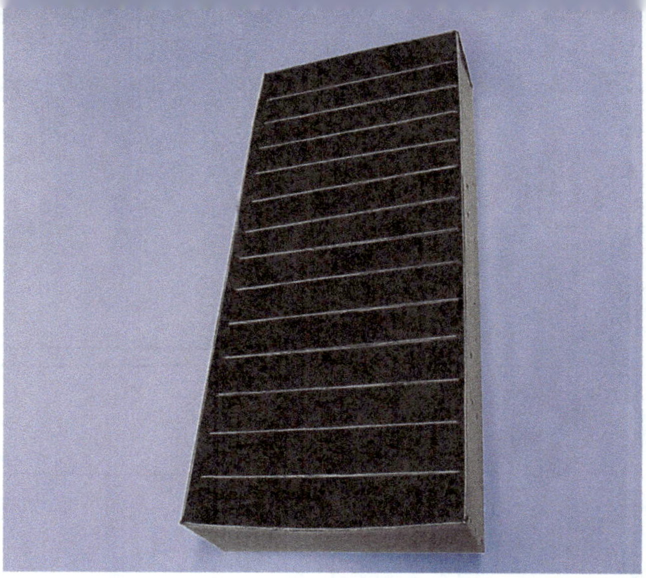

This storage unit/organizer can be divided into any number of sections. It is designed to be mounted on the wall, but can also be placed in a drawer or on a desktop to organize various items. You may even choose to use it to display a collection of baseballs, model cars, or small action figures.

Bill of Materials			
Item	**Quantity**	**Material Description**	**Size**
Body	1	28 gage (or thicker) galvanized sheet metal	16 1/2″ × 29″ (419 mm × 727 mm)
Divider	13	28 gage (or thicker) galvanized sheet metal	3 1/2″ × 12 1/2″ (89 mm × 317 mm)

Procedure

All of the information needed to build this media tower can be found in either the bill of materials or the project pattern. Refer frequently to both of these resources while building this project.

Body

1. Cut out the body material from the sheet metal in the size indicated in the bill of materials.
2. Apply blue layout die over the sheet metal piece.
3. Scribe the pattern layout lines into the dried layout die. Cut out the pattern.
4. Cut out all four corners and notch tabs and hems at 45°.
5. Fold over hems.
6. Fold left and right sides up at 90°.
7. Fold top and bottom sides up 90° or until they touch the left and right sides.

Project Pattern

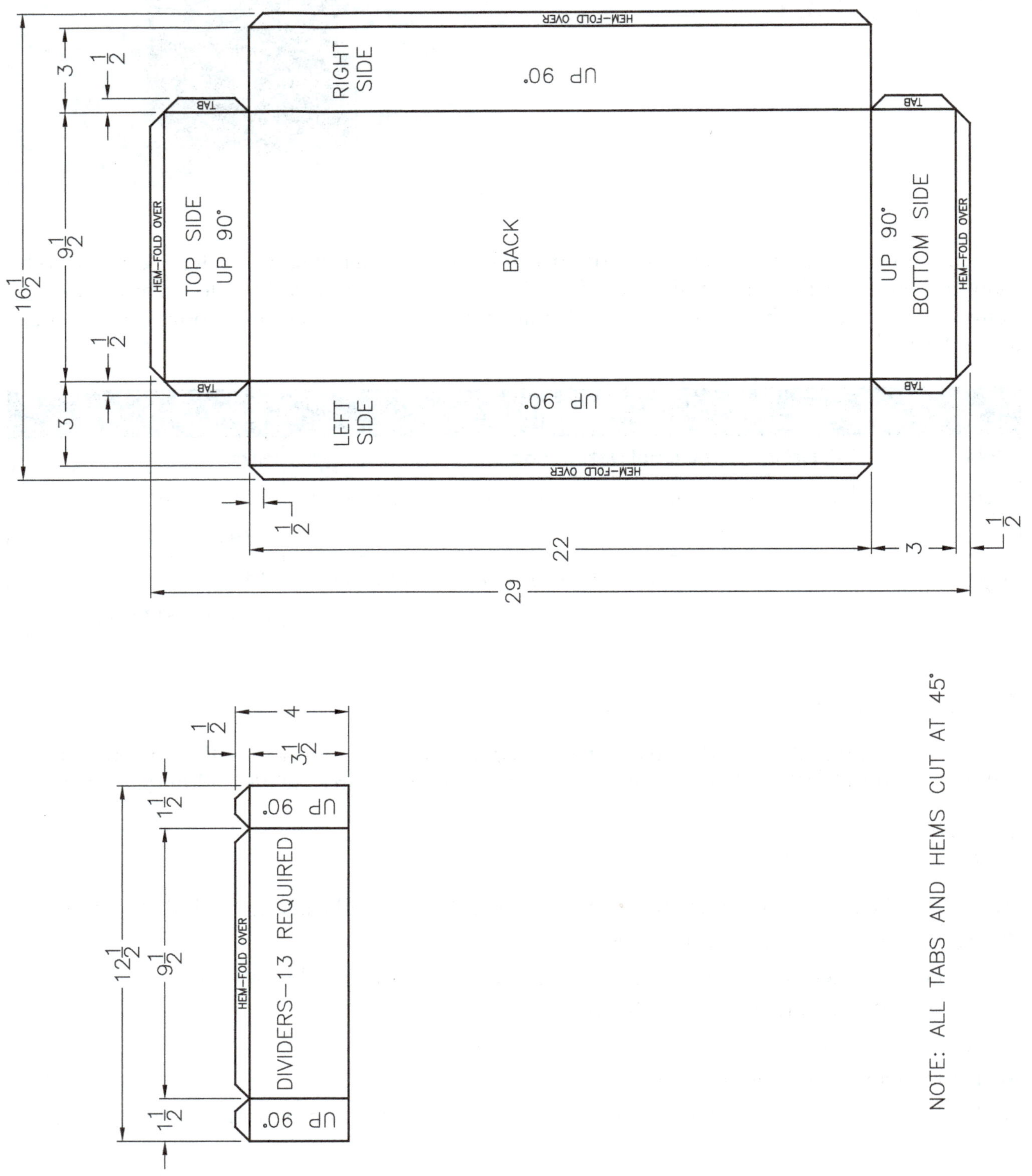

Divider

1. Cut out the divider material from the sheet metal in the size indicated in the bill of materials. The number of dividers and the spacing used is your choice.
2. Apply blue layout die on each of the 13 pieces of sheet metal.
3. Scribe the pattern layout lines into the dried layout die. Cut out the pattern.
4. Notch hems at 45°.
5. Fold over hems.
6. Fold left and right sides up at 90°.

Assembly

1. Spot weld top and bottom side tabs to the left and right sides of the body.
2. Place dividers into the body and spot weld tabs on both sides.
3. Continue to stack the remaining 12 dividers on top of each other and spot weld in place.
4. Prime and paint desired color.

Math for Sheet Metal Workers

Sheet metal workers must be able to take accurate measurements and make correct calculations. Certain mathematical concepts are needed to perform these skills. This appendix will review those basic concepts, as well as give examples of their application to sheet metal work. Most will require the ability to make accurate measurements of distance and volume.

Basic math skills are needed:

- In order to make measurements necessary to cut, fit, and locate members in a structure where air conditioning or heating is being installed.
- In order to calculate the quantities of sheet metal needed or required for a project and to estimate the cost.

This review will include applying basic math skills to fractions, decimals, area, and volume. These applications will cover some of the most common problems met by the sheet metal worker in the shop or at the job site.

Fractions

Fractions are always written with one number above the other with a line separating the two. The top number of a fraction is called the numerator, the bottom number is called the denominator, the line separating them is called the fraction bar.

$$\frac{\text{Numerator}}{\text{Denominator}}$$

← Fraction bar

A fraction is another way of describing some part of a total amount. Imagine a pizza cut into four equal pieces. Each piece represents one part of the total of four parts. This is written as the fraction 1/4, and means one part out of four equal parts.

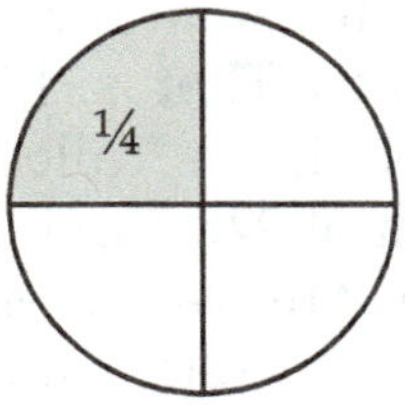

The fraction 3/4 indicates that a whole was divided into 4 equal parts and 3 of these parts are used.

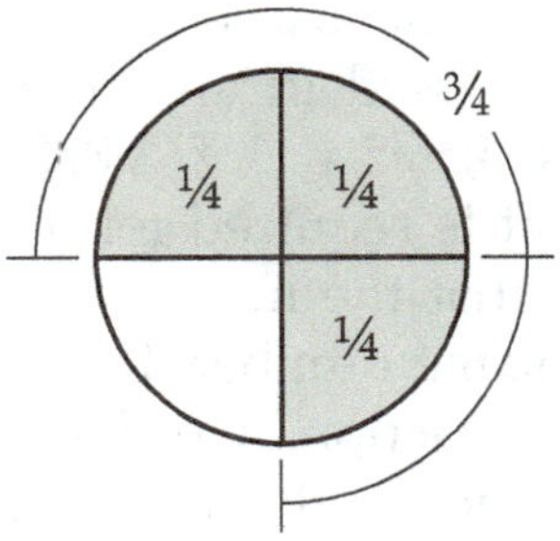

The fraction 1/8 indicates that a whole was divided into 8 equal parts and that 1 of these parts was used.

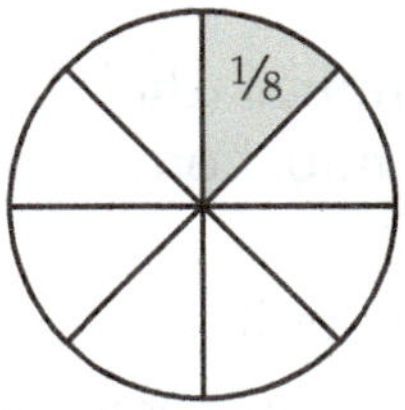

The pie is cut again into 8 equal parts. The shaded portion has increased to 2 sections or 2/8 of the pie. If the pizza had been cut into 4 equal parts, the shaded portion would be 1 part out of 4. Looking at the pie, can you see that 2/8 of the pie is the same as 1/4?

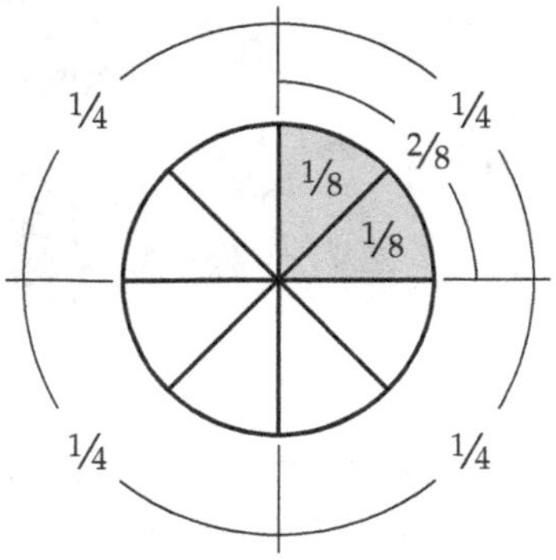

Simplifying Fractions

A portion of the total may be expressed in two different ways. That is, 1/4 is the same as 2/8. The fraction 1/4 is called the simplified, or reduced, form of 2/8. Simplifying a fraction means to change it into another equivalent fraction. When the numerator and denominator of a fraction have no common divisor, except 1, the fraction is said to be simplified, or reduced, to its lowest terms.

If a fraction is multiplied or divided by 1, the value of the fraction is unchanged. The number 1 may be expressed in any number of ways, such as 3/3, 4/4, 5/5, 6/6, 7/7, 8/8, 9/9, 11/11, or 33/33. All that is required for a fraction to be equal to 1 is that the numerator and denominator be the same number. Knowing this, you can simplify a fraction by dividing the fraction by 1, using a fraction that equals 1.

$$\frac{4 \div 4}{12 \div 4} = \frac{1}{3}$$

1. Find any number (the largest is best) that will divide into both the numerator and denominator.
2. Use this number in a fraction that equals 1.
3. Divide both the numerator and denominator of the first fraction by this common divisor.
4. The resulting fraction is the simplified version of the original fraction.

Improper Fractions and Mixed Numbers

An improper fraction has a numerator that is larger than the denominator, such as 5/4. An improper fraction is always equal to a value greater than one. In the case of 5/4, 5 is greater than 4 and the fraction actually represents 1 1/4.

Another way of expressing a value greater than one is as a mixed number, a number with a whole number part and a fractional part, such as 6 5/8. To convert a mixed number to an improper fraction, multiply the whole number by the denominator of the fraction. Using 6 5/8 as an example, multiply 6 by 8. This equals 48. Next, add the numerator of the fraction portion to the product from Step 1. So 48 (product) + 5 (numerator from fraction) = 53. Finally, place this number, 53, over the same denominator used in the improper fraction, or 53/8.

Improper fractions are easier to work with when making calculations. If a mixed number is given in a problem, convert it to an improper fraction. Then do the calculation and convert any improper fractions in your final answer back to mixed numbers so they are more easily understood.

Remember that whole numbers can also be expressed as fractions. The whole number becomes the numerator and 1 is the denominator. For example, the whole number 9 is the same as the fraction 9/1.

Multiplying Fractions

Multiplying fractions is a straightforward process. Simply multiply the numerators for the numerator product. Then multiply the denominators for the denominator product. Work through this multiplication problem: $3/4 \times 1/8$.

1. Multiply the numerators: $3 \times 1 = 3$
2. Multiply the denominators: $4 \times 8 = 32$
   Therefore, $3/4 \times 1/8 = 3/32$

Dividing Fractions

There are three parts to a division equation: the divisor, the dividend, and the quotient. A division equation fraction is actually a division operation because the numerator is the dividend and the denominator is the divisor.

In order to divide fractions, the fraction that is the divisor must be inverted. To invert, switch the numerator and denominator of the fraction. Then multiply the two fractions. For instance, divide 2/3 by 7/4. The fraction will look like this:

$$\frac{2/3}{7/4}$$

1. Invert the divisor from 7/4 to 4/7.
2. Next, multiply the dividend by the inverted divisor.

$$\frac{2 \times 4}{3 \times 7} = \frac{8}{21}$$

Therefore, $2/3 \div 7/4 = 8/21$

Changing Denominators

Fractions can be written in different ways. For example, 1/2 is the same as 4/8 and 12/24. There are situations where it is preferable to change the denominator of the fraction. In order to convert between equivalent (equal) fractions, you simply multiply both the numerator and denominator by the same number. For example, to find an equivalent fraction of 1/3 that has 24 as its denominator, set the equation up as shown.

$$\frac{1}{3} = \frac{?}{24}$$

Determine how many times 3 goes into 24. The answer is 8. Multiply both the numerator (1) and denominator (3) by 8 to find the equivalent fraction.

$$\frac{1 \times 8}{3 \times 8} = \frac{8}{24}$$

Adding Fractions

Fractions must have the same denominators in order to add them. When fractions do have the same, or common, denominator, add the two numerators and place them over the common denominator. For instance, add 3/8 + 1/8. Add the nominators first, for an answer of 4. Place this over the common denominator, or 8. The answer is 4/8. It can be reduced to 1/2.

When fractions have different denominators, one or both of the fractions need to be converted to an equivalent fraction with a common denominator. For instance, in the equation 3/4 + 3/16, the denominators are different. In this case, 3/4 can be converted to an equivalent fraction with 16 as a denominator.

$$\frac{3 \times 4}{4 \times 4} = \frac{12}{16} + \frac{3}{16} = \frac{15}{16}$$

Subtracting Fractions

Subtracting and adding fractions are very similar operations. As with addition, fractions must have common denominators to subtract. The only difference is that you subtract the numerators, rather than add them. For instance, in the example 3/4 – 1/4, the denominators are the same, so simply subtract the numerators. The answer is 2/4. It can be reduced to 1/2.

When denominators are different, you must find the common denominator. In the example 4/5 – 3/15, convert 4/5 to an equivalent fraction that has 15 as the denominator.

$$\frac{4 \times 3}{5 \times 3} = \frac{12}{15} - \frac{3}{15} = \frac{9}{15}$$

The answer can be reduced further to 3/5.

Decimals

A decimal number has three parts: a whole number part, a decimal point, and a decimal part.

Decimal point

19.45

Whole number Decimal

The decimal part is like the numerator in a fraction. The denominator in this fraction is 1, followed by the same number of zeros as there are numbers after the decimal point. A decimal number is another way of expressing a mixed number.

For example, convert 17.45 to a mixed number. There are two digits to the right of the decimal point, so there will be two zeros following the 1 in the denominator. The numerator is 45. Therefore, 17.45 can also be expressed as 17 45/100.

Let's now look at an example with zeros in the decimal portion of the number. Convert 2.004 to a mixed number. There are three digits to the right of the decimal point. That means there will be three zeros following the 1 in the denominator. The numerator is 4. The zeros preceding it are dropped. The mixed number then is 2 4/1000, which can be reduced to 2 1/250.

A place value chart shows the value of numbers to the left and right of a decimal point. As numbers are added to the left of the decimal point their values increase and as numbers are added to the right, their values decrease.

Decimal Places

Ten-Thousands	Thousands	Hundreds	Tens	Ones		Tenths	Hundredths	Thousandths	Ten-Thousandths
1	2,	7	2	6	.	6	5	7	9
10,000	1,000	100	10	1		1/10	1/100	1/1,000	1/10,000

While zeros have no value themselves, they are very important to the value of the complete number. The zero is a placeholder. It is used to keep the other numbers in their places. For example, the decimal number 0.62 is larger than 0.062, the first being 62/100 and the second 62/1000. The zero to the right of the decimal point indicates that the second number is smaller than the first.

Adding and Subtracting Decimal Numbers

To add and subtract decimals, write the numbers in a column with their decimal points aligned. Then add or subtract as if they were whole numbers. Carry the decimal point down into the answer. For instance, add 2.6 + 10.5 + 8.803 + 5 + 0.0155. The first step is to arrange the numbers in a column, with the decimal points aligned.

$$\begin{array}{r} 2.6000 \\ 10.5000 \\ 8.8030 \\ 5.0000 \\ \underline{0.0155} \\ 26.9185 \end{array}$$

The number of zeros at the far right end of the decimal portion does not matter, as long as they are located at the end of the number. For example, 54.78 and 54.780000 are the same number. However, 32.001 and 32.0000001 are different numbers because there is a nonzero digit to the right of the zeros.

Multiplying Decimal Numbers

Multiplying decimal numbers is a two-step process. First, multiply the numbers together, ignoring the decimal points. Then count the total number of decimal places that occur in the numbers being multiplied and use the same number of decimal places in the product. If there are more decimal places than numbers in the product, add zeros to the left to get enough places.

In this example, multiply 2.123 × 2.2. Ignore the decimal points in this first step.

```
 2123
× 22
 4246
4246
46706
```

There are four decimal places in the numbers multiplied, so there must be four decimal places in the product. The answer is 4.6706

Whenever you do any type of calculations, check for a reasonable answer. For instance, in the previous example, round the two factors to the nearest whole numbers (2 and 2), then multiply these together to get 4. An estimate of 4 is reasonably close to 4.6706.

Dividing Decimal Numbers

Using decimals allows division of numbers that do not divide evenly. The division remainder is a fractional part. It can be written as a fraction or the division can continue for as many decimal places as needed.

Follow these steps when neither divisor nor dividend contains decimal points and the division does not have an exact solution. Also refer to the solved problem that is provided.

1. Place a decimal point to the right of the dividend and after the last number in the answer.
2. Add zeros to the dividend.
3. Continue dividing to as many decimal places as needed.

Place decimal in quotient directly above decimal in dividend.

```
    2.666
6 | 16.000
    12
     4 0
     3 6
      40
      36
       40
       36
```

Add decimal point and zeros until answer comes out evenly or necessary accuracy is reached.

When the dividend contains a decimal point and the divisor does not, follow these steps. Also refer to the solved problem that is provided.

1. Divide as usual up to the decimal point.
2. Place a decimal point to the right of the last number in the answer.
3. Continue dividing the rest of the dividend. Add zeros to the dividend until the answer is carried out to the number of decimal places needed.

Place decimal in quotient directly above decimal in dividend.

```
    2.40625
8 | 19.25000
    16
     3 2
     3 2
      050
```

Add zeros until answer comes out evenly or necessary accuracy is reached.

Finally, when the divisor contains a decimal point, follow these steps, while referring to the solved problem that is provided.

1. Move the decimal point to the right of the divisor.
2. Move the decimal point in the dividend to the right the same number of places.
3. Divide, being sure to place the decimal point in the quotient directly above the new location of the decimal point in the dividend.

Decimal point in quotient is placed directly above new decimal location in dividend.

```
         4.25
2.1. | 8.9.25
       8 4
         5 2
         4 2
         1 05
         1 05
            0
```

Converting Between Decimals and Fractions

The system for units of measure in the United States uses both fractions and decimals to indicate parts of a whole. While rulers are marked in inches and fractions, micrometers are marked in decimals. Sometimes it is necessary to convert one to the other.

Converting Decimals to Fractions

1. Use the number to the right of the decimal point as the numerator.
2. Count off the places to the right of the decimal (one place = tenths; two places = hundredths; three places = thousands, and so on). This number is the denominator.
3. Reduce the resulting fraction to its lowest term. For example:
   0.5 = 5/10 = 1/2
   0.05 = 50/100 = 5/10 = 1/2
   0.175 = 175/1000 = 7/40

Converting Fractions to Decimals

1. Divide the numerator by the denominator.
2. If it does not divide exactly, continue the division to as many decimal places as needed.

$$\frac{3}{4} = \begin{array}{r} .75 \\ 4\overline{)3.00} \\ \underline{2\,8} \\ 20 \\ \underline{20} \\ 0 \end{array}$$

Knowing and Using Formulas

Sheet metal workers often need to determine area or volume. They must know what formula to use and how to apply it.

Area

The area of a square or rectangle determines how much material will be needed to make air conditioning and heating duct work. To find the area of either shape, multiply the length of two adjoining sides, or length (L) times width (W). Answers are given in square units, such as square feet or square yards. For example, calculate the area of a room that is 10 feet long and 8 feet wide. 10 (L) × 8 (W) = 80 square feet.

When using any formula, note the unit of measurement being used. When multiplying measurements expressed in feet, the answer will be in square feet. When the measurements are expressed in inches, the answer will be in square inches. Measurements using different units cannot be combined. They must be converted to a single unit.

Volume

Volume is a measurement of length times width times height (L × W × H). The product of dry measure is expressed in cubic inches, cubic feet, or cubic yards.

For example, calculate the amount of air that can be moved through a duct 4 feet wide by 3 feet high by 12 feet long. Using the formula L × W × H, 4 × 3 × 12 = 144 cubic feet. This movement of air is often expressed in cubic feet per minute, or cfm.

Triangulation

When duct changes shape, parallel and radial line development techniques do not work. Triangulation is the best method for forming corners in duct that changes from square to round or some other shape. Using the triangulation method, a series of triangles is formed to make the curve as the piece transitions from one shape to another shape.

The following steps illustrate how to make a pattern for a duct that changes from square to round using the triangulation method. The dimensions in this illustration are smaller than normal so the pattern can be printed on letter-sized paper.

1. Draw a circle and divide it into twelve parts. Label the lower-right quadrant 1–4 as shown.

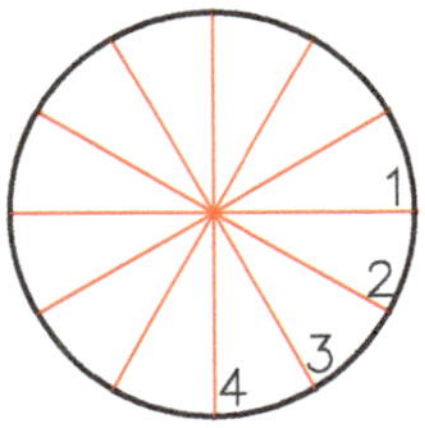

2. Draw a square centered around the circle. Label the bottom-right corner A and the bottom-left corner B. Connect the four numbered points of the circle with the bottom-right corner of the square.

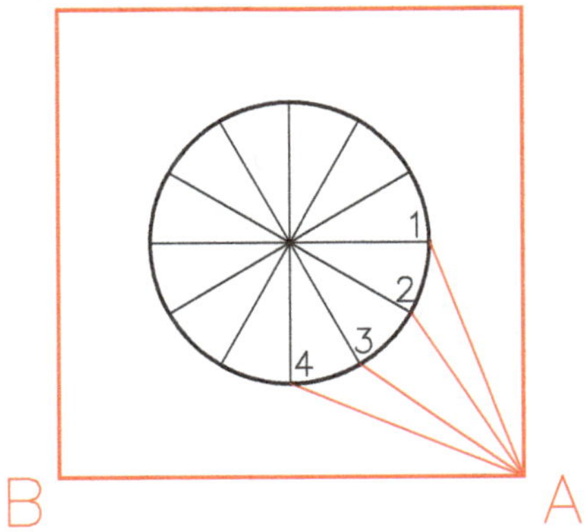

3. Project the corners of the square and the points on the circle down to the front view. Finish drawing the height and cone shape as illustrated.

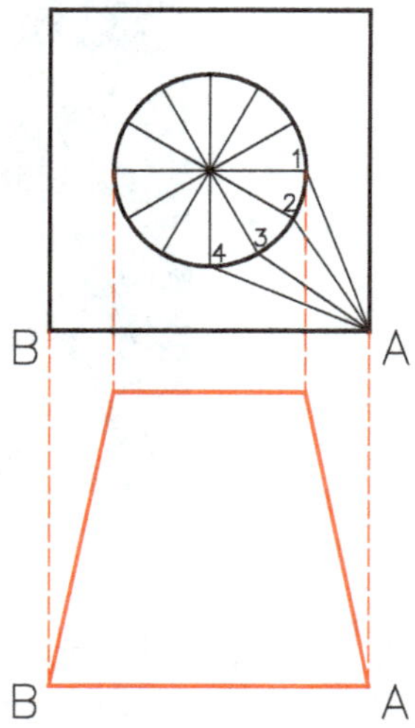

4. To the right of the front view, draw a vertical line that is the same height as the cone. Write A at the top of this vertical line.

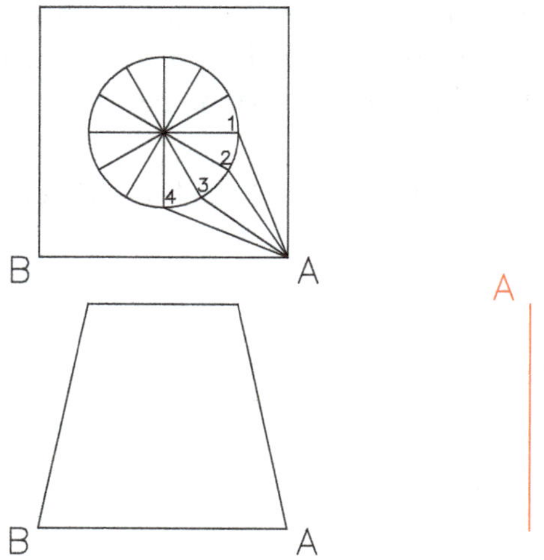

5. Draw a horizontal line from the bottom of this vertical line to the right about the same length as AB. These two lines are the beginning of the true length diagram.

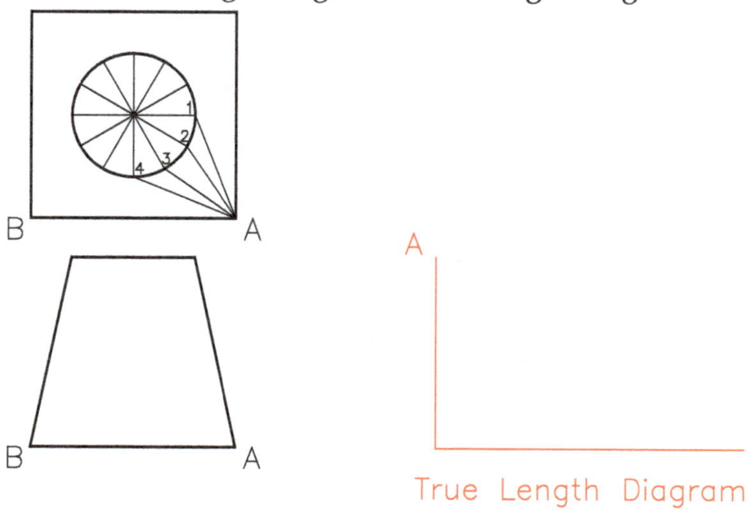

6. Adjust the compass to match the length of A4 in the top view. Place the point of the compass at the intersection of the true length diagram and draw a small mark on the horizontal line. Write 4 at this mark.

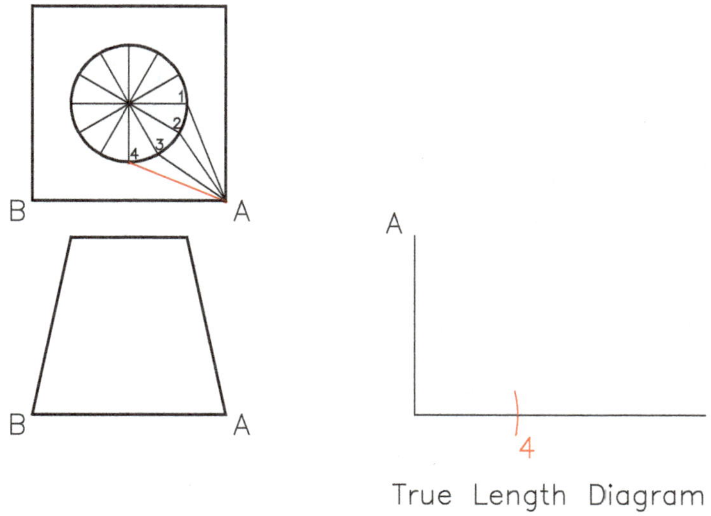

7. Repeat Step 6 with line A3 and identify the mark in the true length diagram with a 3.

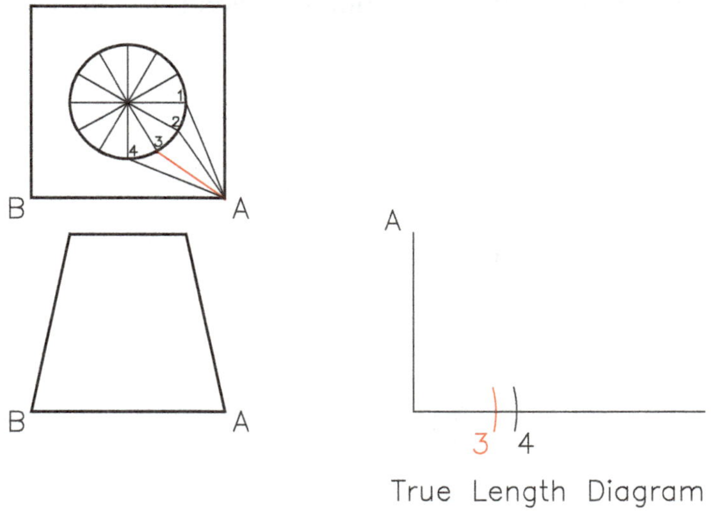

8. In the true length diagram, draw a line from A to 4 and another from A to 3.

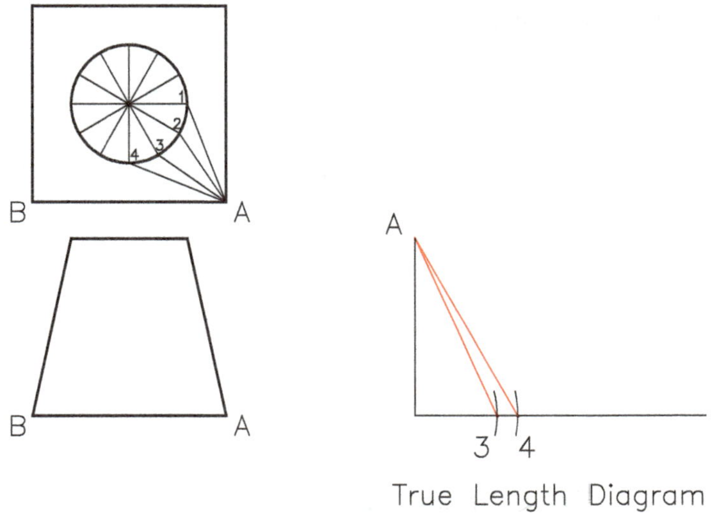

9. To the right of the true length diagram, draw a horizontal line that is the same length as AB in the top view. This is the beginning of the development view.

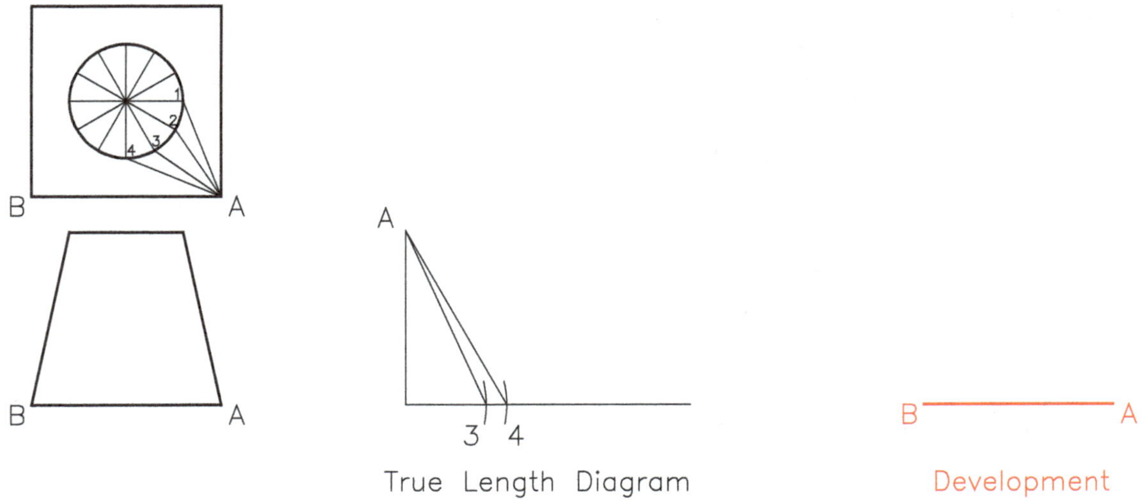

10. Adjust the compass to match the length of A4 in the true length diagram. In the development view, place the compass point on the end of this horizontal line and draw an arc. Draw another arc from the other end of the horizontal line. Write 4 at the intersection of the arcs.

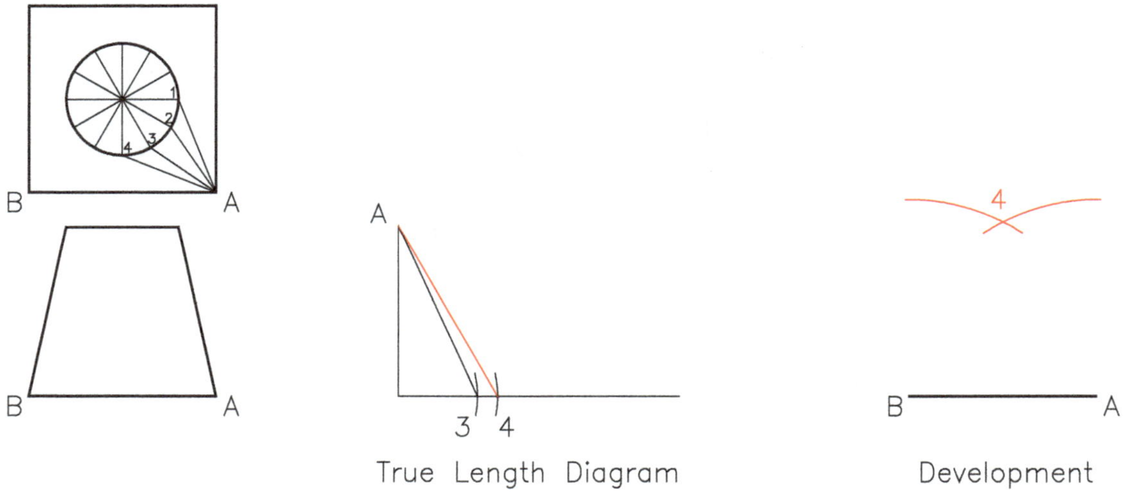

11. Draw a line from the end of the horizontal line to the intersection of these two arcs. Repeat for the other end of the horizontal line.

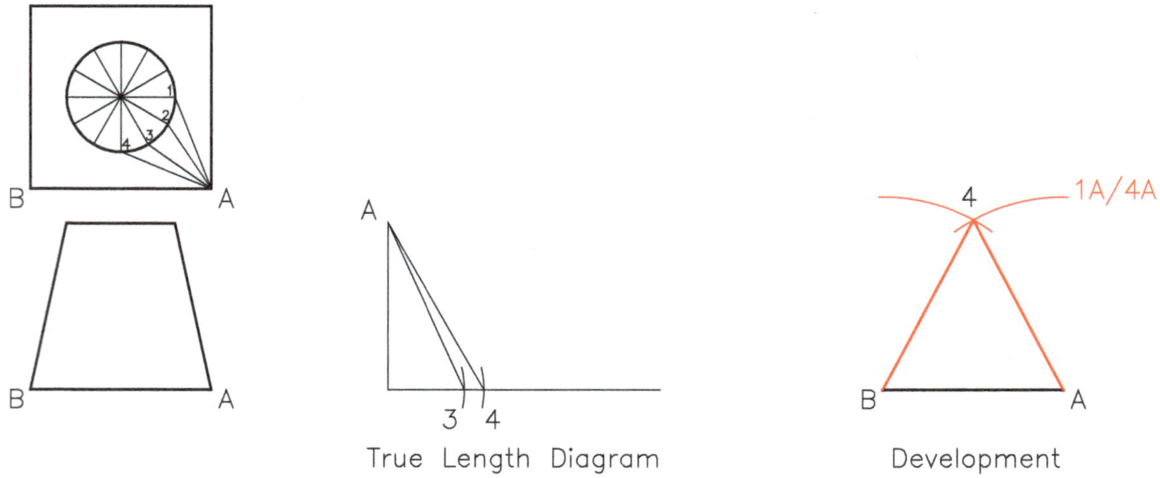

12. Adjust the compass to match line A3 in the true length diagram. Place the compass point on the end of the horizontal line and draw an arc. Draw another arc from the other end of the horizontal line.

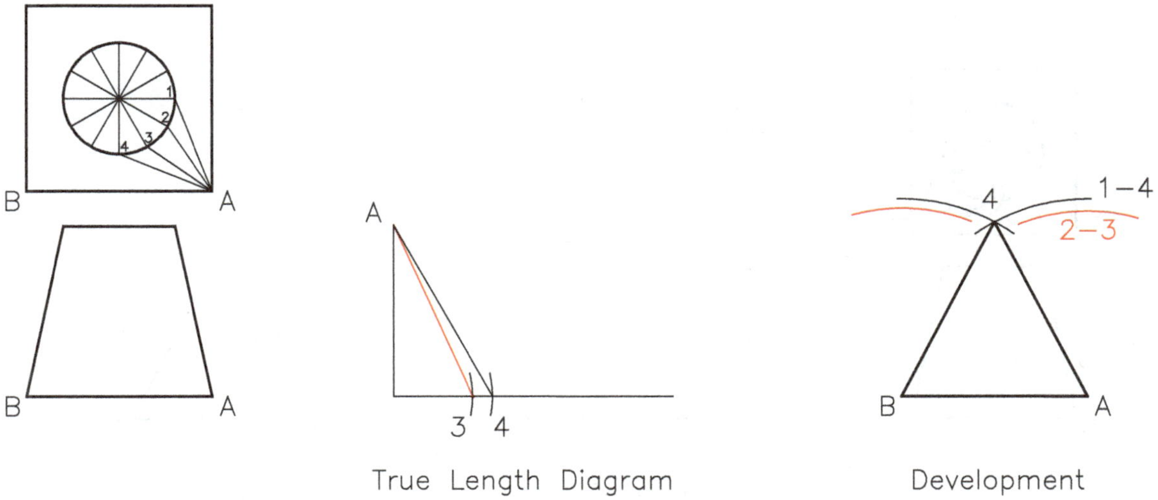

13. Adjust the compass to match the distance from Point 4 to Point 3 in the top view. In the development view, place the point of the compass at Intersection 4 and draw a mark across Arc 2-3.

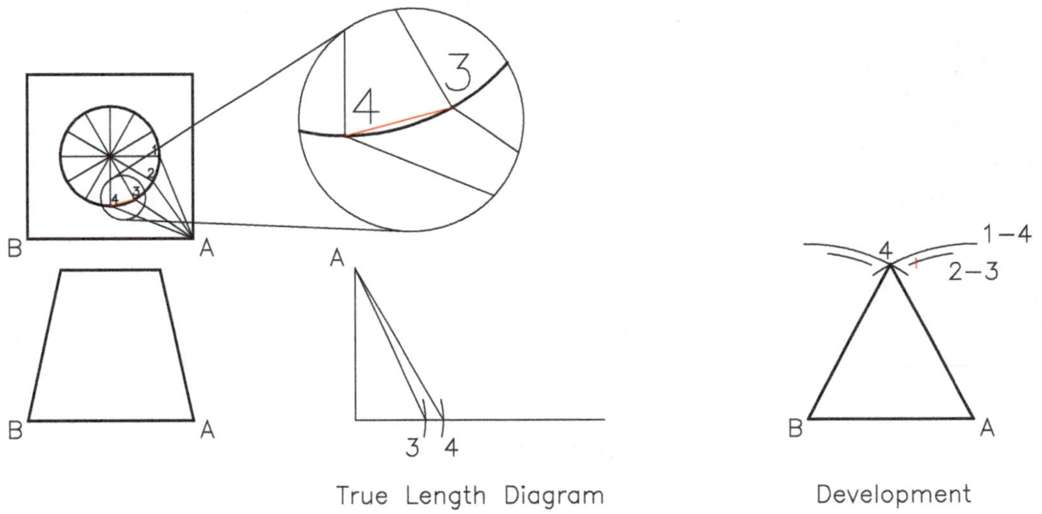

14. Without adjusting the compass, move the compass point to the intersection of these two arcs and draw another mark across Arc 2-3. Move the compass point to this intersection and draw a mark across Arc 1-4. Repeat this process for the other side.

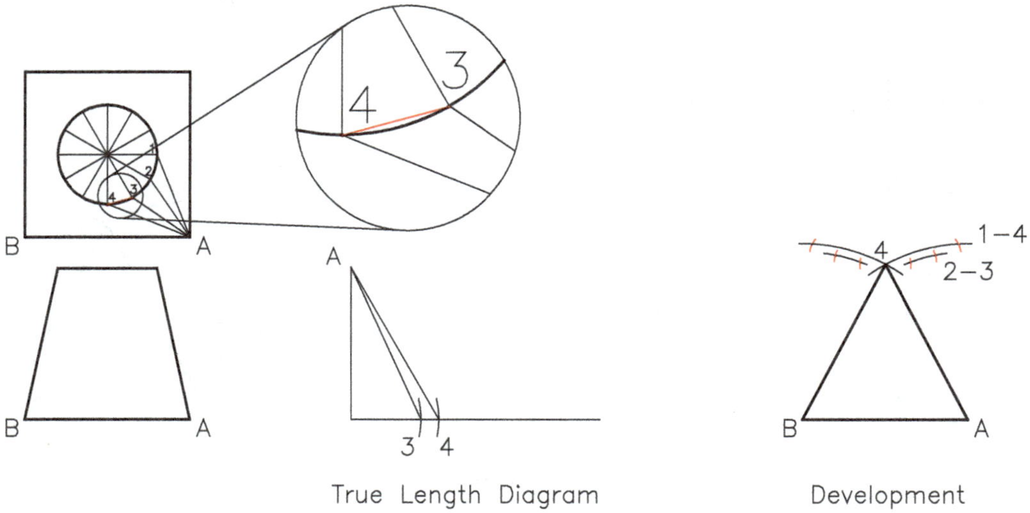

15. Draw lines from the intersections to Point A. Repeat for the other side to Point B.

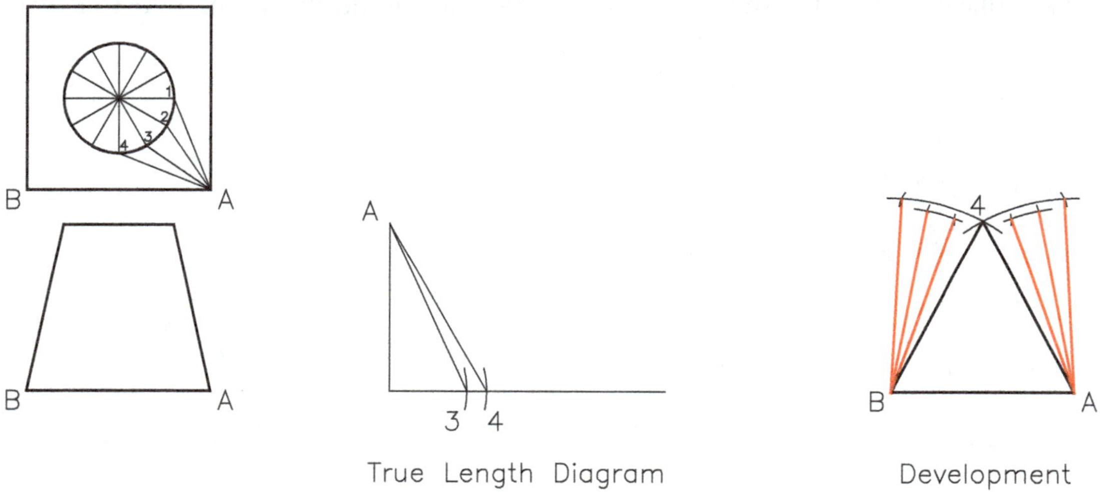

16. Adjust the compass to the length of 4B. Place the compass point at Point 1 and draw a small arc. Adjust the compass to the length of AB. Place the compass point at Point A. Draw a small arc across the previous arc. Repeat for the other side.

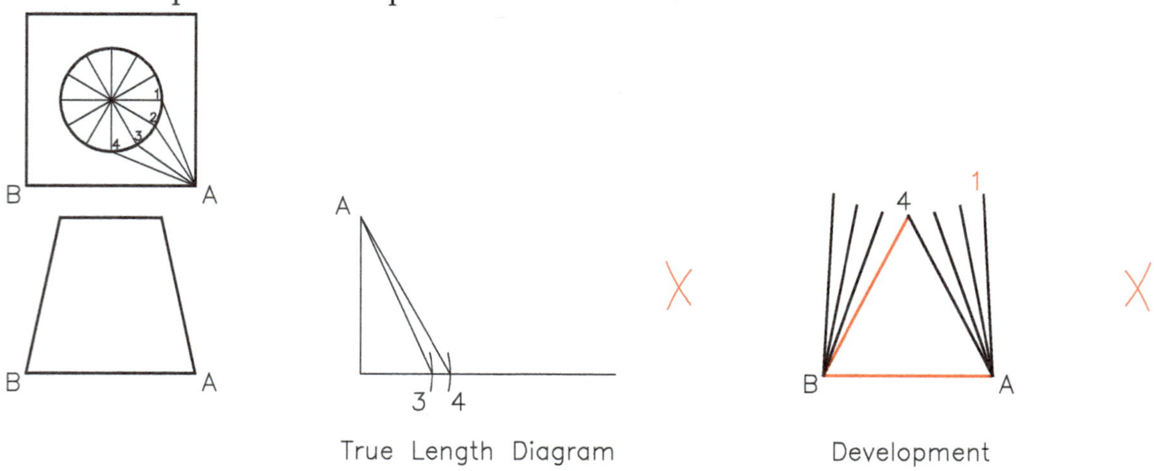

17. Draw lines from the intersection of these arcs to Point 1 and Point A. Repeat for the other side.

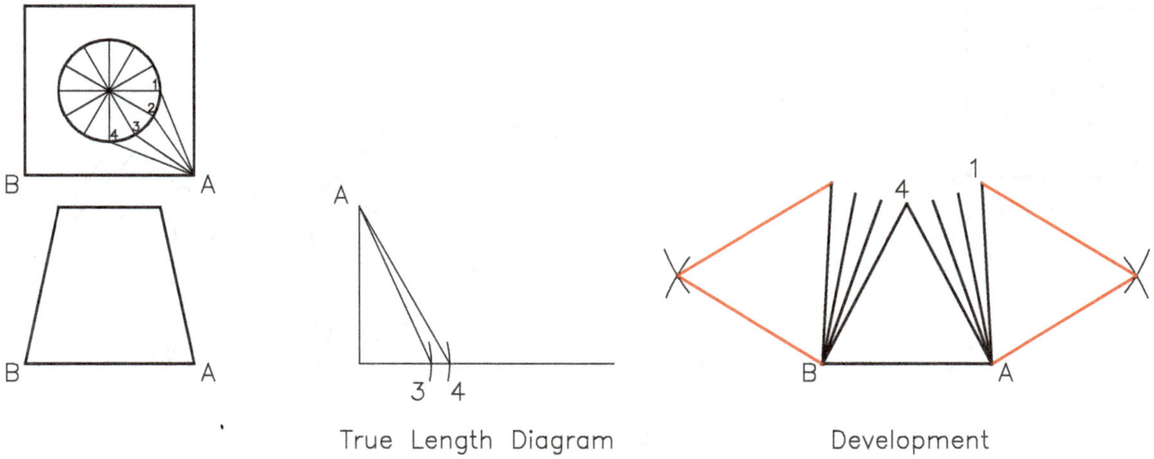

18. Repeat the steps starting at Step 10 to create the lines for each transition corner.

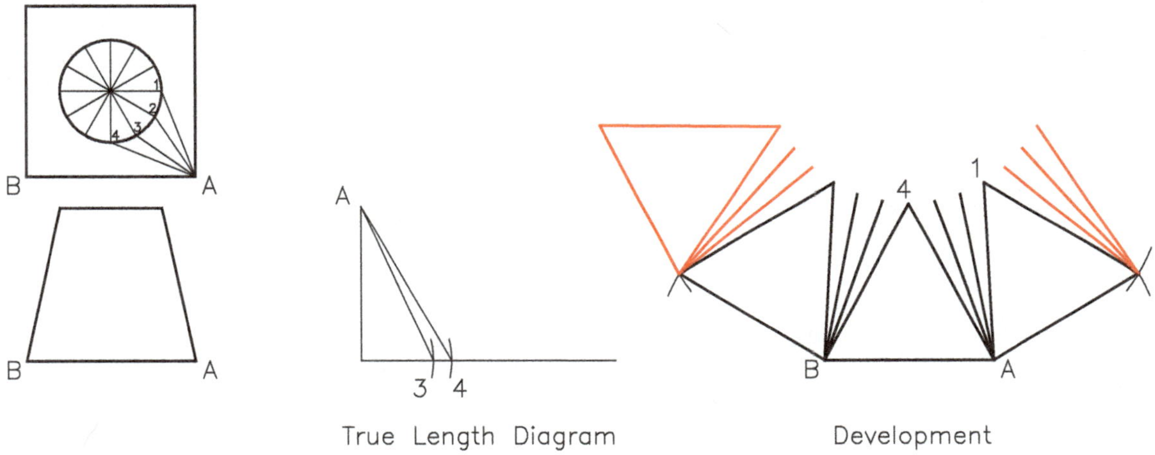

19. Draw a smooth curve across the endpoints of lines.

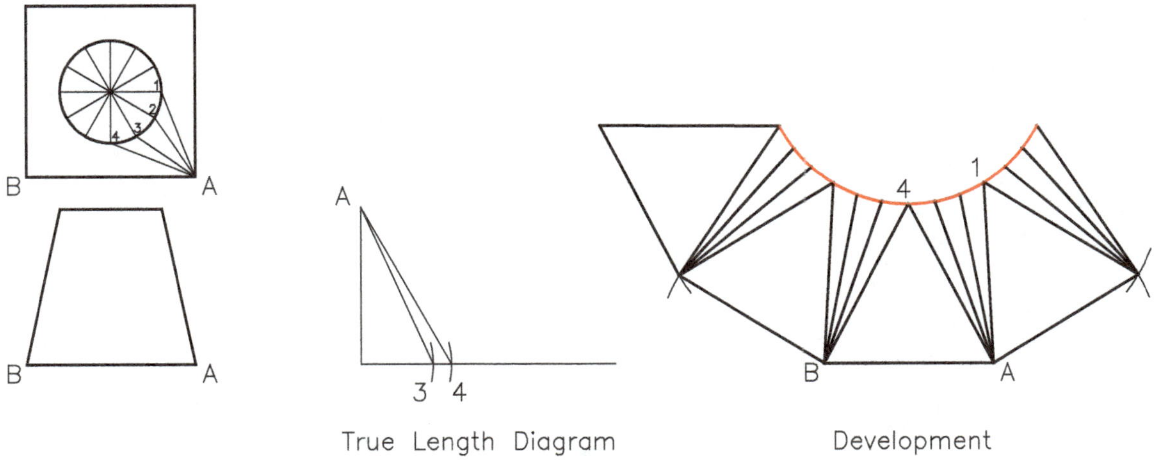

20. Print and then cut out the development. Fold it to check for accuracy.

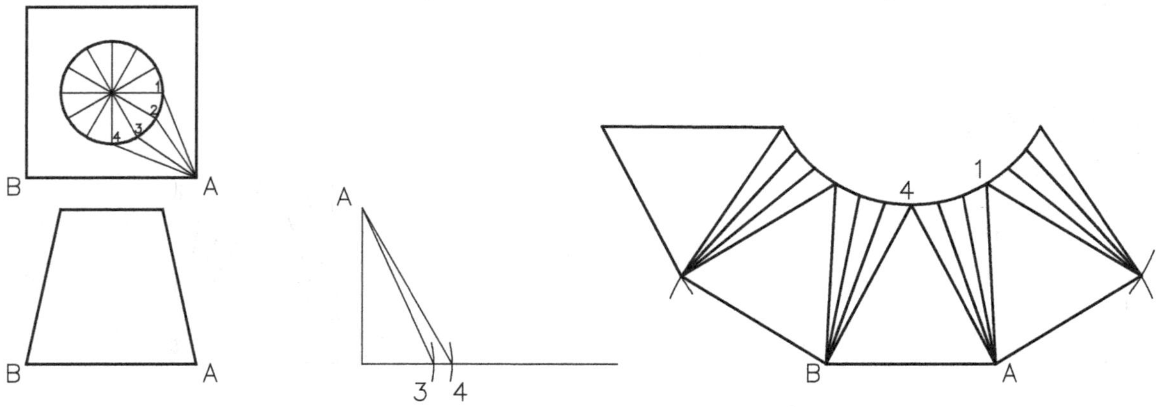

True Length Diagram

Development

Metal Sheet Materials Chart

Material (Sheet less than 1/4″ thick)	How Measured	How Purchased	Characteristics
Copper	Gage number (Brown & Sharpe and Amer. Std.)	24″ × 96″ sheet or 12″ or 18″ by lineal feet on roll	Pure metal
Brass	Gage number (Brown & Sharpe and Amer. Std.)	24″ × 76″ sheet or 12″ or 18″ by lineal feet on roll	Alloy of copper and zinc
Aluminum	Decimal	24″ × 72″ sheet or 12″ or 18″ by lineal feet on roll	Available as commercially pure metal or alloyed for strength, hardness, and ductility
Galvanized steel	Gage number (Amer. Std.)	24″ × 96″ sheet	Mild steel sheet with zinc plating, also available with zinc coating that is part of sheet
Black annealed steel sheet	Gage number (Amer. Std.)	24″ × 96″ sheet	Mild steel with oxide coating, hot-rolled
Cold-rolled steel sheet	Gage number (Amer. Std.)	24″ × 96″ sheet	Oxide removed and cold-rolled to final thickness
Tin plate	Gage number Amer. Std.)	20″ × 28″ sheet 56 or 112 to pkg.	Mild steel with tin coating
Nickel silver	Gage number (Brown & Sharpe)	6″ or 12″ wide by lineal feet on roll	Copper 50%, zinc 30%, nickel 20%
Expanded	Gage number (Amer. Std.)	36″ × 96″ sheet	Metal is pierced and expanded (stretched) to diamond shape; also available rolled to thickness after it has been expanded
Perforated	Gage number (Amer. Std.)	30″ × 36″ sheet 36″ × 48″ sheet	Design is cut in sheet; many designs available

Wire Gages in Decimal Inches

Number of Wire Gage	American or Brown & Sharpe	Washburn & Moen Mfg. Co., A.S.& W. Roebling	Imperial Wire Gage	Stubs' Steel Wire	Birmingham or Stubs' Iron Wire
0000000	...	.4900	.5000	...	...
000000	.5800	.4615	.4640	...	...
00000	.5165	.4305	.4320	...	.500
0000	.460	.3938	.4000	...	.454
000	.40964	.3625	.3720	...	.425
00	.3648	.3310	.3480	...	.380
0	.32486	.3065	.3240	...	.340
1	.2893	.2830	.3000	.227	.300
2	.25763	.2625	.2760	.219	.284
3	.22942	.2437	.2520	.212	.259
4	.20431	.2253	.2320	.207	.238
5	.18194	.2070	.2120	.204	.220
6	.16202	.1920	.1920	.201	.203
7	.14428	.1770	.1760	.199	.180
8	.12849	.1620	.1600	.197	.165
9	.11443	.1483	.1440	.194	.148
10	.10189	.1350	.1280	.191	.134
11	.090742	.1205	.1160	.188	.120
12	.080808	.1055	.1040	.185	.109
13	.071961	.0915	.0920	.182	.095
14	.064084	.0800	.0800	.180	.083

(Continued)

Number of Wire Gage	American or Brown & Sharpe	Washburn & Moen Mfg. Co., A.S.& W. Roebling	Imperial Wire Gage	Stubs' Steel Wire	Birmingham or Stubs' Iron Wire
15	.057068	.0720	.0720	.178	.072
16	.05082	.0625	.0640	.175	.065
17	.045257	.0540	.0560	.172	.058
18	.040303	.0475	.0480	.168	.049
19	.03589	.0410	.0400	.164	.042
20	.031961	.0348	.0360	.161	.035
21	.028462	.0317	.0320	.157	.032
22	.025347	.0286	.0280	.155	.028
23	.022571	.0258	.0240	.153	.025
24	.0201	.0230	.0220	.151	.022
25	.0179	.0204	.0200	.148	.020
26	.01594	.0181	.0180	.146	.018
27	.014195	.0173	.0164	.143	.016
28	.012641	.0162	.0148	.139	.014
29	.011257	.0150	.0136	.134	.013
30	.010025	.0140	.0124	.127	.012
31	.008928	.0132	.0116	.120	.010
32	.00795	.0128	.0108	.115	.009
33	.00708	.0118	.0100	.112	.008
34	.006304	.0104	.0092	.110	.007
35	.005614	.0095	.0084	.108	.005
36	.005	.0090	.0076	.106	.004
37	.004453	.0085	.0068	.103	...
38	.003965	.0080	.0060	.101	...
39	.003531	.0075	.0052	.099	...
40	.003144	.0070	.0048	.097	...

Drill Size Chart

Inch	mm	Drill	Decimals of an Inch
		80	.0135
		79	.0145
1/64			.0156
	.4		.0157
		78	.0160
		77	.0180
	.5		.0197
		76	.0200
		75	.0210
	.55		.0217
		74	.0225
	.6		.0236
		73	.0240
		72	.0250
	.65		.0256
		71	.0260
	.7		.0276
		70	.0280
		69	.0293
	.75		.0295
		68	.0310
1/32			.0313
	.8		.0315
		67	.0320
		66	.0330
	.85		.0335
		65	.0350
	.9		.0354
		64	.0360
		63	.0370
	.95		.0374
		62	.0380
		61	.0390
	1		.0394
		60	.0400
		59	.0410
	1.05		.0413
		58	.0420
		57	.0430
	1.1		.0433
	1.15		.0453
		56	.0465
3/64			.0469
	1.2		.0472
	1.25		.0492
	1.3		.0512
		55	.0520
	1.35		.0531
		54	.0550
	1.4		.0551
	1.45		.0571
	1.5		.0591
		53	.0595
	1.55		.0610
1/16			.0625
	1.6		.0630
		52	.0635
	1.65		.0650
	1.7		.0669
		51	.0670
	1.75		.0689
		50	.0700
	1.8		.0709
	1.85		.0728
		49	.0730
	1.9		.0748
		48	.0760
	1.95		.0768
5/64			.0781
		47	.0785
	2		.0787
	2.05		.0807
		46	.0810
		45	.0820
	2.1		.0827
	2.15		.0846

(Continued)

Inch	mm	Drill	Decimals of an Inch
		44	.0860
	2.2		.0866
	2.25		.0886
		43	.0890
	2.3		.0906
	2.35		.0925
		42	.0935
3/32			.0938
	2.4		.0945
		41	.0960
	2.45		.0966
		40	.0980
	2.5		.0984
		39	.0995
		38	.1015
	2.6		.1024
		37	.1040
	2.7		.1063
		36	.1065
	2.75		.1083
7/64			.1094
		35	.1100
	2.8		.1102
		34	.1110
		33	.1130
	2.9		.1142
		32	.1160
	3		.1181
		31	.1200
	3.1		.1220
1/8			.1250
	3.2		.1260
	3.25		.1280
		30	.1285
	3.3		.1299
	3.4		.1339

Inch	mm	Drill	Decimals of an Inch
		29	.1360
	3.5		.1378
		28	.1405
9/64			.1406
	3.6		.1417
		27	.1440
	3.7		.1457
		26	.1470
	3.75		.1476
		25	.1495
	3.8		.1496
		24	.1520
	3.9		.1535
		23	.1540
5/32			.1563
		22	.1570
	4		.1575
		21	.1590
		20	.1610
	4.1		.1614
	4.2		.1654
		19	.1660
	4.25		.1673
	4.3		.1693
		18	.1695
11/64			.1719
		17	.1730
	4.4		.1732
		16	.1770
	4.5		.1772
		15	.1800
	4.6		.1811
		14	.1820
		13	.1850
	4.7		.1850
	4.75		.1870
3/16			.1875
	4.8		.1890
		12	.1890
		11	.1910

Inch	mm	Drill	Decimals of an Inch
	4.9		.1929
		10	.1935
		9	.1960
	5		.1969
		8	.1990
	5.1		.2008
		7	.2010
13/64			.2031
		6	.2040
	5.2		.2047
		5	.2055
	5.25		.2067
	5.3		.2087
		4	.2090
	5.4		.2126
		3	.2130
	5.5		.2165
7/32			.2188
	5.6		.2205
		2	.2210
	5.7		.2244
	5.75		.2264
		1	.2280
	5.8		.2883
	5.9		.2323

Inch	mm	Letter Sizes	Decimals of an Inch
		A	.2340
15/64			.2344
	6		.2362
		B	.2380
	6.1		.2402
		C	.2420
	6.2		.2441
		D	.2460
	6.25		.2461
	6.3		.2480
1/4		E	.2500
	6.4		.2520
	6.5		.2559

(Continued)

Inch	mm	Letter Sizes	Decimals of an Inch
		F	.2570
	6.6		.2598
		G	.2610
	6.7		.2638
17/64			.2656
	6.75		.2657
		H	.2660
	6.8		.2677
	6.9		.2717
		I	.2720
	7		.2756
		J	.2770
	7.1		.2795
		K	.2810
9/32			.2812
	7.2		.2835
	7.25		.2854
	7.3		.2874
		L	.2900
	7.4		.2913
		M	.2950
	7.5		.2953
19/64			.2969
	7.6		.2992
		N	.3020
	7.7		.3031
	7.75		.3051
	7.8		.3071
	7.9		.3110
5/16			.3125
	8		.3150
		O	.3160
	8.1		.3189
	8.2		.3228
		P	.3230
	8.25		.3248
	8.3		.3268
21/64			.3281
	8.4		.3307
		Q	.3320

Inch	mm	Letter Sizes	Decimals of an Inch
	8.5		.3346
	8.6		.3386
		R	.3390
	8.7		.3425
11/32			.3438
	8.75		.3445
	8.8		.3465
		S	.3480
	8.9		.3504
	9		.3543
		T	.3580
	9.1		.3583
23/64			.3594
	9.2		.3622
	9.25		.3642
	9.3		.3661
		U	.3680
	9.4		.3701
	9.5		.3740
3/8			.3750
		V	.3770
	9.6		.3780
	9.7		.3819
	9.75		.3839
	9.8		.3858
		W	.3860
	9.9		.3898
25/64			.3906
	10		.3937
		X	.3970
		Y	.4040
13/32			.4063
		Z	.4130

Inch	mm	Decimals of an Inch
	10.5	.4134
27/64		.4219
	11	.4331
7/16		.4375
	11.5	.4528

Inch	mm	Decimals of an Inch
29/64		.4531
15/32		.4688
	12	.4724
31/64		.4844
	12.5	.4921
1/2		.5000
	13	.5118
33/64		.5156
17/32		.5313
	13.5	.5315
35/64		.5469
	14	.5512
9/16		.5625
	14.5	.5709
37/64		.5781
	15	.5906
19/32		.5938
39/64		.6094
	15.5	.6102
5/8		.6250
	16	.6299
41/64		.6406
	16.5	.6496
21/32		.6563
	17	.6693
43/64		.6719
11/16		.6875
	17.5	.6890
45/64		.7031
	18	.7087
23/32		.7188
	18.5	.7283
47/64		.7344
	19	.7480
3/4		.7500
49/64		.7656
	19.5	.7677
25/32		.7812
	20	.7874

(Continued)

Inch	mm	Decimals of an Inch
51/64		.7969
	20.5	.8071
13/16		.8125
	21	.8268
53/64		.8281
27/32		.8438
	21.5	.8465
55/64		.8594
	22	.8661
7/8		.8750
	22.5	.8858
57/64		.8906
	23	.9055
29/32		.9063
59/64		.9219
	23.5	.9252
15/16		.9375
	24	.9449
61/64		.9531
	24.5	.9646
31/32		.9688
	25	.9843
63/64		.9844
1		1.000
	25.5	1.0039
1 1/64		1.0156
	26	1.0236
1 1/32		1.0313
	26.5	1.0433
1 3/64		1.0469
1 1/16		1.0625
	27	1.0630
1 5/64		1.0781
	27.5	1.0827
1 3/32		1.0938
	28	1.1024
1 7/64		1.1094

Inch	mm	Decimals of an Inch
	28.5	1.1220
1 1/8		1.1250
1 9/64		1.1406
	29	1.1417
1 5/32		1.1562
	29.5	1.1614
1 11/64		1.1719
	30	1.1811
1 3/16		1.1875
	30.5	1.2008
1 13/64		1.2031
1 7/32		1.2188
	31	1.2205
1 15/64		1.2344
	31.5	1.2402
1 1/4		1.2500
	32	1.2598
1 17/64		1.2656
	32.5	1.2795
1 9/32		1.2813
1 19/64		1.2969
	33	1.2992
1 5/16		1.3125
	33.5	1.3189
1 21/64		1.3281
	34	1.3386
1 11/32		1.3438
	34.5	1.3583
1 23/64		1.3594
1 3/8		1.3750
	35	1.3780
1 25/64		1.3906
	35.5	1.3976
1 13/32		1.4063
	36	1.4173
1 27/64		1.4219
	36.5	1.4370

Conversion Tables

Conversion Table: SI Metric to US Conventional			
When You Know	Multiply By:		To Find
	Very Accurate	Approximate	
Length			
millimeters	0.0393701	0.04	inches
centimeters	0.3937008	0.4	inches
meters	3.280840	3.3	feet
meters	1.093613	1.1	yards
kilometers	0.621371	0.6	miles
Weight			
grains	0.00228571	0.0023	ounces
grams	0.03527396	0.035	ounces
kilograms	2.204623	2.2	pounds
tonnes	1.1023113	1.1	short tons
Volume			
milliliters		0.2	teaspoons
milliliters	0.06667	0.067	tablespoons
milliliters	0.03381402	0.03	fluid ounces
liters	61.02374	61.024	cubic inches
liters	2.113376	2.1	pints
liters	1.056688	1.06	quarts
liters	0.26417205	0.26	gallons
liters	0.03531467	0.035	cubic feet

(Continued)

When You Know	Multiply By: Very Accurate	Multiply By: Approximate	To Find
Volume			
cubic meters	61023.74	61023.7	cubic inches
cubic meters	35.31467	35.0	cubic feet
cubic meters	1.3079506	1.3	cubic yards
cubic meters	264.17205	264.0	gallons
Area			
square centimeters	0.1550003	0.16	square inches
square centimeters	0.00107639	0.001	square feet
square meters	10.76391	10.8	square feet
square meters	1.195990	1.2	square yards
square kilometers		0.4	square miles
hectares	2.471054	2.5	acres
Temperature			
Celsius	*9/5 (then add 32)		Fahrenheit

* = Exact

Conversion Table: US Conventional to SI Metric

When You Know	Multiply By: Very Accurate	Multiply By: Approximate	To Find
Length			
inches	* 25.4		millimeters
inches	* 2.54		centimeters
feet	* 0.3048		meters
feet	* 30.48		centimeters
yards	* 0.9144	0.9	meters
miles	* 1.609344	1.6	kilometers
Weight			
grains	15.43236	15.4	grams
ounces	* 28.349523125	28.0	grams

* = Exact

(Continued)

When You Know	Multiply By:		To Find
	Very Accurate	Approximate	
Weight			
ounces	* 0.028349523125	.028	kilograms
pounds	* 0.45359237	0.45	kilograms
short ton	* 0.90718474	0.9	tonnes
Volume			
teaspoons		5.0	milliliters
tablespoons		15.0	milliliters
fluid ounces	29.57353	30.0	milliliters
cups		0.24	liters
pints	* 0.473176473	0.47	liters
quarts	* 0.946352946	0.95	liters
gallons	* 3.785411784	3.8	liters
cubic inches	* 0.016387064	0.02	liters
cubit feet	* 0.028316846592	0.03	cubic meters
cubic yards	* 0.764554857984	0.76	cubic meters
Area			
square inches	* 6.4516	6.5	square centimeters
square feet	* 0.09290304	0.09	square meters
square yards	* 0.83612736	0.8	square meters
square miles		2.6	square kilometers
acres	* 0.40468564224	0.4	hectares
Temperature			
Fahrenheit	* 5/9 (after subtracting 32)		Celsius

* = Exact

Metric-Inch Equivalents

Inches: Fractions	Inches: Decimals	Millimeters
	.00394	.1
	.00787	.2
	.01181	.3
1/64	.015625	.3969
	.01575	.4
	.01969	.5
	.02362	.6
	.02756	.7
1/32	.03125	.7938
	.0315	.8
	.03543	.9
	.03937	1.00
3/64	.046875	1.1906
1/16	.0625	1.5875
5/64	.078125	1.9844
	.07874	2.00
3/32	.09375	2.3813
7/64	.109375	2.7781
	.11811	3.00
1/8	.125	3.175
9/64	.140625	3.5719
5/32	.15625	3.9688
	.15748	4.00
11/64	.171875	4.3656
3/16	.1875	4.7625
	.19685	5.00
13/64	.203125	5.1594
7/32	.21875	5.5563
15/64	.234375	5.9531
	.23622	6.00
1/4	.2500	6.35
17/64	.265625	6.7469
	.27559	7.00
9/32	.28125	7.1438
19/64	.296875	7.5406
5/16	.3125	7.9375
	.31496	8.00
21/64	.328125	8.3344
11/32	.34375	8.7313
	.35433	9.00
23/64	.359375	9.1281
3/8	.375	9.525
25/64	.390625	9.9219
	.3937	10.00

(Continued)

Inches		Millimeters	Inches		Millimeters
Fractions	Decimals		Fractions	Decimals	
13/32	.40625	10.3188		.70866	18.00
27/64	.421875	10.7156	23/32	.71875	18.2563
	.43307	11.00	47/64	.734375	18.6532
7/16	.4375	11.1125		.74803	19.00
29/64	.453125	11.5094	3/4	.7500	19.05
15/32	.46875	11.9063	49/64	.765625	19.4469
	.47244	12.00	25/32	.78125	19.8438
31/64	.484375	12.3031		.7874	20.00
1/2	.5000	12.70	51/64	.796875	20.2407
	.51181	13.00	13/16	.8125	20.6375
33/64	.515625	13.0969		.82677	21.00
17/32	.53125	13.4938	53/64	.828125	21.0344
35/64	.546875	13.8907	27/32	.84375	21.4313
	.55118	14.00	55/64	.859375	21.8282
9/16	.5625	14.2875		.86614	22.00
37/64	.578125	14.6844	7/8	.875	22.225
	.59055	15.00	57/64	.890625	22.6219
19/32	.59375	15.0813		.90551	23.00
39/64	.609375	15.4782	29/32	.90625	23.0188
5/8	.625	15.875	59/64	.921875	23.4157
	.62992	16.00	15/16	.9375	23.8125
41/64	.640625	16.2719		.94488	24.00
21/32	.65625	16.6688	61/64	.953125	24.2094
	.66929	17.300	31/32	.96875	24.6063
43/64	.671875	17.0657		.98425	25.00
11/16	.6875	17.4625	63/64	.984375	25.0032
45/64	.703125	17.8594	1	1.0000	25.4000

Acknowledgments

The authors wish to thank the following companies for their generous cooperation and assistance for supplying photographs, illustrations, and technical information presented in this book.

American Chain and Cable Company
Butler Manufacturing Company of San Marcos, Texas
Cincinnati Incorporated
Clearing Division of U.S. Industries, Inc.
Cooper Industries
Crescent
General Electric Company
Herlihy & Company
Industrial Fabricators of Corpus Christi, Texas
Justrite Manufacturing Company
Klein Tools
L.S. Starrett
Lodge & Shipley Company
Material Engineering
Milicron Company
Morse Twist Drill Company
Newage Industries
Nicholson
Nicholson File Company, Inc.
Roper Whitney Company
Shore Instrument Company
Stanley Tools
Teco
Tennsmith
The Aluminum Association of America
The Hydraulic Press Manufacturing Company
Tinius Testing Machines
Webster Instruments
Werner Ladders
Wilson Mechanical Instrument Division

Glossary

A

acute angle: A shape formed when the two intersecting lines are less than 90° from each other.

Allen wrench: A hexagonal wrench with a 90° bend. It comes in a number of sizes and fits screw heads that have a similarly shaped hole.

allowance: Extra material added onto the edges of sheet metal patterns to provide for the creation of hems, edges, and seams.

alloy: A material with metallic properties that is a mixture of two or more metals.

alloying element: A substance added to a base substance to improve specific properties.

aluminum: A lightweight, inexpensive, easily worked metal. It is used in alloys with other metals, such as magnesium or manganese.

American Standard Wire Gage: The standard measurement used for nonferrous metals and for wire.

angle changer: An elbow that makes a 45° turn.

annealing: A heat treatment process used to reduce the hardness caused by working a material.

antimony: A hard, brittle nonferrous metal that resembles tin.

apprenticeship: A formal, on-the-job training program in which a trainee learns the skills needed for a trade.

arc: Any section of the curved outside boundary of a circle.

B

bachelor's degree: An academic degree awarded to individuals who complete an undergraduate course of study. It is awarded by a college or university.

bar folder: A forming machine used to bend sheet metal for formed edges. It is also used to form seams.

bead: A decorative impression made on sheet metal that also strengthens and stiffens the metal.

beading machine: A machine used to form beads or impressions on sheet metal.

bismuth: A nonferrous metal whose specific gravity decreases under pressure and expands on cooling.

blank: A piece of sheet metal stock used to make a finished piece.

blind edge: An edge made of a strip of steel, held in place by a nail to create the appearance of a finished hem. The free edge of the metal strip is hammered into place, covering the nail head.

blind rivet: A type of rivet used in situations where only one side of the work is reachable or visible.

body clearance: Part of a drill bit directly behind the margin. It has a smaller diameter than the margin, which decreases the friction between the drill and the wall of the hole so the drill does not bind.

bolt: A fastener with an externally threaded shaft designed to fit through a preformed hole and into an internally threaded nut. A bolt must be used with a nut to do its intended job.

box and pan brake: A machine that produces folds and bends on long pieces of sheet metal. Variation of the cornice brake.

brake: A machine used to bend and fold metal. Can be mechanical or hydraulic.

brass: An alloy of copper and zinc, available in various degrees of hardness. Colors range from copper to nearly white.

brazing: A process of joining metals with a nonferrous filler at temperatures exceeding 840°F (450°C).

Brinell hardness test: A type of indentation test used to determine the hardness of a material. A load is applied and held for a predetermined time to an indenter. The resulting indentation is measured in two directions and averaged.

brittleness: A characteristic of a material that allows it to break suddenly.

bronze: A nonferrous metal that is an alloy of copper and tin, or copper and alloying elements other than zinc.

burred edge: An edge created by rolling two pieces of metal together providing a watertight seal. It is made on a burring machine and is found on cans, buckets, and similar objects.

burring machine: A machine used to make small flanges on the edges of sheet metal discs. Flanges are used as part of a seam.

C

caliper: A measuring tool used to take internal and external measurements of diameters, thicknesses, hole depth, and distances between pieces.

cape chisel: Tapered chisel used to cut grooves and keyways in corners.

capillary action: A physical reaction between a solid and liquid that causes liquid solder to flow into air spaces found between pieces of tightly fitted metal.

capped edge: An edge made from a strip of metal to cap a raw edge of metal. The metal strip is bent and slipped onto the raw edge. It is held in place by rivets, screws, or soldering.

center punch: A punch used to make indentations in metal and on 18-gage and thicker sheet metal.

centerline: A thin, broken line that indicates the location of the middle of an object.

chisel edge: The portion of a drill bit that connects the lips.

chord: A straight line located on the interior of a circle that ends at two points on the outside edge of that circle.

chuck: Portion of a drill that both holds and drives the drill bit. Drill bit is held using a movable jaw.

circle: A closed-plane curve in which every point is equidistant from the center.

circumference (perimeter): The length of the outside boundary of a circle.

clinch-tee: A duct with a straight or curved throat. It provides a separate path for airflow anywhere along the pipe.

cold-rolled steel: Steel that is not processed until it is cooled. Its finished appearance is silver gray, with a smooth surface.

combination bevel: A measuring device used to transfer and to set angles.

combination set: A combination square that contains a center head, a square head, and a protractor head. Different heads are used for different tasks.

combination square: A tool used to check angles, measure lengths, heights, and depths. A scriber contained in the square can be used to scribe lines on work.

compressive strength: A material's ability to resist the force created by opposite objects pushing against each other.

computer-aided drafting: Software used to create prints and all other types of drawings used in construction.

conventional spinning: A type of sheet metal spinning in which the blank is pressed against the mandrel, forcing it into the shape of the form. A conventionally spun piece has a smaller diameter than the blank, but the thickness remains the same. Several passes may be made, resulting in hardening of the metal and subsequent annealing.

copper sheet: Reddish brown metal. An excellent conductor of electricity that contains a small quantity of alloy.

cornice brake: A forming machine used to form large pieces of sheet metal. Contains forming blocks that can be arranged in different ways to form different shapes.

corrosive flux: A type of flux that contains acid. The work piece must always be cleaned after soldering with corrosive flux to prevent residue from corroding the work.

crimper: A machine that reduces the diameter of the end of a sheet metal pipe, allowing it to slide into a pipe of the same diameter.

curved elbow: An elbow that changes ductrun direction gradually.

cutting speed (peripheral speed): The speed difference between the drill bit and the surface of the workpiece it is drilling into. It is measured in surface feet per minute.

D

deep drawing: A stretching process in which a blank is placed in a draw-press die. The blank holder places pressure on the blank, holding it tight enough that it slides into the die, while keeping the metal flat and free from wrinkles.

depth gage: A gage used to measure the depth of holes, slots, and other hard-to-reach places.

diagonal pliers: Pliers used to cut wires, nails, pins, and screws.

dial indicator: A precision instrument used to check the roundness of a piece of work or to center work on a metal lathe.

diameter: The length of a straight line through the center of a circle, ending at each side.

diamond-point chisel: Chisel used to cut grooves and to square corners. Cutting face is diamond shaped.

dimension line: A line that displays the measurement of a particular feature in it and terminates at each end with thin arrowheads.

divider: A tool used to transfer measurements from one surface to another and to scribe arcs and circles in layout work.

double seam: A seam used on irregular fittings and to join cylindrical shaped objects to their bottom pieces.

double-seaming machine: A forming machine that folds a joint over, securing the seam.

dovetail seam: A seam used to join a flange and a collar. It is used almost exclusively on round or elliptical pipes.

draw filing: Filing method in which the file is pushed and pulled across the work and at a right angle to the work. Produces a finer finish that straight filing.

drill bit: A metal cutting tool that makes cylindrical holes.

drill press speed: Measurement of the revolutions per minute made by the drill press spindle.

drill walking: Tendency for an uncentered drill bit to move away from a marked hole location.

drive clip: A piece of metal that joins folded edges together.

ductility: The ability of material to be drawn out, stretched, twisted, or bent when cold, without breaking.

ductrun: A complete system of ducts and fittings.

ductwork: Combination of pipes and fittings in the HVAC system that contains the circulating air.

E

edge: Surface of a file that is adjacent to the face. Shape is determined by shape of the file.

elastic limit: The maximum load a material can hold without being permanently damaged.

elasticity: A material's ability to return to its original shape after a load is removed.

elbow: A duct used to make a turn in a ductrun.

extension line: A line drawn perpendicular to dimension line that projects out from an object's feature and is connected by the dimension line.

F

fabrication software: A computer program that contains ready-made patterns for common sheet metal pieces.

face: The surface of a file that contains the teeth.

fatigue strength: The ability of a material to withstand repeated stress.

feed: Downward motion of the drill bit into the material made in each revolution of the drill press spindle.

firm-joint caliper: A caliper that is held in place by friction at the joint of the two legs.

fitting: Hardware used to change airflow direction, the size of the duct ends, or both.

flat chisel: Chisel used for general cutting work.

flex strength: The ability of a material to withstand bending forces.

flexible curve: A thin piece of bendable plastic that contains a thin metal wire. It is bent to go through all the plotted points and then the points are connected. The wire allows the curve to retain its shape once it is bent.

flute: Groove that runs the length of the drill body.

flux core solder: A solder that has a flux core. As the solder melts, the flux core cleans the surface of the joint.

flux: Material (acid or rosin) used in soldering that prevents oxidization of the soldering iron tip and the heated metal. It can also remove the oxide film that forms on metals.

French curve: A type of irregular curve that has a variety of curves on the outside and inside edges.

G

gage: The thickness of sheet metal. Each gage represents a specific metal thickness.

galvanized steel: Sheet steel that is coated with a thin layer of zinc to prevent corrosion. Commonly used to make roofing, siding, gutters, and downspouts.

galvanizing: Process that coats steel with a thin layer of zinc to prevent corrosion.

general note: An instruction or comment that provides information that applies to the entire drawing.

graduation: A line on a steel rule that represents points of measurement.

grooved seam: A seam with two folded edges locked together with a hand groover. Used on light- and medium-gage sheet metal.

H

hand punch: A blunt tool used to pierce a solid material.

hand snips: A cutting tool used to cut sheet metal. Another name for *shears.*

hard solder: Solder that melts at temperatures above 750°F (400°C). It is usually a copper/zinc or copper/silver alloy.

hardness: The ability of a material to resist penetration or scratching by another material.

heating, ventilation, and air-conditioning (HVAC) system: System that circulates either warm or cooled air throughout a building.

heel: Bottom portion of a file, nearest the tang.

hem: An edge used to strengthen sheet metal objects. The edge has a rolled-over or rounded appearance.

hermaphrodite caliper: A caliper that has a straight, pointed leg and a curved leg. It is used to scribe lines on surfaces parallel to an edge and to locate center points on round stock.

hexagon: A six-sided figure with sides of equal length.

hidden line: A thin, broken line that represents internal features of a part that are not visible from its exterior.

high-energy-rate forming (HERF): A group of sheet metal forming methods that shape a work piece using the force created by a sudden blast of extremely intense energy.

hook rule: A steel rule with an arm attached to one end that grabs the end of the work piece and allows for hands-free measuring.

I

impact strength: The ability of a material to withstand a sudden shock or load.

initiative: The ability to start and finish a task or job.

irregular curve: A plastic drafting tool used to connect plotted dots that designate the outline of a curve.

isometric drawing: A pictorial representation of an object in which all three dimensions are drawn full scale rather than foreshortened, or compressed.

K

kerf: The slot produced by a saw blade.

L

lands: The outermost portion of the drill bit body between adjacent flutes.

lathe: A machine that holds a work piece while it rotates and is machined by a stationary tool.

lip: The cutting edges of the drill bit.

local note: An instruction or comment that provides information about a specific process or feature; placed near the view to which it refers.

lock washer: A type of washer used to secure nuts and bolts to an assembled part.

M

machine screw: Fastening device that has threads made to fit into a nut or into a tapped hole.

manager: An employee who is responsible for assigning and directing work for other employees.

mandrel: A cylindrical piece of metal used to support work pieces being formed.

margin: Part of a drill bit that determines the diameter of the hole. It runs the entire length of the flute.

materials: The physical substances used in production or manufacturing.

micrometer: A precision instrument capable of measuring to 1/1000″ (0.001″) and 1/100 mm (0.001 mm). The outside micrometer is the type most commonly used in sheet metal work.

mitered duct: A sheet metal object with a beveled end.

multiview drawing: An exact representation of an object that usually consists of at least two views in order to include all sides, visible and invisible.

N

needle-nose pliers: Pliers with a long nose and flat gripping surface, used to hold work in tight spaces and to shape light metal and wire.

noncorrosive flux: A type of flux that will not continue to corrode metal after soldering.

nonferrous metal: Metal that does not contain iron.

note: An instruction or comment that provides information that is not part of the normal dimensioning system. It is stated as briefly and concisely as possible and is lettered horizontally.

nut: A fastener used with bolts and some screws to fasten and lock parts together.

O

object line: Solid, thick line that represents the visible edges and details of a part.

obtuse angle: A shape formed when two intersecting lines are more than 90° from each other.

Occupational Safety and Health Administration (OSHA): A government agency responsible for setting and enforcing work safety standards.

octagon: An eight-sided figure with sides of equal length.

offset: A flexible duct used to move ductwork around an obstacle.

overtime: Time spent working beyond a typical 40-hour work week.

P

parallel line development: Method of creating patterns for objects in which the sides run parallel to each other.

pattern: A flat outline of an object that has only width and depth and in which all sides of the object can be seen.

pentagon: A five-sided figure with sides of equal length.

peripheral speed: See *cutting speed.*

personal protective equipment (PPE): Clothing, hats, face shields, shoes, and other items that protect a person's body from danger. PPE must meet OSHA standards.

pickling: A cleaning process that removes oxide scale formed during the annealing process. The process involves submerging the work piece into a diluted sulfuric acid bath.

pin punch: A hand punch used to drive pins out of holes.

Pittsburgh seam: A seam in which a single lock is inserted into a pocket lock, after which the edge of the pocket lock is folded down to seal the seam.

plasticity: The ability of material to permanently deform but not break.

plate: Any type of metal produced in flat sheets or rolls more than 1/4″ (7 mm) thick.

plug gage: A measuring tool that tests tolerances in two steps. The piece being tested must pass one test (go) and fail the other (no-go). Also called a go no-go gage.

point: Cone-shaped end of the drill bit that does the actual cutting.

point: Tip of the file, at the opposite end of the tang.

press brake: A powered machine used to bend heavy sheet metal and steel plate. Dies used on the machine produce a large number of different bends and shapes.

prick punch: A punch used to make small indentations in metal and to locate points for dividers and fold lines in sheet metal work.

print: An original, detailed drawing that illustrates how an object is to be built. It includes information on the size, shape, and tolerances of objects, materials to be used, and finishes needed.

properties: The characteristics or qualities of a material when that material is placed in certain conditions or combined with other materials.

R

radial line development: Method of creating patterns for objects that radiate from a common center. Arcs are used to create the pattern.

radius: A measurement that is half the distance of the circle diameter or the length of a straight line from the center of a circle to an outside edge.

rectangle: A four-sided figure whose sides are at right angles, with only opposing sides of equal length.

resistance spot welding: A welding technique. The pieces of sheet metal to be welded are held in contact between two copper electrodes. Current from the electrodes passes through the metal to fuse the pieces together.

right angle: A shape formed when the two intersecting lines are 90° from each other.

rivet: A metal fastener made up of a head and a shaft. It is attached to a work piece using a riveting hammer, other riveting hand tools, or a riveting machine.

rivet set: A tool used to seat and lock a rivet into place.

Rockwell hardness test: A type of indentation test used to determine the hardness of a material. An indenter is forced into the material being tested. The resulting dent is measured and given a hardness value.

rotary machine: Forming machine used to form edges on round and cylindrical objects. Comes with interchangeable rollers.

round-nose chisel: Chisel used to cut grooves and to chip inside corners.

S

scaffolding: A temporary platform used to reach elevated work areas.

screw pitch gage: A gage that measures the pitch of screw threads.

seam: A method of joining sheet metal pieces together. Ends of sheet metal are folded together and then fastened, usually with rivets or solder.

self-drilling sheet metal screw: A machine screw that has a point capable of drilling a hole into the material before the threads engage.

self-locking nut: Type of nut with a mechanical feature that prevents the nut from loosening. It is used in assemblies that experience a great deal of vibration or stress.

set: The tooth pattern on a saw. The set determines the width of the cut and prevents blades from binding in the cuts.

setting-down machine: A forming machine that prepares the seams in cans and other vessels for double seams.

shank: Portion of the drill bit that fits into the spindle or the chuck.

shear forming: A simplified variation of the spinning process in which the roller bends the blank and also applies downward force.

shear forming: A simplified variation of the spinning process in which the roller bends the blank and also applies downward force. This stretches the blank over the mandrel, resulting in a piece with the same diameter as the blank, but a different thickness. Usually only one pass is required in shear forming, eliminating the need for annealing.

shears: A cutting tool or machine used to cut out sheet metal patterns.

shear strength: A material's ability to resist sliding forces.

sheet metal forming: The process of applying force to a metal sheet to change its surface shape. No material is removed. The metal sheet is deformed but not weakened by the applied force.

sheet metal screw: A type of machine screw that has a fully threaded shaft, ending in a point. This screw is used on sheet metal that has existing holes.

sheet metal: Any type of metal produced in flat sheets or rolls less than 1/4″ (7 mm) thick.

sheet steel: An uncoated sheet of steel having a bluish-black surface. After being assembled into a finished product, the steel is coated to prevent corrosion.

shore scleroscope hardness test: A rebound test used to determine the hardness of a material.

shrink rule: A rule made slightly longer than ordinary. Extra length allows for shrinkage of cooled metal.

sleeve: A drill press holding tool used to hold drill bits that are too small for the spindle.

slide caliper rule: A measuring tool used to make inside and outside measurements on round or flat stock.

slip joint seam: A seam that is held together with a separate piece of metal called a drive clip.

slip-joint pliers: Pliers with an adjustable hinge, useful for holding different sizes of work.

slip-rolls forming machine: A forming machine used on light gage sheet metal to bend stock into cylindrical shapes.

snap gage: A fixed limit gage that has two sets of jaws. The outside jaws test the upper limit tolerance and the inside jaws test the lower limit tolerance.

socket: A drill press holding tool used to hold drill bits that are too large to fit in the spindle.

soft solder: Solder that melts at temperatures below 750°F (400°C).

solder: Filler material used in the soldering process. It is an alloy of several materials.

soldering: The process of joining metals by heating a filler material (solder). The filler material melts, joining the base metals without melting them.

solid punch: A hand punch used to make small diameter holes in light-gage metal.

solvent: A liquid that dissolves another substance and forms a solution.

spot welding: A welding technique. The pieces of sheet metal to be welded are held in contact between two copper electrodes. Current from the electrodes passes through the metal to fuse the pieces together.

springback: Partial rebounding of formed material caused by its elasticity.

spring-joint caliper: A caliper that is held in place by the tension of the spring against the nut.

square: A four-sided figure with equal-length sides that are at right angles to each other.

square elbow: An elbow that makes a 90° turn.

squaring shears: A foot-operated machine used to trim or square large metal sheets. It can cut metal up to a maximum gage.

stainless steel: A durable, somewhat expensive metal that contains 12–25 percent chromium and resists rust and corrosion.

standing seam: An upright seam that provides its own reinforcement. Used on large expanses of metal, where additional strength is needed.

steel rule: A metal measuring tool divided into graduations that represent measurement points.

straight filing: Filing method in which the file is pushed lengthwise straight ahead or at a slight diagonal to the work.

straight-fluted drill bit: A drill bit with unbent flutes.

strain: Deformation caused by stress that causes elongation, compression, or distortion.

strength: A material's ability to resist stress.

stress: The result of a force or load applied to a material.

stretch forming: Shaping a metal sheet by first applying suitable tension or stretch and then wrapping it around a form of the desired shape.

stretchout: The distance from each end of a flat pattern development. It also refers to the line projecting from the right of the front view when drawing a pattern.

surface gage: A gage used to accurately scribe lines and transfer measurements.

swinging-blade protractor: A thin plastic or metal tool on which degrees zero through 180 are marked. It is used to draw angles and to check angles.

T

TAB technician: A sheet metal worker who specializes in the testing, adjusting, and balancing of existing HVAC systems to ensure they are working properly.

tang: Tapered portion of a file designed to fit into a handle.

tangent: A straight line that touches the outside edge of a circle at one point.

taper mount: Portion of a drill press that both holds and drives the drill bit. Tapered opening holds the drill bit in place through friction.

temper: Hardness and strength of metal.

template: A pattern developed on paper or metal, that is then transferred onto a piece of metal.

tensile strength: A material's ability to resist pulling forces.

tensioning: The proper adjustment of a saw blade. Assures the blade will cut properly, cleanly, safely.

thickness gage: A gage used to make adjustments and check clearances.

thread pitch: Measurement on a threaded fastener that identifies the number of threads per inch.

throatless shears: A cutting machine used to make straight, curved, and irregular cuts in sheet metal. It is designed so that the piece of metal being cut can be moved into different positions for difficult cuts.

tin plate: Sheet metal coated with a layer of pure tin, making it corrosion resistant.

tinning: A process done before soldering. It involves rubbing the tip of a heated soldering iron on a sal ammoniac block and then covering it with a thin layer of solder.

torsion strength: Related to shear strength; the ability of a material to resist a rotating force.

triangle: A three-sided figure whose interior angles add up to 180°.

triangulation: Method of creating patterns for objects that gradually transition from one shape to the other. Patterns for these pieces are created using triangles.

true length: A line on a plane that is perpendicular to a viewer's line of vision.

try square: A tool used to check the accuracy of 90° angles and to mark 90° angles when laying out patterns.

turning machine: A forming machine that creates a rounded edge for holding wire.

twist drill bit: A drill bit with helically grooved flutes.

two-way Y branch: A duct that splits the main airstream into two separate streams.

U

Underwriters Laboratories (UL): An independent, not-for-profit product safety certification organization. Products that pass their testing are said to be UL approved and a seal is affixed to the product.

United States Standard Gage for Sheet and Plate Iron and Steel: The standard measurement used for plate iron, steel, and uncoated steel. It is based on weight in ounces per square foot.

utility pliers: Pliers with parallel jaws which can hold large work. Useful for heavy-duty work.

V

velocity: Speed of air through ductwork.

Vernier scale: A graduated scale used to measure lengths and angles.

vertex: The point at which two straight lines forming an angle meet.

visualization: The ability to see an object as a collection of shapes and how the shapes fit together.

W

washer: A flat disc used to distribute the load created by a screw, to protect the finish of a part, or to create a finished look to a part.

web: The center part of the drill bit body where the lands meet.

welding: A nonmechanical fastening method that permanently fuses metal pieces together by melting the metal.

wired edge: An edge that is additionally strengthened through the use of wire, angle iron, band iron, or other material. The strengthened edge aids the object in keeping its shape.

wiring machine: A forming machine that closes the edge surrounding a wire.

working drawing: A copy of a print made for use in the shop or on the job site.

Z

zinc: A nonferrous metal used to galvanize iron and steel against corrosion.

Index

E

F

G

H

I

Q

R

S